Comparative Physiology of the Vertebrate Digestive System is useful for any-
one interested in the basic structural and functional characteristics of the digestive system and how these vary among vertebrate groups and species. It shows
how fish, amphibians, reptiles, birds, and mammals have adopted various diets
that range from other animals to plant fiber, and habitats that vary from fresh
or salt water to extremely arid terrestrial environments.

This book is unique in several respects. It discusses all of the major aspects
of nutrition, anatomy, and physiology in all of the major groups of vertebrates.
It contains numerous figures and tables to aid in their comparison, including
seventy-six illustrations of gastrointestinal tracts of amphibians, reptiles, birds,
and mammals, prepared to allow for their direct comparison and labeled with
animal silhouettes. The terminologies of scientific specialties and subspecial-
ties are defined, and animals are referred to by both their common and scientific names. Relationships between digestive strategies and the diet and environment are discussed throughout the text and brought together in Chapter 11
on the evolution of the digestive system. Chapter 12 offers a brief summary of
the major concepts covered in this book and suggests future directions for research.

The first edition of this book, published in 1988, was prepared as a response to the needs of veterinary, physiology, zoology, and nutrition graduate
students enrolled in an NIH-sponsored program on comparative gastroenterology. This second edition takes advantage of the considerable research conducted in the intervening years. It contains three additional chapters, on nutrition, digesta transit, and evolution. Each of the other chapters has been
updated and expanded. These changes provide a more complete discussion of
the subjects and a broader interpretation of the information.

The authors have collectively studied the nutrition and digestive physiology
of a wide range of vertebrates in North and South America, Australia, Europe,
Africa, Indonesia, and Israel, and they have taught these subjects to undergraduate, graduate, and veterinary students for over two decades.

COMPARATIVE PHYSIOLOGY OF THE VERTEBRATE DIGESTIVE SYSTEM

COMPARATIVE PHYSIOLOGY OF THE VERTEBRATE DIGESTIVE SYSTEM

Second Edition

C. EDWARD STEVENS
College of Veterinary Medicine
North Carolina State University

IAN D. HUME
School of Biological Sciences
University of Sydney

CAMBRIDGE
UNIVERSITY PRESS

PUBLISHED BY THE PRESS SYNDICATE OF THE UNIVERSITY OF CAMBRIDGE
The Pitt Building, Trumpington Street, Cambridge, United Kingdom

CAMBRIDGE UNIVERSITY PRESS
The Edinburgh Building, Cambridge CB2 2RU, UK
40 West 20th Street, New York NY 10011–4211, USA
477 Williamstown Road, Port Melbourne, VIC 3207, Australia
Ruiz de Alarcón 13, 28014 Madrid, Spain
Dock House, The Waterfront, Cape Town 8001, South Africa

http://www.cambridge.org

First edition published 1988
Second edition published 1995
First paperback edition 2004

Typeset in Times

A catalogue record for this book is available from the British Library

ISBN 0 521 44418 7 hardback
ISBN 0 521 61714 6 paperback

For ALVIN SELLERS and REG MOIR,
who introduced us to the wonders of the
digestive system and the value of the comparative approach

Contents

Contents   xiii

Preface

Studies of comparative physiology provide us with the information needed for the understanding of physiological mechanisms, the management of domesticated animals, and the preservation of wild species. This edition of *Comparative Physiology of the Vertebrate Digestive System*, like the previous edition, is designed for use by nutritionists, physiologists, zoologists, veterinarians, and others interested in the vertebrate digestive system and its many variations. Its objectives are to describe the anatomical and physiological characteristics of this system in relation to nutritional niches and other characteristics of the environment.

Most concepts and theories of functional mechanisms are predicted from the study of animals that perform these functions in a different or simpler manner. Many vertebrates are propagated for the production of food or fiber, raised as pets or companion animals, or maintained as captive (zoo) animals. The numbers of many wild species and even classes of vertebrates, such as the amphibians, are decreasing at rates that have been compared with those seen during the demise of the dinosaurs. Therefore, studies of the vertebrate digestive system provide information needed for the understanding of basic physiological mechanisms, the proper maintenance of domesticated and captive animals, and the preservation of endangered species.

In this edition chapters that deal with the general characteristics of the vertebrate digestive system and its enormous range of structural and functional variations were updated and expanded to include new material and recent interpretations. A chapter on energy and nutrient requirements was added to provide information about metabolic and nutrient requirements and how these vary across vertebrate groups and accommodate to physiological states and changing needs. The related topics of motor activity and digesta transit have been separated into two chapters, reflecting significant advances in the past seven years. Chapter 11, on evolution of the digestive system, summarizes current thoughts and several themes that run through earlier chapters. The final chapter has been expanded to include both general conclusions and suggestions for future studies.

Research on the digestive system has resulted in a veritable explosion of new information and concepts. New techniques for the examination of mechanisms at the cellular and molecular levels have expanded knowledge on both the basic mechanisms and their adaptations to the diet and environment. The expansion in scientific specialties and techniques makes it impossible to discuss the significance of all of this new information. This book does not attempt to provide an encyclopedic coverage or exhaustive review of the available literature. However, earlier classical studies of comparative anatomy and physiology are discussed for their contributions, and reference is made to the authors of important discoveries or ideas as often as possible. Readers are directed to recent reviews and other sources that give more detailed or extensive coverage of many of these subjects.

We are grateful to the many people who helped in the preparation of this book. Many colleagues have provided specimens or photographs for the series of line drawings comparing the gastrointestinal tracts of amphibians, reptiles, birds, and mammals. Erica Melack again deserves special recognition for the preparation of most of the earlier drawings. We are grateful to Frank Knight for his preparation of the animal silhouettes that grace these figures and the book cover and to Brenda Bunch for the preparation or modification of many of the new illustrations. We greatly appreciate the help of Robert Argenzio in reviewing portions of the manuscript, Kendall Clements for information and insights on herbivory in fish and, Judith Caton for information and insights on folivorous primates. We appreciate the contributions of Sylvia Bennett for her assistance in editing the manuscript and Rebecca Bladon for her aid in its preparation. Completion of this project would have been difficult without the encouragement of our respective wives, Barbara and Desley – to them our heartfelt thanks.

C. Edward Stevens
Ian D. Hume

1

Energy and nutrient requirements

The major purpose of the digestive system is to provide for the assimilation of nutrients required for energy, maintenance, growth, and reproduction. Digestion consists of a number of physical and chemical processes. Food is ingested, broken down into smaller particles, macerated, mixed with digestive enzymes, and propelled through the digestive tract by the motor or muscular activities of the tract. Salivary, gastric, pancreatic, biliary, and intestinal secretions collectively provide mucus for protection and lubrication of the tract, enzymes that aid in digestion, and the watery medium and optimal pH required for digestion. Digestive enzymes aid in the hydrolysis of carbohydrates, protein, and lipids into a limited number of much smaller compounds suitable for absorption. Microorganisms indigenous to the digestive tract can provide additional nutrients by breaking down structural carbohydrates that are not subject to attack by endogenous enzymes and by synthesizing amino acids and vitamins essential to the host animal.

Variations in the digestive system among vertebrates can often be related to the animal's nutritional niche. The nutritional niche of an animal can be defined by two basic parameters: (1) the energy and nutrients it needs; and (2) how it harvests and extracts what it needs from its nutritional environment. This book is primarily about the system that extracts energy and nutrients from food once it is harvested, or ingested, that is, the digestive system. However, the digestive system cannot be adequately discussed without some knowledge of what the animal needs in the way of energy and nutrients. These needs, in turn, are set largely by the animal's metabolic rate, which differs widely across taxons. Even within taxons, there can be significant differences in the rate of metabolism, depending on such things as the degree of homeothermy, the physiological state (maintenance, growth, reproduction), the environment (aquatic versus terrestrial), or whether an animal is arboreal (lives in trees) or fossorial (lives underground) (McNab 1978). The success of an animal is a function of how well the digestive tract can process food and extract energy and nutrients at rates that match rates of use in the body under these conditions. In quantitative terms, energy is the most important requirement of the body, and it is almost self-evident that animals eat to meet their energy requirements first. Energy can be derived by catabolism of carbohydrates, lipids, and proteins, which also provide the building blocks needed for a range of biosynthetic pathways. The other nutrients required by all vertebrates are water, vitamins, and minerals.

Metabolic rates

Despite the wide range in the rates of energy consumption (metabolic rates) observed among animal groups, rates of metabolism can be related to body size (M) by the equation $R = aM^b$, in which R is the metabolic rate measured under "basal" or standard conditions, a is a proportionality coefficient that differs among species, and b is an empirically determined exponent that expresses the rate of change of R with changes in body mass (M). Basal metabolic rate (BMR) is the rate of metabolism of an endothermic animal that is resting in a postabsorptive state and thermoneutral environment and is not experiencing any physical or psychological stress. Standard metabolic rate is the equivalent minimal metabolic rate of an ectothermic animal at a particular temperature (Withers 1992). Processes that utilize a significant percentage of the BMR include the maintenance of ionic gradients across cell membranes (ion pumping) and the synthesis and turnover of protein. The relationship described by the equation $R = aM^b$ holds for unicellular organisms through ectothermic invertebrates and vertebrates, to the endothermic birds and mammals (Fig. 1.1).

The true value for b, the slope of the three lines in Fig. 1.1, is the subject of great debate (Schmidt-Nielsen 1984; Withers 1992). One argument is that, theoretically, b should be close to 0.67, because many physiological processes, such as uptake of oxygen by the lungs, gills, or across body surfaces, are functions of surface area. Heat loss takes place across the body surface. The surface area of a body of given density and constant proportions increases as a function of the square of the linear dimension, but mass increases as a function of its cube. Thus the metabolic rate should be proportional to the 0.67 (2/3) power of the body mass.

Alternatively, McMahon and Bonner (1983) argued that, based on engineering principles, the power output of any particular muscle depends only on its cross-sectional area, which is proportional to $M^{0.75}$. If the power output of a single muscle is applied to all metabolic variables involved in supplying all muscles with energy and oxygen, the complete metabolic system should be scaled to body size in the same way (Schmidt-Nielsen 1984).

The slopes of all of the curves for the three groups in Fig. 1.1 are close to 0.75.

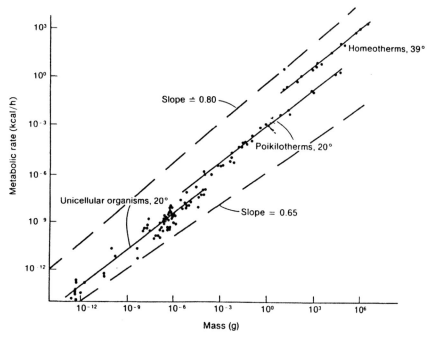

Fig. 1.1. Log–log relationship between metabolic rate, in kcal/L and body mass in three groups of animals, ranging from unicellular organisms to large homeotherms (endotherms). Note that the slope for all three groups is close to 0.75. (From Hemmingsen 1960).

However, within taxonomic groups, the exponent may differ. For example, "*b*" is 0.67 for the mammalian order Rodentia but 0.75 for the Primates (Haysson and Lacey 1985). Whatever the value for the exponent, the relationship enables us to compare the metabolic rates of animals of vastly different sizes, from unicellular organisms to the largest vertebrates (Hemmingsen 1960). In these comparisons, it is the magnitude of the coefficient, "*a*," that is important, as shown in Table 1.1 for various vertebrate groups.

One of the most important influences on "*a*" is temperature; this is because enzymatic reactions are highly temperature dependent. A temperature difference of 10°C has become a standard span over which to determine the temperature sensitivity of biological functions; this is the Q_{10} value. In general, chemical processes have Q_{10} values of 2 to 3, while purely physical processes, such as diffusion, have Q_{10} values closer to 1. Metabolic rates in most ectotherms increase twofold to threefold for every 10°C increase in ambient temperature. If the same Q_{10} is applied to endotherms, values of the coefficient "a" can be recalculated to "correct" for differences in resting body temperatures. These converted rates are given in Table 1.1, which shows that the difference in standard or resting metabolic rates

Table 1.1. *Basal or Standard Metabolic Rates of Higher Vertebrates at Normal Body Temperatures and Recalculated to a Uniform Body Temperature of 38° C*

Parameter	Reptiles (lizard)	Mammals			Birds	
		Monotreme	Marsupial	Eutherian	Nonpasserine	Passerine
Body temperature (°C)	30.0	30.0	35.5	38.0	39.5	40.5
BMR at Body temp.	23	143	201	290	313	515
Q^{10}	3.3	2.1	2.5	2.5	2.5	2.5
BMR at 38° C	61	259	252	290	272	411

Source: Modified from Schmidt-Nielsen (1984). Metabolic rate in $kJ \cdot kg^{-0.75} \cdot day^{-1}$.

between eutherian mammals and nonpasserine birds, such as domestic fowl and sea birds, is purely a Q_{10} effect, but that rates for passerines (the perching birds) are 60 percent higher than rates for nonpasserines. On the other hand, metabolic rates of ratites (the large flightless birds) are 37 percent lower than the nonpasserine rate (Withers 1992), no doubt because of the lower energy cost of bipedal locomotion.

At the same ambient temperature of 38°C, the metabolic rate of the ectothermic lizard in Table 1.1 is 21 percent of that of the basal eutherian mammal, but at 30°C, it is only 8 percent because of the Q_{10} effect. Withers (1992) prepared a comprehensive table of metabolic rates of ectotherms corrected to a common temperature of 20°C. This table showed that although the metabolic rates of fish range over more than an order of magnitude, the metabolic rates of amphibians and reptiles are of the same order as that of the lizards in Table 1.1. Invertebrates such as crustaceans, spiders, and insects also exhibit metabolic rates within the same range as the amphibians and reptiles.

Metabolic rate also can be expressed in terms of unit mass of tissue; for example, ml of oxygen consumed per kg mass per hour. This is the mass-specific metabolic rate, or metabolic intensity, of the animal. Mass-specific metabolic rate declines with increasing species body size, as illustrated in Table 1.2 and Fig. 1.2. If this is plotted with both the mass-specific metabolic rate and the body mass on logarithmic coordinates, the regression line has a slope of −0.25 (the reciprocal of 0.75). Other physiological parameters, such as heart rate, respiratory frequency, and rate of digesta passage, also scale as $M^{-0.25}$ (Karasov and Hume in press).

Table 1.2. *Standard Rates of Oxygen Consumption in Eutherian Mammals of Various Body Sizes*

Animal	Body mass (g)	Metabolic rate		
		Whole-body (ml $O_2 \cdot h^{-1}$)	Mass-specific (ml $O_2 \cdot g^{-1} \cdot h^{-1}$)	BMR (ml $O_2 \cdot kg^{-0.75} \cdot h^{-1}$)
Shrew	5	36	7.40	45.95
Harvest mouse	9	23	1.50	18.89
Kangaroo mouse	15	27	1.80	15.12
House mouse	25	41	1.65	15.65
Ground squirrel	96	99	1.03	13.78
Rat	290	250	0.87	15.18
Cat	2,500	1,700	0.68	20.52
Dog	11,700	3,870	0.33	14.68
Sheep	42,700	9,590	0.22	13.78
Human	70,000	14,760	0.21	26.03
Horse	650,000	71,100	0.11	13.26
Elephant	3,833,000	268,000	0.07	13.20

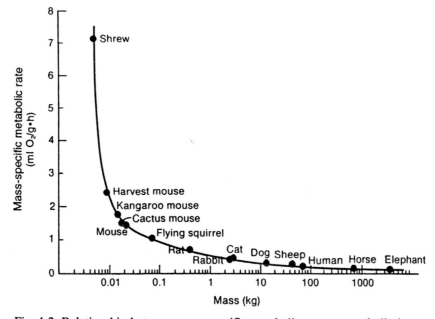

Fig. 1.2. Relationship between mass-specific metabolic rate or metabolic intensity for eutherian mammals ranging in body size from 6-g shrews to 1,300-kg elephants. Note the inverse relationship between mass-specific metabolic rate and body mass. (From Schmidt-Nielsen 1984).

Mass-specific metabolic rates have important consequences in animal nutrition and digestive tract design and function. In general, smaller animals are restricted to more digestible diets of higher quality because of the need to process energy (and nutrients) at higher rates than larger animals (Parra 1978; Demment and Van Soest 1985). This topic is pursued in Chapter 6.

Energy and food requirements

Notwithstanding the strong influence of body temperature on standard metabolic rates, Table 1.1 illustrates that there are sometimes large phylogenetic differences in rates of metabolism even within vertebrate classes. These differences often translate into markedly different requirements for both energy and nutrients. Not surprisingly, there is usually a strong correlation between basal metabolic rates and the energy required for longer term energy balance.

In housed domestic and captive wild animals, energy additional to basal requirements is needed for feeding, drinking, digestion, absorption, and metabolism, but little is needed for thermoregulation or other activities. Under these conditions, maintenance energy requirements are often approximately double the BMR for the species. For instance, Hume (1974) determined that sheep and captive red kangaroos *Macropus rufus* and euros, or hill kangaroos *M. robustus*, needed 569, 456 and 414 $kJ \cdot kg^{-0.75} \cdot day^{-1}$ respectively for maintenance. These values are approximately double the respective BMR of 293 $kJ \cdot kg^{-0.75} \cdot day^{-1}$ for eutherians (Kleiber 1961) and 201 $kJ \cdot kg^{-0.75} \cdot day^{-1}$ for marsupials (Dawson and Hulbert 1970). Note that in both BMR and maintenance energy requirements, the kangaroos are 30 percent below the eutherian level; thus the basic phylogenetic difference in BMR is retained in the maintenance energy requirements of these herbivores. Similarly, in a series of studies in which adult kangaroos and merino sheep were fed the same diet, three species of kangaroos ate 26–27 percent less than the sheep (Hume 1982).

In free-ranging domestic stock and free-living wild animals, the total energy cost of free existence, or field metabolic rate (FMR), is usually a higher multiple of BMR, because it includes additional energy expenditures associated with foraging, predator avoidance, thermoregulation, territorial and other social behaviors, and sometimes growth and/or reproduction (Nagy 1987). FMRs are also more variable than captive maintenance energy requirements, being dependent on the relative intensity of each of the factors listed. For instance, the published FMRs of marsupials range from 1.2 times BMR in the sedentary koala *Phascolarctos cinereus* to 4.8 times BMR in the small, omnivorous Leadbeater's possum *Gymnobelideus leadbeateri* (Hume 1982). Karasov (1992) reported a range in the ratio of FMR to BMR of 1.3 to 5.3 for seventeen species of terrestrial eutherians.

The highest ratios reported to date are 6.9 in marmots (Salsbury and Armitage 1994) and 7.2 in lactating ground squirrels (Kenagy et al. 1990). Koteja (1991) found the correlation between FMR and BMR to be very high in eutherians (mainly rodents) but weak in birds.

Nagy (1987) compared the FMRs of thirty-six mammals and twenty-five birds with those of reptiles. The energy cost of free existence of 250-g endotherms was about seventeen times that of the ectothermic iguana of similar size. The difference is explained by two main factors. First is that endotherms have a threefold to sixfold greater capacity to process energy at the cellular level because of greater enzyme activities (Bennett 1972), mitochondrial densities, relative membrane surface areas, sodium transport (Else and Hulbert 1981), and thyroid activities (Hulbert and Else 1981). The second is the result of differences in metabolic responses to daily ambient temperature regimes. A lizard's body temperature declines along with ambient temperature at night, with a drop in energy metabolism (the Q_{10} effect). On the other hand, as the difference between the endotherm's body temperature and ambient temperature increases, metabolic heat production must increase to maintain a constant body temperature. Thus at night, the resting metabolic rate of the 250-g endotherm is about 200 times that of the 250-g lizard (Nagy 1987). Many small endotherms avoid such a huge metabolic drain by entering torpor when their body temperature falls toward some minimal level, in response to the low ambient temperature and/or food shortages.

By calculating the mean metabolizable energy contents of various natural food categories, Nagy (1987) generated equations for food requirements of free-living mammals, birds, and lizards. These equations indicated that, among herbivores, 250-g iguanas require only 6 percent of the dry matter per day that is required to fuel the FMR of eutherian mammals, and 15 percent of that required by marsupials. Thus the adaptive significance of differences in BMR and, more directly FMR, can be enormous. Amphibians and reptiles can survive much longer periods without food than birds and mammals can. On the other hand, the reproductive output of many small mammals can be much greater than that of reptiles of similar size.

These great differences in rates of metabolism raise the question of what is the maximal metabolic rate, fueled by food intake rather than by transient depletion of energy reserves, that can be sustained by animals. Peterson, Nagy, and Diamond (1990) addressed this question by comparing BMR with sustained metabolic rate of nineteen mammals, thirteen birds, and five reptiles (all lizards), using values of sustained metabolic rate from studies in which body mass changed by less than 1 percent per day. Most values for the ratio of sustained to basal metabolic rate (sustained metabolic scope) fell between 1.5 and 5; the highest was 6.9 for a small marsupial. There were no differences between ectotherms and endotherms. They concluded that metabolic rates higher than seven times basal values, which constituted a "metabolic ceiling," cannot be sustained in the long-term.

Peterson et al. (1990) aired two alternative hypotheses for the proximate physiological causes for metabolic ceilings. The first is that the capacities of clusters of peripheral systems such as lactation, exercise, and heat production, are coadjusted to operate at similar ceilings. This is the concept of *symmorphosis* advanced by Taylor and Weibel (1981) and disputed by Garland and Huey (1987). The second is that the ceiling is imposed by one central system, most likely the digestive system. The capacity of the intestine to absorb nutrients, or the liver to process absorbed nutrients, might set a limit on the supply of metabolites to the entire body, regardless of the purposes for which the metabolic energy is used (Hammond and Diamond 1994). Whatever the proximate cause, the existence of metabolic ceilings may well set limits to animals' reproductive output, foraging behavior, and geographic distribution (Peterson et al. 1990).

Water turnover and requirements

Water also limits animals' geographic distributions, being of central importance to all physiological processes. It has often been stated that the rate of water usage by an animal (water turnover) is related to its metabolic rate (Macfarlane et al. 1971), since the various avenues of water loss from the body are all indirectly influenced by the metabolic rate. Although fecal water loss is determined largely by electrolyte absorption in the colon, it also depends in part on the amount of food eaten, which in turn can be predicted from the animal's FMR (Nagy 1987). Urinary water loss is partly determined by glomerular filtration rate (GFR), which is lower in desert-adapted animals than in their more mesic counterparts. Insensible water loss via the skin is directly related to the animals' body temperature, and respiratory water loss is directly related to rates of oxygen consumption. Thus, rates of water turnover are much lower in reptiles than in mammals and birds (Nagy and Peterson 1988). Among mammals, water turnover rates are lower in macropod marsupials than in terrestrial herbivorous eutherians from similar environments, as would be expected on the basis of the lower BMRs of macropods (McNab 1978). Similarly, water turnover rates measured in free-living animals are higher than those measured under standard conditions (Hume 1982), as might be predicted from the previous discussion on FMR versus BMR.

Because of the many factors that can influence water loss, water turnover rates should be measured under standardized conditions for comparisons between species. Nicol (1978) suggested that the ambient temperature used for comparisons should be at the lower end of the thermoneutral zone because higher temperatures may increase water used for evaporative cooling, and lower temperatures raise metabolic rate and thus water turnover. Water must be available ad libitum since water deprivation reduces metabolic rate (Schmidt-Nielsen et al. 1967) and water turnover rate (Hulbert and Dawson 1974; Nagy, Shoemaker, and Costa 1976). Water turnover measured under these conditions can be described as the Standard Water Turnover Rate of the species (Nicol 1978).

Protein requirements

In contrast to energy and water turnover rates, requirements for protein or nitrogen are much less affected by additional costs of free existence, such as activity and thermoregulation (Hume 1982). Thus, estimates of the maintenance requirement of captive animals for nitrogen are likely to reflect the needs of nonreproductive adult animals in the wild. Only growth and reproduction impose significant increments on the total nitrogen required.

Because of the major contribution to BMR made by whole-body protein turnover (Milligan and McBride 1985; Reeds, Fuller, and Nicholson 1985), we should expect a direct relationship between BMR and the animal's protein requirement. Indeed, Smuts (1935) found that about 2 mg of endogenous urinary nitrogen (EUN)–the urinary nitrogen excretion on a protein-free diet–was excreted for each kcal of basal heat production. This is equivalent to 0.5 mg/kJ in eutherian mammals. Available data support the notion that animals with a low BMR have low maintenance nitrogen requirements. For example, the rock hyrax *Procavia habessinica*, which has a BMR significantly below the eutherian mean (Taylor and Sale 1969), also has an unusually low maintenance nitrogen requirement (Hume, Rübsamen, and Engelhardt 1980). Most marsupials also have low maintenance nitrogen requirements, in line with their low BMRs (Hume 1982). Most exceptions can be related to differences in nutritional habitat among species. White, Hume, and Nolan (1988) found that the lower maintenance nitrogen requirement of the tammar wallaby *Macropus eugenii*, compared with the parma wallaby *M. parma* and red-necked pademelon *Thylogale thetis*, correlated with lower rates of whole-body protein turnover and energy metabolism. The tammar regularly copes with the low nitrogen levels of vegetation in its semiarid habitat, but in their preferred moist forest habitat, the parma and pademelon rarely face such nutritional stresses.

Energy and protein requirements of the digestive system

The gastrointestinal tract is responsible for a disproportionately high fraction of whole-body protein turnover and energy utilization (McBride and Kelly 1990). Estimates of the gut's contribution to the total oxygen consumption of the animal range from 12 percent in rats to 25 percent in pigs. Rates of protein synthesis are particularly high in actively proliferating and secreting tissues. In ruminants, the gut wall constitutes 6 percent of the protein pool of the body but accounts for 28–46 percent of whole-body protein synthesis. In contrast, skeletal muscle, which constitutes up to 60 percent of the protein pool of the body, accounts for only 25 percent of whole-body protein synthesis (Reeds 1988). Thus the gastrointestinal tract is by far the most metabolically intense organ of the vertebrate body in terms of both protein synthesis and energy utilization, and is expensive to maintain. The sea cucumber (*Parastichopus*) sheds its gut in autumn and regains it in the spring. Evidently, the cost of maintaining the gut over the winter, when only poor quality food is available, exceeds the cost of its entire replacement (Self, Jumars and Mayer 1995).

Summary

This chapter introduces the idea that the nutritional niche of an animal can be defined by what it needs in the way of energy and nutrients and how it harvests and extracts what it needs from its nutritional environment. The animal's needs for nutrients, such as water and protein, are usually directly related to its basal metabolic rate (BMR). In turn, BMR is usually closely correlated with body mass, from unicellular organisms to the largest vertebrates. This allows us to compare the metabolic rates of animals of vastly different sizes.

Of more direct relevance to the total energy requirements of free-living animals is their field metabolic rate (FMR), or energy cost of free existence. FMR is not a constant multiple of BMR. The highest value for sustained metabolic scope (the ratio of the maximal rate of metabolism fueled by food intake, to basal values) appears to be about 7 in ectotherms and endotherms alike. This metabolic ceiling may be set by the rate at which the digestive system can process food and absorb nutrients, regardless of the purpose for which the nutrients are to be used in the body.

2

General characteristics of the vertebrate digestive system

Phylum Chordata, which includes animals with pharyngeal clefts and either a persistent notochord or a segmented spinal column, contains approximately 5 percent of species in the animal kingdom. The earliest members of this phylum were the protochordates, such as *Amphioxus* (Subphylum Cephalochordata) and the sea squirts (Subphylum Tunicata), from which the vertebrates are believed to have evolved approximately 500 million years ago. Our discussion will be limited to the Subphylum Vertebrata, which contains approximately 45,000 species. The major groups of vertebrates are listed in Table 2.1. These can be divided into two superclasses, the Agnatha (cyclostomes) and the Gnathostomata (vertebrates with jaws), which includes two classes of fish and four classes of tetrapods. The two classes of fish consist of Chondrichthyes, which have a cartilaginous skeleton, and Osteichthyes, which have a bony skeleton. The amphibians, reptiles, birds, and mammals make up the four remaining classes.

Many structural and functional characteristics of the digestive system are common to all vertebrates and are found in the digestive system of some invertebrates as well (see Chapter 11). Others have resulted from adaptations to the diet or environment either through divergence from a common or more primitive form or by convergence, the appearance of similar structures or functions in completely unrelated species. These similarities and variations will be discussed in greater detail in the following chapters. However, it is useful to first consider the general characteristics of the vertebrate digestive system and the major variations among classes.

Although all vertebrates have a digestive tract and accessory glands, various parts of this system are not necessarily homologous, analogous, or even present in all species. Therefore, broad comparisons can be best made under the listings of *headgut, foregut, midgut, pancreas and biliary system, hindgut.*

Table 2.1. *Classification of Vertebrates*

SUPERCLASS: Agnatha
    CLASS: Cephalaspidomorphi. Lampreys
    CLASS: Myxini. Hagfish
SUPERCLASS: Gnathostomata
  GRADE: Pisces
    CLASS: Chondrichthyes. Cartilagenous fishes
        SUBCLASS: Holocephali. Chimaeras (one order)
        SUBCLASS: Elasmobranchii. Sharks, skates, and rays (five orders)
    CLASS: Osteichthyes. Bony fishes
        SUBCLASS: Dipneusti. Lungfish (two orders)
        SUBCLASS: Crossopterygii. Lobe-finned fishes (one order)
        SUBCLASS: Brachiopterygii. Gar, sturgeon, paddlefish, and bowfin (one order)
        SUBCLASS: Actinopterygii. Ray-finned fishes (thirty-eight orders)
  GRADE: Tetrapoda
    CLASS: Amphibia
        ORDER: Gymnophiona (Apoda). Limbless, short-tailed amphibians
        ORDER: Caudata (Urodeles). Salamanders, newts, conger eels, and related species
        ORDER: Salientia (Anura). Frogs and toads
    CLASS: Reptilia
        ORDER: Crocodilia. Crocodiles, alligators, caimens, and gavials
        ORDER: Testudinata (Chelonia). Turtles, tortoises, and terrapins
        ORDER: Squamata
          SUBORDER: Ophidia (Serpentes). Snakes
          SUBORDER: Sphenodontia. Tuatara
          SUBORDER: Lacertilia (Sauria). Lizards
          SUBORDER: Amphisbaenia. Worm lizards
    CLASS: Aves (twenty-eight orders)
    CLASS: Mammalia
        SUBCLASS: Prototheria. Egg-laying mammals
          ORDER: Monotremata. Echidnas and platypus
        SUBCLASS: Theria. Mammals bearing their young live
          INFRACLASS: Metatheria. Pouched mammals
          ORDER: Marsupialia. Kangaroos, koala, etc.
          INFRACLASS: Eutheria. Placental mammals
          ORDER: Insectivora. Shrews, moles, etc.
          ORDER: Scandentia. Tree shrews
          ORDER: Macroscelidea. Elephant shrews
          ORDER: Dermoptera. Flying lemurs
          ORDER: Chiroptera. Bats
          ORDER: Primates. Lemurs, monkeys, apes, humans
          ORDER: Carnivora. Cats, dogs, weasels, bears, etc.
          ORDER: Hyracoidea. Hyraxes
          ORDER: Proboscidea. Elephants
          ORDER: Sirenia. Manatees, dugongs
          ORDER: Cetacea. Whales, porpoise, dolphins

Table 2.1. *(cont.)*

ORDER: Perissodactyla. Horses, tapirs, rhinoceros
ORDER: Artiodactyla. Cattle, sheep, camels, pigs, etc.
ORDER: Edentata. Sloths, armadillos, anteaters
ORDER: Tubulidentata. Aardvarks
ORDER: Lagomorpha. Hares, rabbits, pika
ORDER: Rodentia. Rats, mice, squirrels, etc.
ORDER: Pholidota. Pangolins

NOTE: Extinct groups of animals are excluded. Fish are classified according to Nelson (1984), amphibians according to Duellman and Trueb (1986), reptiles according to Evans (1986), and mammals according to Vaughan (1986). For classification of birds, see Storer (1971a).

Headgut

The headgut is the cranial portion of the digestive tract that includes the oral, or buccal cavity, and the throat, or pharynx. In fish and larval amphibians, it includes the gill cavity and is referred to as the *orobranchial cavity*. Aside from its respiratory functions in fish and amphibian larvae, the headgut serves principally for the capture and preparation of food. The headgut shows a wide range of divergence and convergence with the diet and feeding habits of animals. Articulated jaws are present in all vertebrates other than the cyclostomes (lampreys and hagfish). Lips, tongue, a beak, or teeth may be used for the prehension of food. The chelonians (turtles, tortoises, and terrapins) and birds use their beaks, and most other vertebrates use their teeth, to grasp, position, puncture, sort, tear and/or triturate food.

The tongue of most vertebrates is attached to the floor of the mouth throughout most of its length. It aids in the positioning of food for mastication and deglutition (swallowing) of food. However, it may be used by some species of fish, adult amphibians, birds, and mammals for other purposes such as the capture of prey or in reptiles as a sensing organ. The headgut of some fish, larval amphibians, flamingos, and whales contains elaborate filtering mechanisms that separate small aquatic plants and animals from water and larger particles. The gill flaps, mouth, or tongue provide the pump for these filters.

Lubrication of the oral cavity for the deglutition of food is provided by mucus-secreting cells in the mouth of fish and multicellular glands in more advanced vertebrates. Oral glands of reptiles, birds, and mammals can be very complex in structure and can perform additional functions as well. The glands of adult frogs and toads, swifts, woodpeckers, and mammalian anteaters secrete an adhesive material, which aids swifts in building their nests or, when applied to the tongue, aids the other species in capturing prey. Digestive enzymes are found in the oral glandular secretions of some adult amphibians, reptiles, birds, and mammals. Toxins and spreading agents are secreted by some species. The serous (watery) component of mammalian saliva is used as an aid to evaporative heat loss in the dog and cat and provides the large quantities of bicarbonate and phosphate required for the buffering of fermentation end products in the stomach of foregut-fermenting herbivores.

Mammals show a wide variation in their feeding apparatus in relation to diet (Cloudsley-Thompson 1972). Those that feed primarily on ants or termites tend to have weak jaws, simple teeth, and a tongue adapted for this purpose. Examples are seen in the orders Pholidota, Tubulidentata, Edentata, and Marsupialia. Baleen whales, which include the largest of all fossil and living animals, are filter feeders. The filter apparatus consists of two rows of horny baleen plate, less than 0.5 cm thick, which hang from each side of the upper jaws (Ridgeway 1972). These may be long and narrow (right whales) or much shorter (rorquals), and can number from 250 to 400 in various species. However, even the sei whale *Balaenoptera borealis*, which has a fine baleen strainer, will eat cuttlefish in areas where krill is less profuse, and some baleen whales feed on fish in part (rorqual and humpback) or almost exclusively (Brydes whale). Other carnivorous mammals tend to have well-developed incisors and canine teeth. However, the canine teeth of herbivores are often absent, and the incisors are usually adapted for cropping of vegetation. The upper incisors of most ruminants are absent, and the area is covered by a horny pad. The grinding surface of molars in herbivorous mammals tends to have a square and elaborate surface (Walker 1987).

One characteristic of mammals that is generally absent in other vertebrates is their ability to masticate, or chew, their food (Crompton and Parker 1978). The teeth of reptiles are used for the prehension, puncturing, and tearing of food, which is then swallowed whole or in relatively large pieces. New teeth erupt between earlier ones, as replacements throughout the life of the animal. Most mammals differ from reptiles and other vertebrates in having large premolars and molars in the upper and lower jaw. These grow opposite one another with uneven surfaces that fit together during occlusion. The presence of cheek muscles and a more muscular and mobile tongue provide the suction required for nursing and aid in the placement of food between the shearing, grinding surfaces of the premolar and molar teeth. Furthermore, the muscles attached to the mandible (lower jaw) form a complex sling, which allows horizontal as well as vertical movement. This provides either the lateral grinding action seen in most mammals or the anterior-posterior action seen in rodents and elephants.

Foregut

Esophagus

The esophagus transfers ingesta from the mouth to the stomach, or to the intestine of those species that lack a stomach. It also serves as a temporary site for food storage in some reptiles, and the esophagus of many birds includes a crop, which provides a major site for the storage of food. The esophagus is organized for controlled regurgitation in some vertebrates. Some birds regurgitate crop contents for the feeding of their young, and others regurgitate undigested parts of their prey from the gizzard. Regurgitation of forestomach digesta and gas is an important function of the ruminant esophagus.

The esophagus of vertebrates is generally lined with nonglandular stratified squamous epithelium, which contains goblet (mucus-secreting) cells in some species. However, it can be lined with columnar epithelium in marine and euryhaline fish, and glands that secrete pepsinogen and/or HCl have been reported in some fish and amphibians, and in the bat *Plecotus auritus* (Botha 1958). Aberrant areas of cardiac glandular, proper gastric glandular, ciliated epithelium, and intestinal-like columnar epithelium also have been reported in the esophagus of some humans (Burns et al. 1970).

Esophageal epithelium contains ciliated cells in some fish, adult amphibians, and reptiles, but, with the exception of the ciliated esophagus of embryonic fish and larval amphibians, ingesta are transferred through the esophagus principally or entirely by muscular activity. Generally, the esophagus is invested with an inner layer of circular muscle and an outer layer of longitudinal muscle. However, this arrangement can be reversed at the terminal esophagus of fish, and the longitudinal layer may be complete only near the gastroesophageal junction in adult amphibians and reptiles. Furthermore, the circular muscle is actually oblique in segments of the esophagus of some mammals. The circular muscle forms a well-defined sphincter at the termination of the esophagus of some fish, adult amphibians, and reptiles. A similar sphincter can be found in many mammalian species. However, in other mammals this area is delineated only by a constriction, with no anatomical sphincter. Esophageal muscle is striated in fish but is smooth in amphibians, reptiles, and birds. Mammals show considerable species variation in the presence and distribution of these two types of muscle. The remainder of the digestive tract consists of smooth muscle. The only apparent exceptions to this are stomachless fish, which may have a short segment of striated muscle in the upper intestine.

Stomach

With the exception of some fish and the larval toads, all vertebrates have a stomach or analogous organs. The stomach serves as a site for storage and maceration of food and for the trituration of food in most birds and some reptiles and mammals. With the exception of some amphibian larvae and a few mammalian species, the stomach secretes pepsinogen and HCl, which initiate protein digestion. The stomach of fish, amphibians, reptiles, and most mammals is a relatively simple tubular or asymmetric expansion of the digestive tract. However, these functions are carried out by the crop (storage), proventriculus (pepsinogen and HCl secretion) and ventriculus, or gizzard, (trituration) of birds, and a complex, voluminous stomach is found in some bats, whales, and a variety of mammalian herbivores.

Different regions of the human stomach have been characterized as the *cardia, body, fundus*, and *pylorus*. The cardiac region is the segment of stomach nearest the gastroesophageal junction, and thus the heart. The body is the distended portion of the stomach and the term fundus (bottom) refers to the outpocketed area. The pyloric region is the terminal segment, which often has a thicker layer of circular muscle and includes the pyloric valve. However, these terms are useless in describing the external characteristics of the more complex stomachs of birds and many mammals and show no correlation with the functional characteristics of the gastric epithelium in many species. Therefore, for comparative purposes, the regions of the stomach are best described by their epithelial lining of proper gastric, cardiac, pyloric glandular mucosa, and nonglandular stratified squamous epithelium.

The stomach of most vertebrates contains a region of proper gastric glandular mucosa. In this region, pepsinogen and HCl are secreted by the same glandular cell in fish, adult amphibians, reptiles, and birds but are secreted by separate cells in mammals. A cardiac glandular region is absent in fish but is present near the gastroesophageal junction in reptiles, some adult amphibians, and most mammals. A pyloric glandular region is described in fish, adult amphibians, reptiles, and mammals. Mucus is secreted by glandular and surface epithelium in all three of these regions. Although neither cardiac nor pyloric glandular regions are described in birds, Ziswiler and Farner (1972) found that the proventriculus contained centrally located glands that secrete only mucus and fluid. The avian gizzard also contains tubular glands that secrete a substance that provides the horny material lining this gastric region.

The gross structure and epithelial lining of the mammalian stomach show extremely wide variations among species and demonstrate some features that are not found in other vertebrates (Fig. 2.1). Starting at the gastroesophageal junction, the stomach of humans, dogs, and many other mammals shows the same successive regions of cardiac, proper gastric, and pyloric glandular mucosa that were noted

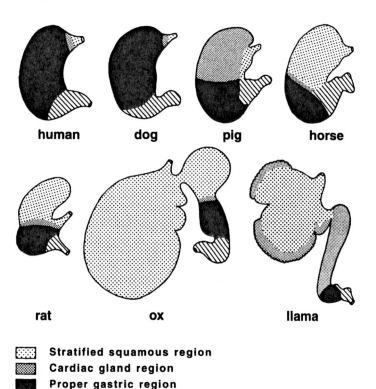

human   dog   pig   horse

rat   ox   llama

▢ Stratified squamous region
▨ Cardiac gland region
■ Proper gastric region
⧅ Pyloric gland region

Fig. 2.1. Some variations in the distribution of gastric mucosa. The stomach of each species demonstrates regions of cardiac, proper gastric, and pyloric glandular mucosa. The pig and llama stomachs contain a relatively large region of cardiac glandular mucosa, and the pig, horse, rat, llama, and cow also show increasing areas of stratified squamous, nonglandular epithelium. Stomachs are not represented on same scale, for example, the volume capacity of the adult bovine stomach is approximately seventy times that of the human stomach, or fourteen times the capacity per kg of body weight (modified from Stevens 1973).

in reptiles. However, the stomach of some mammals shows an additional region of nonglandular stratified squamous epithelium, which extends to varying degrees from the gastroesophageal junction. Furthermore, the cardiac glandular mucosa, which forms only a narrow band in the stomach of reptiles and most mammals, occupies large areas of the stomach of some mammals and forms islands within the region of stratified squamous epithelium in others. In addition to this, the stomach of many species is both enlarged and sacculated or compartmentalized.

The musculature of the vertebrate stomach generally consists of an inner layer of circular muscle, which forms a valve or sphincter at the junction with the intestine, and an outer layer of longitudinal muscle. A third oblique layer of muscle is present near the gastroesophageal junction of some mammals. The outer longitudinal layer is almost lacking from the gizzard of birds and is concentrated into a few bands (taeniae) on the stomach of macropod marsupials and some herbivorous primates, where it draws the stomach into a series of haustrations (nonpermanent sacculations). The arrangement of these muscle layers is extremely complex in the forestomach of ruminants.

Midgut

The vertebrate midgut is the major site for the digestion of carbohydrate, fat, and protein. It is also the major site for the absorption of the end products of this digestion as well as the vitamins and minerals required as nutrients. The luminal surface of the midgut is increased by the presence of microvilli (Fig. 2.2) on the apical border of absorptive cells and is further expanded by pockets, folds, or ridges. This includes pyloric ceca or a spiral valve in some species of fish (Harder 1975a), intricate folds and ridges in some reptiles (Parsons and Cameron 1977), and orad-facing pouches in beaked whales (Flower 1872).

In salamanders, birds, and mammals, the intestinal absorptive surface is increased by villi, macroscopic projections of epithelial and subepithelial tissue (Fig. 2.2). These contain capillaries and, in salamanders and mammals, lacteal vessels for the removal of absorbed fat. The villi of mammals are closely associated with tubular glands called the *crypts of Lieberkühn*, which contain a variety of endocrine secretory cells, mucus-secreting goblet cells, and undifferentiated cells. The latter undergo rapid mitosis and are the precursors of cells that migrate up the crypt and onto the villi as they develop the microvilli and other characteristics of mature absorptive cells. Both the goblet and absorptive cells migrate to the tip of the villus where they are sloughed into the lumen.

Although most fish do not have intestinal glands that extend into the submucosa, the Gadidae (Jacobshagen 1937) and Macrouridae (Geisterdoerfer 1973) have glands at the base of surface folds throughout the length of the intestine. They have been called crypts of Lieberkühn, but they contain no cell types different from those of the surface epithelium (Harder 1975a). Crypts of Lieberkühn have been described in the midgut of salamanders (Reeder 1964), some reptiles (Luppa 1977), and some birds (Ziswiler and Farner 1972). The crypts found in reptiles are less developed than those of birds and mammals, and their epithelium is similar to that of the surface. The crypts found in birds vary among species, from those that contain only absorptive and goblet cells to those that contain cells with basophilic granules.

Clearly defined zones of cell proliferation such as those at the base of the mam-

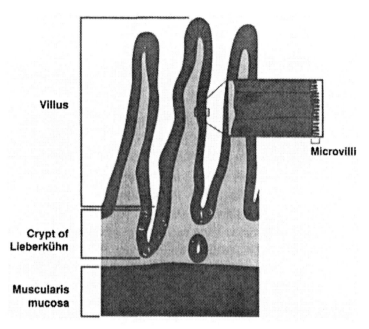

Villus

Microvilli

Crypt of
Lieberkühn

Muscularis
mucosa

Fig. 2.2. Intestinal villus and crypt. The crypts contain cells that secrete mucus and fluids into the intestine, endocrine cells that secrete hormones, and cells that migrate up the villi and become the absorptive-digestive cells. The inset shows an enlarged absorptive-digestive cell, with its microvilli, or brush border.

malian crypts were not seen in the midgut of larval and adult lampreys (Youson and Horbert 1982; Youson and Langille 1981), larval amphibians (Marshall and Dixon 1978), or the fresh water painted terrapin *Chrysemys scripta* (Wurth and Musacchia 1964). Zones of proliferation at the base of intestinal folds have been described in more advanced species of fish (Hyodo-Taguchi 1970; Gas and Noaillac-Depeyre 1974; Stroband and Debets 1978) and adult amphibians (Andrew 1963; Martin 1971; McAvoy and Dixon 1977).

The mammalian midgut, or small intestine, can be subdivided into the duodenum, jejunum, and ileum. The proximal small intestine of many mammals contains subepithelial glands called *Brunner's glands*, which are believed to secrete an alkaline fluid and mucus. They extend a short distance along the duodenum of carnivores, a longer distance in omnivores, and the greatest distance in herbivores. Brunner's glands are said to be absent in nonmammalian vertebrates (Andrew 1959), but Ziswiler and Farner (1972) noted glands of similar structure at the gastroduodenal junction of some birds.

The cyclostome intestine has a very thin musculature, consisting of an oblique muscle in lampreys, and digesta are mixed and transported with the aid of cilia.

However, in other vertebrates, these functions are carried out by inner circular and outer longitudinal layers of muscle.

Pancreas and biliary system

The pancreas and liver are embryonically derived from the midgut epithelium, and their contributions to digestion center on this segment of the tract. The exocrine pancreas secretes enzymes important to the digestion of carbohydrate, fat, and protein. The pancreas of cyclostomes appears to be represented in a primitive stage as ceca, which are closely associated with follicles that are believed to represent the islets of Langerhans of the endocrine pancreas (Barrington 1957). The form and distribution of pancreatic tissue vary in the more advanced classes of fish. It is diffusely distributed along the intestinal wall of many fish and may even extend into the liver. However, the pancreas is a compact organ in the elasmobranchs, lungfish, some Siluridae (freshwater catfish), and all higher classes of vertebrates. The avian pancreas is relatively large in insectivores, piscivores (fish eaters), and omnivores, but it is smaller in carnivores. It usually consists of three lobes with individual ducts–an arrangement that has proven useful in experiments designed to study partial pancreatic fibrosis (Bensadoun and Rothfeld 1972).

The vertebrate liver is a distinct, compact organ in all species. It serves a number of functions, but the one primarily associated with digestion is bile secretion. Bile is stored in the gallbladder in most vertebrates, which allows for its release when high concentrations of lipid are present in the midgut. The gallbladder is absent in some fish, and both the gallbladder and hepatic ducts disappear in lampreys after completion of metamorphosis to the adult feeding stage. A gallbladder also is absent in horses, deer, seals, rats and a few other rodent species.

Hindgut

The terms *midgut* and *hindgut* are preferable to those of *small intestine* and *large intestine* because these two gut segments are of similar diameter in many vertebrates. Although different regions have different embryological origins, the term *hindgut* is generally used to define the entire large intestine of mammals. The hindgut serves as the final site for storage of digesta and retrieval of dietary or endogenous electrolytes and water. It is also the principal site of microbial fermentation in herbivorous reptiles and most herbivorous birds and mammals.

The musculature of the hindgut is similar to that of the midgut in most species. It is thin in cyclostomes and has a double layer of circular and longitudinal muscle in other vertebrates. The circular layer is heavier than that of the midgut in some species. In many mammals, the longitudinal muscle of the cecum and varying lengths of the colon form taeniae similar to those described for the stomach of some mammalian species. These bands serve the similar purpose of drawing the included segment into haustra, which can help delay the rate of digesta passage. The colon is further divided into permanent compartments in a few mammalian herbivores.

The hindgut is difficult to distinguish from the midgut in most fish and amphibian larvae with respect to epithelial morphology, change in diameter, or the presence of a sphincter or valve. The hindgut of adult amphibians can usually be distinguished by its larger diameter, but valvular separation of the midgut and hindgut seems to be limited to frogs. The reptilian hindgut is separated from the midgut by a valve or sphincter and is enlarged in many species. The initial segment contains a cecum in a few species, and this, plus a segment of the proximal colon, are partially compartmentalized by mucosal folds in some herbivores (Iverson 1980). The avian hindgut is separated from the midgut by a sphincter or valve. It generally consists of paired ceca and a relatively short, straight, enlarged extension of the intestine, which is often called the *rectum* (because it is usually straight).

The hindgut of mammals shows a wide range of structural variation and tends to be longer than that of other vertebrates. In some species it lacks a cecum or even a valvular or other distinction that separates it from the midgut. However, most mammals have a distinct hindgut, which consists of a colon, rectum, and often a cecum (which is paired in a few species) at its valvular junction with the small intestine. Some herbivores have an extremely large cecum, and others have an extremely large proximal colon. The mammalian hindgut is lined with crypts and a surface epithelium containing goblet cells and columnar cells with microvilli, but villi are absent. In some species, the cecum and varying lengths of the colon are haustrated due to the presence of longitudinal bands of muscle (taeniae). The hindgut of adult amphibians, reptiles, and birds, and some mammals terminates, along with the renal and reproductive systems, in a cloaca.

As with the stomach, comparative descriptions of the gross structure of the large intestine have suffered from attempts to apply terminology used in human anatomy. The human large intestine can be divided into a cecum, colon, and rectum. As the colon leaves its junction with the small intestine and cecum, it can be further subdivided into ascending, transverse, and descending segments, according to the direction it takes in the abdominal cavity. The transverse colon can be compared with similar segments in other species, on the basis of its mesenteric attachment and the loop that it and the duodenum form around the cranial mesenteric artery. However, the "ascending" colon of other species can vary considerably in its length, volume, and the course it takes in the abdominal cavity (Fig. 2.3). Furthermore, as can be seen in Chapters 5 and 8, the functional divisions of the colon show little or no correlation with these anatomical divisions.

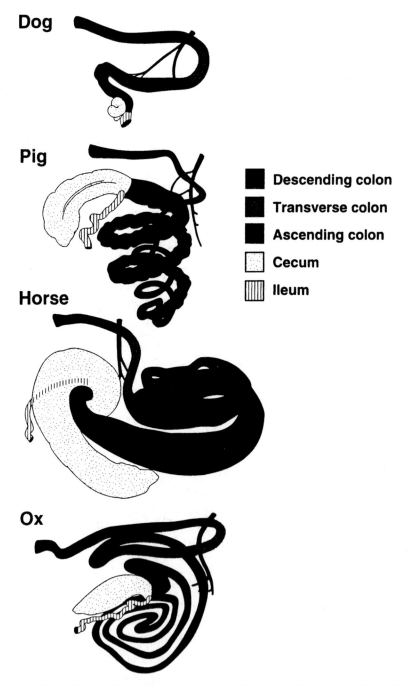

Fig. 2.3. The hindgut, or large intestine, of the dog, pig, horse, and ox. Note that both the cecum and the segments homologous to the ascending, transverse, and descending colon of humans vary in their relative length, shape, and volume. (Modified from de Lahunta and Habel 1986).

Cloaca

Ducts from the urinary and genital tracts of fish usually exit from the body at a point separate from that of the digestive tract. However, those three tracts terminate in a cloaca in embryos of most species and in adults of some teleosts, the lungfish, the Elasmobranchii, the crossopterygian *Latimeria*, and the hagfish. A cloaca also is present in adult amphibians, reptiles, birds, and some species of mammals.

The reptilian cloaca consists of a coprodeum, urodeum, and proctodeum. The first two compartments are lined with an epithelium similar to that of the rest of the hindgut, and the proctodeum is lined with stratified squamous epithelium. The urinary and reproductive tracts enter the urodeum. The cloaca of some turtles is capable of acting as a pump, and bilateral cloacal bursae aid in the adjustments for buoyancy (Jackson 1971). These turtles also can regulate their resting lung volume, and changes in lung volume are accompanied by an equal, but opposite, uptake or release of water by the cloacal bursae.

The avian cloaca consists of three chambers similar to those of reptiles. The coprodeum is often represented only as an enlargement of the intestinal lumen. It is usually partially separated from the urodeum by a septum and receives the ducts from the ovaries or testes, and the kidney. The urodeum is less distinctly separated from the proctodeum, which terminates at the anus. Stratified squamous epithelium is limited to the caudal compartments in young birds. The dorsal wall of the cloaca also contains the cloacal bursa, a relatively solid lymphoid organ that has provided significant contributions to our present understanding of immunology.

The hindgut terminates in a cloaca in fetal mammals, and adult montremes, marsupials, some Insectivora, and at least one species of rodent (the mountain beaver, *Aplodontia rufa*) retain the cloaca. However, the outlets for the renal and digestive systems eventually separate prior to the birth of most mammals.

Summary

Although the digestive system of all vertebrates share a number of common structural and functional characteristics, it also shows major differences both between and within vertebrate classes. The major structural variations are discussed in Chapters 3 and 4.

3

The digestive system of fish, amphibians, reptiles, and birds

The digestive tract of fish, amphibians, reptiles, and birds shows a considerable degree of structural variation both between and within the different vertebrate classes. Some of these variations can be attributed to adaptations to the diet or feeding strategy. Others can be attributed to the habitat, such as aquatic versus terrestrial, or other physiological characteristics, such as the ability to control body temperature. For general references, we suggest Andrew (1959), Reeder (1964), Sturkie (1970), Ziswiler and Farner (1972), Harder (1975a,b), Luppa (1977), and Skoczylas (1978).

Although there are many excellent illustrations of the gastrointestinal tract of vertebrates, it is difficult to obtain photographs or drawings that depict the entire tract in a manner that allows comparison of dimensions and gross structural characteristics. Therefore, the series of illustrations of the amphibian, reptilian, avian, and mammalian gastrointestinal tract in this and the following chapter were prepared to allow comparison under similar conditions with special attention to details of the stomach and hindgut structure. Because the shape and dimensions of the gut can be affected to a large degree by the time after feeding and by postmortem changes, wherever possible, the gastrointestinal tract was collected from animals that had been feeding and immediately after death. Arrangement of the intestine in the pattern chosen removes some anatomical characteristics such as duodenal loops or spirals of the colon. However, these structural characteristics generally show little correlation with function.

Fish

The four classes of fish are the Cephalaspidomorphi, Myxini, Chondrichthyes, and Osteichthyes (see Table 2-1). The first two classes contain the lowest craniate vertebrates, the cyclostomes (lamprey and hagfish). Cartilaginous fish can be classified into two subclasses: the Elasmobranchii, which includes sharks, skates, and rays, and the Holocephali (*Chimaera* sp.)–fish with a blunt snout and thread-like tail. The Osteichthyes (teleosts, or bony fish) can be subdivided into four sub-classes: Crossopterygii (lobe-finned fish), Dipneusti (lungfish), Brachiopterygii (gar, sturgeon, paddlefish, and bowfin), and Actinopterygii. The Dipneusti have both gills and lungs with a pulmonary circulation. The Actinopterygii comprise a large number of the present day fishes. Family Cyprinidae is of special interest because its members are distributed throughout the world (1,000 sp.), and it con-tains the majority of the freshwater fish of North America including the carp, min-nows, chubs, and barbs. The anatomy of the fish digestive system has been re-viewed by Andrew (1959); Harder (1975a,b); Kapoor, Smit, and Verighina (1975); Kapoor and Khawna (1993); and Zihler (1982). The physiology of the digestive system is reviewed in its various aspects by Barrington (1957), Campbell and Burnstock (1968), Barnard and Prosser (1973), Hoar (1983), and Vonk and West-ern (1984).

Fish include species that are carnivores, omnivores, and herbivores. They may feed on dead or living material and may be microphagous or macrophagous. Some species can adapt their diet to almost any of these forms or, they may, as Steven (1930) stated, ". . . eat what they can get." For example, the roach *Rutilus rutilus* can be a carnivore or an herbivore, according to necessity, and some fish can vary their diet from plankton in the summer to fish in the winter, even utilizing bac-teria and algae as a source of food (Barrington 1957).

The cyclostomes have no jaws. Species belonging to more advanced classes have movable jaws, but muscles are absent in their lips and cheeks. The tongue usually lacks muscle and shows little mobility. Exceptions include the archerfish *Toxotes jaculator*, which uses a muscular tongue to squirt streams of water at in-sect prey, and several other species that have a mobile, tooth-bearing tongue. The orobranchial cavity of fish shows a wide variety of arrangements for the capture and preparation of food. Teeth, used for capture, tearing, or crushing are present in most species. These may consist of sharp teeth for the capture of prey (sharks and piranhas) or grazing on algae attached to rocks or coral (Scaridae). Teeth may be located on the jaws, tongue, pharynx, or almost any surface of the orobranchial cavity, but mandibular or pharyngeal teeth are the most common types.

Many herbivorous fish scrape algae off coral reefs and rocks with a row of small, closely-spaced mandibular teeth. In some fish, such as the Scaridae (par-rotfish) and Odacidae (butterfish), the mandibular teeth are fused into a parrot-like beak (Clements and Bellwood 1988). However, the grinding, or trituration, of food is generally limited to the pharyngeal teeth of those fish that have them. Pharyngeal teeth are situated on the modified fifth gill arch, which does not carry gills. In cyprinids, the pharyngeal teeth are used to break up hard food by cutting in some species and crushing in others. In some species, the pharyngeal apparatus has been modified to form a second set of functional jaws (Liem and Greenwood 1981). In odacids, the pharyngeal apparatus consists of opposable, ridged bones, which can chop ingested algae into small pieces (Clements and Bellwood 1988). In scarids, the pharyngeal teeth act as a grinding mill, and ingested algae are triturated into a fine paste with the aid of ingested inorganic material (Bellwood and Choat 1990).

Because continuous flow of water through the mouth and gills is required for respiratory and osmoregulatory functions, escape of food past the gills may be prevented by gill rakers, which are projections that serve as a sieve. These may be spaced quite far apart, for fish that feed on large prey, or they may form a fine meshwork for microphagous feeders such as the menhaden, shad, or basking shark *Cetorhinus maximus*. A number of microphagous species such as the parrotfish *Scarus radicans* have pharyngeal pockets, which collect algae during grazing.

The esophagus of fish is generally short, wide, and straight. Ducts to the swim bladder, a hydrostatic organ for the adjustment of specific weight, open into the esophagus of many species. The esophagus of freshwater fish is lined with a multilayer of squamous epithelium containing a large number of mucous cells. However the esophagus of some marine teleosts is lined with complex, highly vascularized mucosal folds of columnar epithelium and a few mucosal cells (Al-Hussaini 1946, 1947). Yamamoto and Hirano (1978) suggested that this may be a common feature in marine teleosts, associated with the osmotic regulatory function of their esophagus (see Chapter 9). The distal esophagus of some species contains glands that secrete pepsinogen and HCl (Kapoor and Khawna 1993). Ciliated epithelium is found in the esophagus of embryonic fish and adult cyclostomes, perch, and some elasmobranchs. The esophagus of teleosts terminates in a sphincter, which may serve to prevent excessive swallowing of freshwater.

The stomach is absent in cyclostomes (Fig. 3.1) and a number of other species belonging to more advanced orders and families. Some of these fish are microphagous, with well-developed pharyngeal teeth for trituration of food (Fig. 3.2). When present, the general form of the stomach can be classified as *straight*, *siphon (U-shaped)*, or *Y-shaped* with a gastric cecum. The straight stomach, as seen in pike (Fig. 3.3), is rare. The siphon-shaped stomach, depicted in the sturgeon (Fig. 3.4) and trout (Fig. 3.5), is common to Elasmobranchii and Osteichthyes. The Y-shaped stomach, such as that of the eel (Fig. 3.6), has a blind sac on its greater curvature. This blind sac, which is well-developed in many macrophagous predatory fishes, is believed to be an adaptation for the storage of very large food particles (typically other fish) in food-limited, midwater environments (Wassersug and Johnson 1976). The circular muscle of the stomach can be well developed, as in mullet (*Mugil*), menhaden (*Brevoortia*), and shad (*Dorosoma*), where this, plus a tough epithelial lining, provide a gizzardlike action (Horn

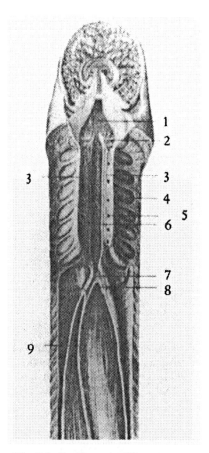

Fig. 3.1. Sea lamprey (*Petromyzon marinus*). *1*, Prebranchial foregut, buccal (pharyngeal) cavity; *2*, velum, valve between buccal cavity and branchial foregut; *3*, branchial foregut; *4*, gill-sac; *5*, ductus branchialis internus; *6*, epibranchial foregut; *7*, metabranchial foregut; *8*, anterior end of midgut; *9*, spiral valve, typhlosolis. (After Harder 1975b; from Pernkopf 1930.)

1992). A valve, or sphincter, of circular muscle, a mucous membrane fold, or both are present at the junction of the stomach and intestine.

The intestinal tract of fish can vary from one that is relatively short and straight to one that is long and arranged in spirals and loops. The length does not necessarily correlate with feeding habits, but it tends to be longest in herbivorous species and especially in fish that ingest large amounts of indigestible mud or plant material (Horn 1989). The mucosal surface of the intestine is increased by folds of various design, and by a spiral valve in species belonging to all groups except the hagfish and the teleosts. The spiral valve is present only in fish with a short, straight intestine and usually those in which the stomach is small or absent. It

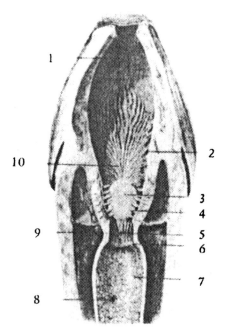

Fig. 3.2. Chub (*Leuciscus cephalus*). Note the absence of a stomach in this cyprinid fish. *1*, Buccal cavity; *2*, gill slits and arches; *3*, masticating plate; *4*, pharyngeal teeth (*3* and *4* form the chewing apparatus of the branchial foregut); *5*, ductus pneumaticus; *6*, pylorus; *7*, midgut; *8*, opening of ductus choledochus; *9*, metabranchial foregut; *10*, branchial foregut. (Harder 1975b.)

consists of folds of mucosa and submucosa, which project into the intestinal lumen and vary in their number of turns (Fig. 3.7). In some species, these leave a space free for direct passage of digesta. In others, they come in contact at the center, forming a drill-bit arrangement, or fold over in the center to form a conelike structure. This remarkable structure, which is unique among vertebrates, serves to both delay the passage of digesta and increase the absorptive surface.

Another arrangement of the upper intestinal tract that is unique among vertebrates, but found in a wide range of fish that have a stomach, is the presence of pyloric ceca. These can vary in size and shape from small evaginations of the intestinal wall to tubular, branched, or tuftlike structures (Figs. 3.8 and 3.9). Their number can vary among species from 1 to 1,000. In alepisauroid fishes, which are large, open-ocean predators such as tuna, a single pyloric cecum extends forward from the ventral-anterior margin of the intestine, beneath the heart. In one genus, *Coccorella*, it extends into the head (Wassersug and Johnson 1976). The pyloric ceca are lined with cells similar to those of the intestinal wall, and their presence does not seem to correlate with the length of the intestinal tract or the diet. However, they appear late in ontogenic development and could act both to

delay passage and increase the absorptive surface. Therefore, they have been suggested as a site for microbial digestion. However, Buddington and Diamond (1987)
found that the ceca of four species of teleosts absorbed glucose and amino acids
at rates similar to those of the rest of the proximal intestine, and concluded that
they served principally to increase the absorptive area. The pyloric ceca may also
secrete fluids that buffer the intestinal contents (Montgomery and Pollak 1988).

The hindgut of most fish cannot be distinguished from the midgut by the relative diameter or changes in epithelium. Lungfish, and at least some Crossopterygii, lack a valvular separation between midgut and hindgut, but an ileorectal valve
is present in many teleosts. A small cecum has been described in catfish (*Bagarius bagarius*), knifefish (*Notopterus notopterus*), and cod (*Raniceps raniceps*).

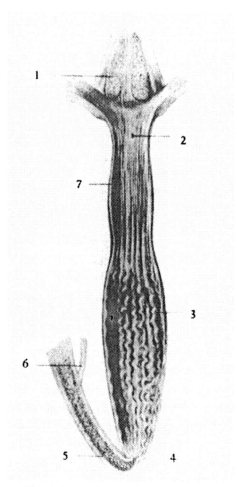

Fig. 3.3. Northern pike (*Esox lucius*). *1*, Branchial foregut; *2*, ductus pneumaticus; *3*, stomach; *4*, pylorus; *5*, midgut; *6*, ductus choledochus; *7*, esophagus (From
Harder 1975b.)

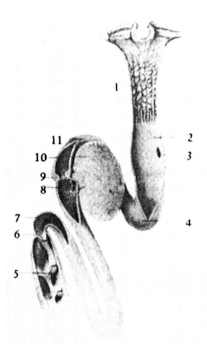

Fig. 3.4. Sturgeon (*Acipenser sturio*). Foregut, intermediate gut, and anterior portion of spiral valve. *1*, Papillae of esophagus; *2*, stomach; *3*, opening of ductus pneumaticus; *4*, stomach bend; *5*, spiral valve; *6*, valve between intermediate gut and spiral valve; *7*, intermediate gut, without spiral valve; *8*, opening of pyloric appendages; *9*, pylorus; *10*, appendices pyloricae; *11*, pars pylorica of stomach. (From Harder 1975b.)

Some herbivores, such as the sea chub (Fig. 3.10, *D*), have paired ceca that may be separated from the rest of the intestine by a sphincter (Rimmer and Wiebe 1987). An annulospiral septum consisting of circular muscle and glandular epithelium is found in the hindgut of the trout *Salmo facia*, and an ileorectal valve is present in some teleosts. However, the hindgut muscle is seldom increased in thickness.

The hindgut epithelium is usually similar to that of the adjacent midgut, except for an increase in goblet cells and sometimes cilia. The terminal gut of the cyprinids, a family that includes carp and minnows, has no distinguishing characteristics, other than an increase in zymogen-secreting cells (Al-Hussaini 1949a,b). However, the terminal intestine of elasmobranchs is lined with stratified squamous epithelium, and studies of the scup *Stenotomus chrysops* indicated that cells in the posterior midgut have special phagocytic capabilities (Strauss and Ito 1969). Barrington (1957) also cited the work of Young and Fox, showing that the cells of the hindgut of surfperch (Embiotocidae) can concentrate and possibly excrete pigments contained in their diet of shrimp. Therefore, it appears that the hindgut has special excretory, absorptive, and motor functions in some species of fish.

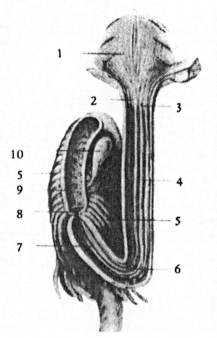

Fig. 3.5. Trout (*Salmo fario*). *1*, Branchial foregut; *2*, esophagus; *3*, ductus pneumaticus; *4*, stomach; *5*, appendices pyloricae; *6*, stomach bend; *7*, pars pylorica of stomach; *8*, pylorus; *9*, anterior end of midgut; *10*, gallbladder. (From Harder 1975b.)

Horn (1989) classified the digestive tracts of marine herbivorous fish into four types (Fig. 3.10), based on adaptations designed to degrade cell walls of algae and/or bacteria by acid lysis, mechanical trituration, or microbial fermentation. Type I, which also includes freshwater cyclids, consists of fish with a thin-walled stomach and relatively long intestine, but no mechanisms for the trituration of food. Type II, which includes the mullets, features a thick-walled, gizzardlike segment of the stomach and an intestine of variable length. Type III, which includes the scarids, odacids, and freshwater grass carp, have no stomach and an intestine shorter than that of types I and II. However, their pharynx contains teeth (see Fig. 3.2), which grind food to a small particle size. Type IV, which is seen in the sea chubs, has a long intestine that includes a distinct hindgut.

Amphibians

As implied by its name, class Amphibia represents the transition of vertebrates from water to land. Most amphibians begin life as free-living, aquatic larvae, which transform into terrestrial adults. The degree of transformation may differ among the three orders: Gymnophiona (wormlike burrowing amphibians), Caudata (salamanders, newts, and conger eels), and Salientia (frogs and toads). The

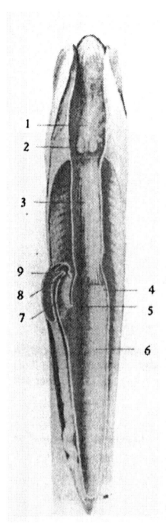

Fig. 3.6. Eel (*Anguilla anguilla*). *1*, Branchial foregut; *2*, horny pads; *3*, esophagus; *4*, stomach, showing the circular muscles; *5*, pars angularis; *6*, gastric cecum with independent circular muscles; *7*, pars pylorica of stomach; *8*, first pyloric valve; *9*, second pyloric valve. (After Pernkopf 1930; from Pernkopf and Lehner 1937.)

Salientia are the most highly specialized and undergo the greatest metamorphosis. General information on the digestive system of amphibians is provided by Reeder (1964).

Amphibian larvae include carnivores, omnivores, and herbivores. Many feed on plankton, bacteria, or detritus. Others graze on plants. Salientian larvae have a horny beak and labial teeth. Water is pumped by the muscular activity of the mouth through an elaborate pharyngeal filtering system, gill slits, and spiracle

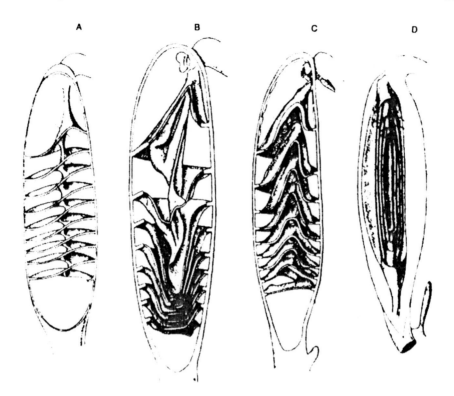

Fig. 3.7. Different types of spiral valves in Elasmobranchii. **A,** Spiral valve with medial column; **B,** with caudad directed funnels; **C,** with craniad directed funnels; **D,** with cylindrically wound-up mucosal fold, scroll-type valve (*Sphyrna*). Dorsal wall of gut removed; intermediate intestine at upper right. (After Parker 1885; from Grassé 1955.)

(breathing tube). Food particles are trapped in mucus, covering a set of filter plates, and transferred by cilia to and through the esophagus. A stomach serves for proteolytic digestion in salamander larvae but provides only a storage function in the larvae of frogs and is absent in toad larvae. The intestine of amphibian larvae is longer than the body and is often coiled with no distinct division into the midgut and hindgut.

Metamorphosis of amphibian larvae into adults is associated with major changes in structure and function of the digestive system over a relatively short time. There are marked changes in the buccal cavity and feeding apparatus. Glands develop in the stomach and in the esophagus of some species (Janes 1934). Intestinal epithelium degenerates and is replaced from subepithelial nests of cells. This process is associated with an increase in proteolytic enzyme activity, which is believed to aid in the breakdown of cells, followed by a disappearance and then a reappear-

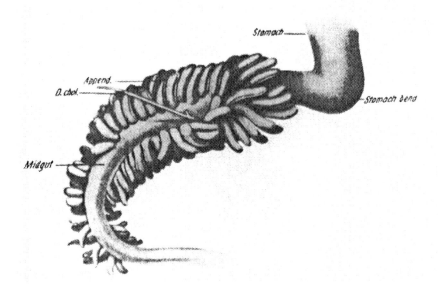

Fig. 3.8. Stomach of a salmonid (*Hucho*) with about 200 appendices pyloricae. *Append.*, Appendices pyloricae; *D. chol.*, ductus choledochus. (After Pernkopf 1930; from Pernkopf and Lehner 1937.)

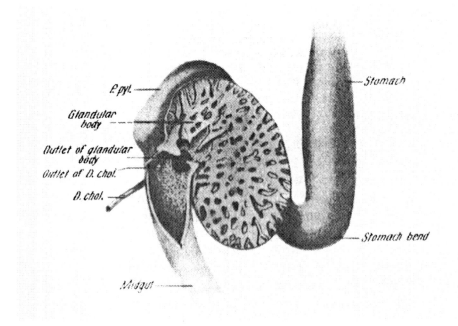

Fig. 3.9. Stomach of the sturgeon (*Acipenser sturio*) with homogenous glandular body of appendices pyloricae, cut open. *P. pyl.*, Pars pylorica; *D. chol.*, ductus choledochus. (After Pernkopf 1930; from Pernkopf and Lehner 1937.)

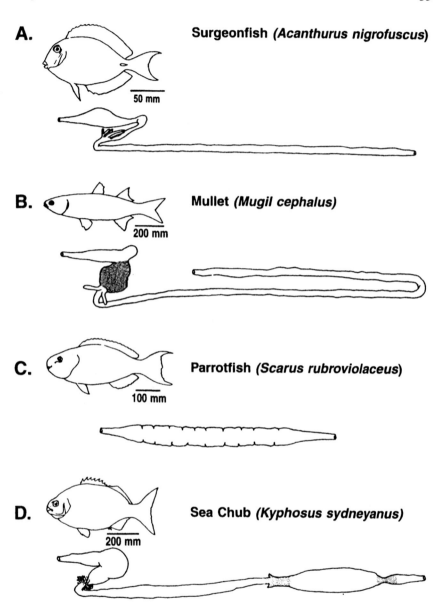

A. Surgeonfish *(Acanthurus nigrofuscus)*

50 mm

B. Mullet *(Mugil cephalus)*

200 mm

C. Parrotfish *(Scarus rubroviolaceus)*

100 mm

D. Sea Chub *(Kyphosus sydneyanus)*

200 mm

Fig. 3.10. Gastrointestinal tracts of four species of marine herbivorous fish. The surgeonfish **(A)** is a browser with pyloric ceca and a relatively long intestine. The mullet **(B)** is a grazer with a thick-walled, gizzardlike stomach, *(shaded area)* and pyloric ceca. The parrotfish **(C)** is a grazer with pharyngeal teeth, no stomach or pyloric ceca, and an intestine that is relatively short but large in diameter. The sea chub **(D)** is a browser with pyloric ceca and a distinct hindgut that contains sphincters or valves at its junction with the midgut and at a more distal location *(shaded areas)*, and small paired ceca. (Modified from Horn 1989).

ance of this activity. The terminal segment of the intestine develops into the enlarged hindgut in many species.

All adult amphibians are carnivores that swallow their prey whole (Erdman and Cundall 1984; Larsen 1984). Dentition tends to be weak and used only for grasping and positioning of prey, although caecilians (tropical burrowing amphibians) have sharp and relatively long teeth that are directed toward the rear of the mouth. The tongue is generally well developed and acts in frogs and toads as an organ for seizure of prey. Multicellular glands in the mouth produce mucus, which aids in deglutition and provides an adhesive surface to the tongue. These glands appear to produce no enzymes except for weak amylolytic activity. The mouth and esophagus are usually lined with ciliated and mucus-secreting cells. The esophagus is relatively short but wide. Esophageal glands that secrete pepsinogen have been described in adult frogs and toads (Reeder 1964), and those of the red-legged pan frog *Kassina maculata* are said to have secreted more pepsinogen than the stomach (Hirji 1982).

The stomach of adult amphibians tends to be tubular (Fig. 3.11). The stomach of many amphibians shuts down its secretory activity during hibernation. That of the gastric brooding frog ceases both its secretory and motor activity and serves as a "uterus" for the hatching and development of larvae until they are juvenile frogs (Fanning, Tyler, and Shearman 1982). The intestine is relatively short in comparison to that of amphibian larvae. However, the hindgut is generally differentiated from the midgut by its larger diameter. They are separated by a valve in the frog, but there appears to be no valve or sphincter at this point in most amphibians. The hindgut is lined with a columnar epithelium containing goblet cells. Although the amphibian skin, kidney, and bladder have been carefully examined for their contributions to water absorption, the contribution of the large intestine has received less attention. This may be due to the fact that amphibians do not usually drink water, and it is assumed that much of the electrolyte–water balance is effected by the skin, kidney, and urinary bladder.

Reptiles

The reptiles are air breathing, scaled, ectothermic vertebrates, which first appeared in the Paleozoic Era and reached their greatest height (in both senses) in the Mesozoic Era. If the insect invasion of land was the most spectacular in numbers, the reptilian dinosaurs established the record for size of individual terrestrial animals, but few large species of reptiles remain. Class Reptilia (see Table 2.1, pp. 12–13) contains three orders: Crocodilia, Testudinata, and Squamata. Crocodilia includes the crocodiles, alligators, caimans, and gavials. The Testudinata, often referred to as *chelonians*, can be subdivided according to habitat into turtles (marine), ter-

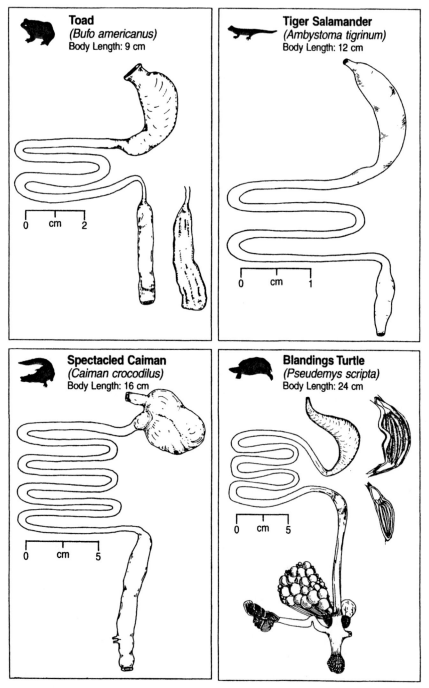

Fig. 3.11. Gastrointestinal tracts of a toad, tiger-salamander, spectacled caiman, and Blandings turtle. The illustration of the turtle gut shows sagittal sections of the stomach and midgut–hindgut junction, as well as the joining of the digestive, reproductive, and urinary tracts at the cloaca.

rapins (freshwater), and tortoises (terrestrial) species. The Squamata consists of the snakes, lizards, worm lizards, and tuatara, a primitive New Zealand reptile. The anatomy of the reptilian digestive system has been reviewed by Parsons and Cameron (1977), Luppa (1977), and Ottaviani and Tazzi (1977). Skoczylas (1978) reviewed information on the physiology of this system.

Most reptiles are either carnivores or omnivores, and many species subsist on insects during their early or entire life. Although many dinosaurs were herbivores, only about forty of the present 2,500 species of lizards are herbivorous (Pough 1973). Some tortoises and the marine green sea turtle are also herbivorous. With the exception of the chelonians, which obtain their food by means of a horny beak, all reptiles have teeth. These are generally used for grasping and tearing, and they are continuously replaced and often added to as the jaw lengthens. The jaws of some snakes are arranged for distension and even disarticulation to allow ingestion of prey, and the fang teeth of some snakes are used to inject venom or digestive enzymes. The teeth of mollusk-eating lizards are modified for crushing (Edmund 1969), and herbivorous lizards, such as the iguanids, have cusplike teeth (Hotton 1955). However, the upper and lower jaws of reptiles are of equal width, and their articulation allows only a scissorslike closure, which is unsuitable for the grinding of food.

The oral cavity of reptiles contains ciliated and mucus-secreting cells. A number of species have a distensible tongue, similar to that of amphibians, but which serves as a sensory organ. As the first group of vertebrates to generally adopt completely terrestrial living and subsistence on dry food, many species demonstrate complex oral glands. Salivary glands are usually absent and, when present, secrete only mucus. However, complex glands that secrete venoms and digestive enzymes are found in snakes and lizards (Elliot 1978; Kochva 1978).

The esophageal epithelium is often ciliated and, like that of amphibians, contains goblet cells, which increase in number along its length, to the point of almost covering the terminal mucosa in alligators and snakes. Esophageal glands may be present, and pepsinogen-secreting glands have been reported in some species. These are most frequently seen in chelonians, and in some species they appear to secrete pepsinogen (Luppa 1977). The reptilian esophagus can be very distensible, serving as a storage area during gastric digestion of large prey. In some egg-eating turtles and snakes, the ventral surface of the vertebral column is attached to the dorsal wall of the esophagus and provides a surface for the crushing of egg shells.

The gastrointestinal tracts of six reptiles are illustrated in Figs. 3.11 and 3.12. The stomach of reptiles tends to be tubular, but the stomach of Crocodilia is outpocketed, with a very muscular pylorus (Fig. 3.11). The pylorus of the alligator stomach is separated from the remainder of the stomach by a constriction and

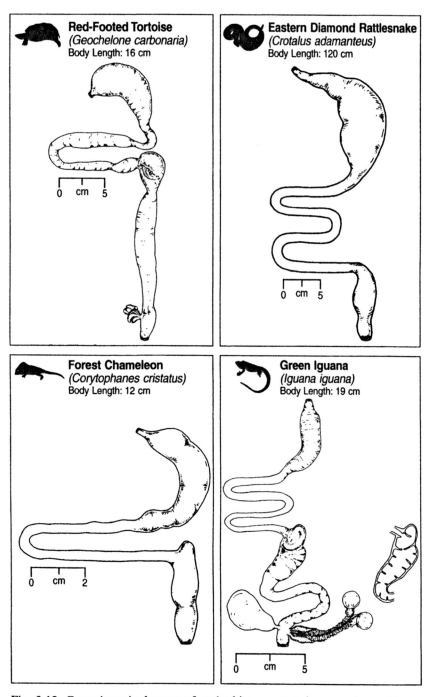

Fig. 3.12. Gastrointestinal tracts of an herbivorous tortoise, a snake, and insectivorous and herbivorous lizards. A sagittal section of the proximal hindgut of the iguana shows the mucosal folds extending into the cecum and proximal colon.

from the intestine by semilunar valves. Pyloric muscle also is extremely well developed in the boa constrictor and Florida indigo snake (Blain and Campbell 1942). Gastroliths (stones, gravel, or sand) have been reported in the stomach of crocodilians, chelonians, and both insectivorous and herbivorous lizards. These appear to be the most prevalent and most studied in the crocodilians. In *Crocodylus nyloticus*, Corbet (1960) reported a 73 percent incidence of gastroliths in individuals up to 1 m in body length, 88 percent in those between 1 and 2 m in length, and 100 percent in larger specimens. It has been suggested that they may serve as ballast, but their presence suggests a triturating function similar to that seen in the gizzard of birds.

The midgut of reptiles tends to be longest in carnivores and shortest in herbivores, whereas the opposite is true for both the length and volume of the hindgut. Skoczylas (1978) cited the study of forty species by Lönnberg (1902), which found the midgut and hindgut of herbivores to be about two times versus one time the body length, respectively, compared with 0.87 versus 0.34 times the body length in carnivores. Parsons and Cameron (1977) described the internal relief of the midgut of a wide range of reptiles. An ileocolic valve or sphincter separates the midgut and hindgut. These are of equal diameter in some species, but the hindgut is enlarged in most reptiles, and in some herbivores it includes a cecum at its junction with the midgut. The cecum and proximal colon of herbivorous lizards in the families Agamidae, Scincidae, and Iguanidae are compartmentalized by mucosal folds (Fig. 3.12). Iverson (1980) counted from one to eleven folds in the proximal colon of iguanine lizards. Two kinds of folds, or "valves", were described (sometimes with a sphincter), which may slow the passage of digesta and increase the absorptive surface. The number of colonic folds increased with body size.

Birds

Birds differ from other nonmammalian vertebrates in their ability to control their body temperature, and they differ from all extant vertebrates in their cover of feathers and (with the exception of bats) the modification of their forelimbs for flight. Modifications for flight are accompanied by a redistribution of weight, which includes an absence of teeth and decreased weight of the jaw skeleton and its muscles and acquisition of the gizzard as the organ for trituration. Storer (1971a,b) reviewed the problems associated with the taxonomic classification of birds and listed twenty-eight orders of existing species. Many of the common species are found in the orders Galliformes (cocklike) and Passeriformes (sparrowlike). Galliformes include the pheasant, partridge, grouse, quail, and common domestic fowl. Passeriformes are the songbirds and include a high percentage of avian species. The general characteristics of the avian digestive system have been described by Ziswiler and Farner (1972) and Duke (1986). Barnard and Prosser (1973) provide additional comparative information.

The horny beak, or bill, and associated mouth parts show a wide range of modifications in carnivores, piscivores, insectivores, and carrion eaters, as well as in birds that feed primarily on fruit, seeds, pollen, nectar, leaves, or roots (Storer 1971b; Kear 1972; Cloudsley-Thompson 1972). The bill of flamingos contains marginal projections, which provide a filter for feeding on small invertebrates or, in the lesser flamingo, blue-green algae and diatoms. In the pelican and some passerine species, the floor of the mouth cavity is distensible and serves for food storage. The tongue shows as many modifications as the bill, but it contains muscle only in a few birds such as the parrots. Although salivary glands were reported to be absent in the snakebird *Anhinga anhinga* (Antony 1920), they usually are present and highly developed. They function primarily for mucigenous lubrication. However, as noted in Chapter 2, they secrete an adhesive material in swifts and woodpeckers and are said to secrete amylase in some species.

The gastrointestinal tracts of a hawk, budgerigar, chicken, grouse, hoatzin, rhea, emu, and an ostrich are illustrated in Figs. 3.13 and 3.14. The gastrointestinal tract of the goose *Anser anser* resembles that of the chicken, except that it has a longer and more tubular crop, a longer midgut, and longer ceca (Clemens, Stevens, and Southworth 1975b). The avian esophagus tends to be long and wide and is usually dilated into a unilateral, bilateral, or spindle-shaped crop. The crop serves for storage of food, which is regurgitated to feed the young in some species. These variations, and the special secretion of "milk" by the pigeon crop, are discussed by Ziswiler and Farner (1972). In the chicken, the crop is isolated by a sphincter that only opens after the gizzard is filled. The crop then fills and undergoes periodic, rhythmic contractions that empty it over the next few days. The crop and distal esophagus of the hoatzin (Fig. 3.14) function as a fermentation organ (Grajal et al. 1989).

Other functions of the conventional vertebrate stomach are carried out by the proventriculus (gastric secretion) and the ventriculus (trituration, maceration, and pumping). The proventriculus is a spindle- or cone-shaped structure, whose size can vary. For example, predacious carnivores, such as the hawk (Fig. 3.13), petrel, heron, or gull, tend to have a highly distensible proventriculus, but in galliform, passeriform, and certain other species, the proventriculus appears to function only for the secretion of gastric juice during the passage of food into the gizzard.

The ventriculus, or gizzard, is a muscular organ lined with koilen, a horny material consisting of protein and carbohydrates, which is periodically molted by many species. The gizzard generally serves the primary function of trituration, and swallowed grit increases its efficiency, although it is not essential. It can vary a great deal in its relative size and degree of musculature. The single layer of circular mus-

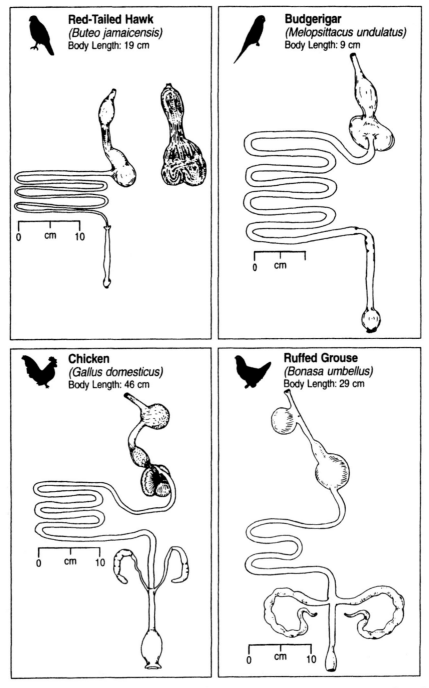

Fig. 3.13. Gastrointestinal tracts of the red-tailed hawk (with sagittal section of foregut), budgerigar, chicken, and ruffed grouse.

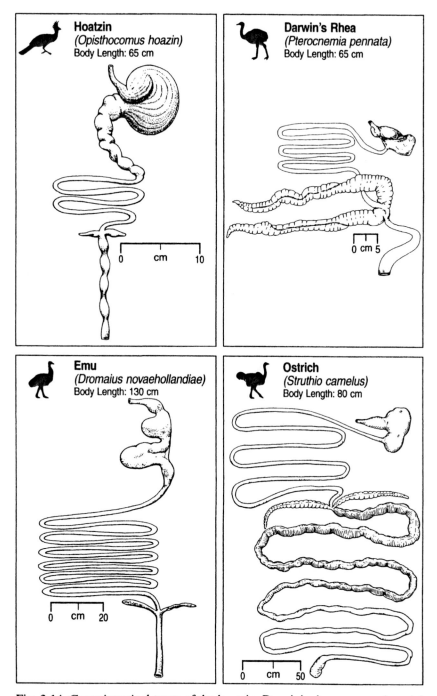

Fig. 3.14. Gastrointestinal tracts of the hoatzin, Darwin's rhea, emu, and ostrich.

cle is thin walled in most carnivores and frugivores. In the fruit-eating tanager, the gizzard has been reduced to an insignificant band. Among Australian parrots, the gizzard of lorikeets, which feed largely on nectar and pollen, is much less muscular than that of granivorous and frugivorous parrots (Richardson and Wooller 1990). In buzzards, carrion-eating birds, the gizzard may generate only 8–26 mm Hg intraluminal pressure. Conversely, it shows the greatest development in the granivorous and herbivorous species, and it is capable of producing intraluminal pressures of 200 mm Hg in the domestic fowl and 280 mm Hg in the goose.

Although the gizzard is generally considered to be the compensatory organ for lack of teeth, it also serves as a site for food storage, acid-pepsin digestion of protein, and propulsion of food into the intestine (Ziswiler and Farner 1972). There is even evidence of pepsinogen secretion by the gizzard of the common kestrel *Falco tinnunculus*, a European falcon, and the buzzard, *Buteo buteo*. In some birds, such as the owls, hawks, kingfishers, and shrikes, the gizzard also serves as a filter for less digestible parts of the prey, which are then formed into pellets and regurgitated. An additional chamber, sometimes called the *pyloric stomach*, is present in penguins, grebes, pelicans, and many storks, as well as in some ducks, geese, and rails. Its function is uncertain, but its presence appears to correlate with diets high in water and, at least in the grebe, it may serve as a filter.

The avian midgut, or small intestine, is usually considered to be composed of a duodenum and ileum. The wide variation in the number and arrangement of intestinal loops was illustrated by Mitchell (1901). Leopold (1953) discussed the relationship between intestinal morphology and diet in galliform birds.

The hindgut of birds generally consists of the ceca and a relatively short, straight extension of the intestine. Its origin is distinguished by a sphincter or valve, and the ceca may be separated from the remainder of the hindgut by valves in many species. The ceca are usually paired, although some birds such as the heron have only one, and others (hummingbirds, swifts, and some pigeons and woodpeckers) have none. They are absent in both the embryo and adult parrot and budgerigar (Fig. 3.13) and small in predators such as the hawk (Fig. 3.13). Ceca are most highly developed in omnivores and herbivores (but not necessarily granivores), and are especially well developed in the grouse (Fig. 3.13) and rhea (Fig. 3.14). Mitchell (1901) classified avian ceca into four different types: (1) *Primitive*, which are well developed and thin-walled (for example, duck, goose, grebe, and loon); (2) *Enlarged*, which are characterized by a greater amount of lymphoid tissue and apparently a greater role in cellulose digestion (for example, galliform species); (3) *Lymphoepithelial*, which are characterized by apparently little digestive function (for example, passeriform species); and (4) *Functionless vestigial* (or absent), which are seen in the penguin, hawk, and budgerigar. The ceca are usually histologically similar to the intestine, except for lymphoid tissue. The remainder of the avian hindgut is generally short and straight and therefore often referred to as the *rectum*. However, it includes a long segment of colon in the horned screamer *Aakima cornuta* (Mitchell 1901) and ostrich (Fig. 3.14). The colon of the ostrich is over ten times the length of the ceca (Skadhauge et al. 1984).

Summary

The digestive systems of fish, amphibians, reptiles, and birds show a number of structural and functional characteristics that are common to all classes. This is particularly true for the midgut. However, the headgut, foregut, and hindgut can demonstrate considerable differences both among and within the vertebrate classes. Some adaptations of the digestive system can be related to other physiological characteristics of the taxon or group of related animals. Passage of large volumes of water through the oral cavity and gills of fish, for respiratory purposes, limits the alternatives by which food can be ingested in small particles. Termination of the urinary, reproductive, and digestive tracts in the cloaca of amphibians, reptiles, and birds provides limitations and opportunities for adaptations that are not shared by most fish and mammals. Control of body temperature in birds decreases the influence of ambient temperature on feeding, but it requires a more regular and efficient system for the assimilation of nutrients. Flight restricts the distribution and weight of gut contents.

Many adaptations of the digestive system that are seen in different species show a closer correlation with their diet or environment than with their taxonomic classification. Therefore, knowledge of the diet and habitat of an animal is necessary for an understanding of its digestive system. Members within each class have developed highly specific, and often elaborate, adaptations to their diet or environment. This is especially true for buccal cavity and pharynx, and it includes adaptations of the lips, teeth, or tongue for procurement of food and reducing its particle size. Microphagous animals often have mechanisms that allow for the separation and ingestion of small food particles. Some animals feed almost continuously. Others require adaptations for the intermittent storage and digestion of food. The hindgut shows numerous adaptations that allow for retention of digesta. These may be related to the diet, as is the case with the hindgut of herbivorous reptiles, birds, and mammals. However, an increase in absorptive surface area or digesta retention time also allows for the conservation of electrolytes, water, and nitrogen.

4

The mammalian gastrointestinal tract

Mammals can be subdivided into two subclasses and twenty orders (see Table 2.1, p. 12). They are endothermic like birds, but differ from other vertebrates in their hair covering and the fact that they are born at an early stage of development and suckle their young. The length and capacity of the mammalian digestive tract increase during prenatal development. Musculature develops in a craniocaudal sequence, circular muscle first. In species with complex stomachs, such as ruminants, compartmentalization may be evident quite early in gestation. The hindgut terminates in a cloaca in all fetal mammals, and the cloaca is retained by adult monotremes, marsupials, some insectivores, and at least one species of rodent, the mountain beaver *Aplodontia rufa*. However, the outlets for the renal, reproductive, and digestive systems separate prior to the birth of most mammals. Birth at an early stage of development and conversion from milk to other diets result in some major changes in the digestive system during the period between birth and weaning. Teeth do not generally erupt until after birth, and the digestive system continues to develop. The stomach and hindgut of some species continue to increase in size and complexity, and the composition of digestive enzymes and mechanisms of nutrient absorption undergo major changes.

Eisenberg (1981) divided mammals into sixteen categories of dietary specialization, and Langer and Chivers (1994) discussed other classifications of animal foods. However, Table 4.1 divides the mammalian orders more simply into three categories based on the general type of diet of inclusive species. The first category includes all animals that feed principally on other vertebrates or invertebrates; thirteen of the twenty orders contain species in this category. Ten orders include omnivores that feed on both plants and animals, or species that feed principally on plant concentrates, such as seeds, fruit, pollen, or roots. The third category contains herbivorous species, defined here as animals that can subsist largely on a diet containing the fibrous portion of plants. Eleven orders contain these

Table 4.1. *Mammalian orders listed according to diets of inclusive species*

Order	Animal	Animal and plant or plant concentrates	Plant fiber
Monotremata	+		
Pholidota	+		
Tubulidentata	+		
Cetacea	+		
Macroscelidea	+		
Insectivora	+	+	
Scandentia	+	+	
Chiroptera	+	+	
Carnivora	+	+	+
Marsupialia	+	+	+
Edentata	+	+	+
Rodentia	+	+	+
Primates	+	+	+
Dermoptera		+	
Artiodactyla		+	+
Lagomorpha			+
Perissodactyla			+
Proboscidea			+
Sirenia			+
Hyracoidea			+

Source: Modified from Stevens (1980).

species, most of which demonstrate major adaptations of the stomach and/or hindgut in association with this capability.

The following discussion deals with the gross structural characteristics of the gastrointestinal tract. The early studies of Flower (1872), Mitchell (1905), and Bensley (1902–03) provide information on a wide range of mammals. The major structural differences between the gastrointestinal tract of lower vertebrates and many mammals are seen in the stomach and hindgut. The series of comparative drawings of the gastrointestinal tract have concentrated on these segments of the tract. The shape and dimensions of the stomach and hindgut can be affected to a large degree by the time between feeding and examination, as demonstrated for the rat (Fig. 4.1). Therefore, when possible, the gastrointestinal tract was collected four hours after a meal in animals fed at twelve-hour intervals. Discussion of the gastrointestinal tract proceeds by mammalian orders, as listed in Table 4.1, and, except where otherwise indicated, the taxonomic classifications listed by Vaughan

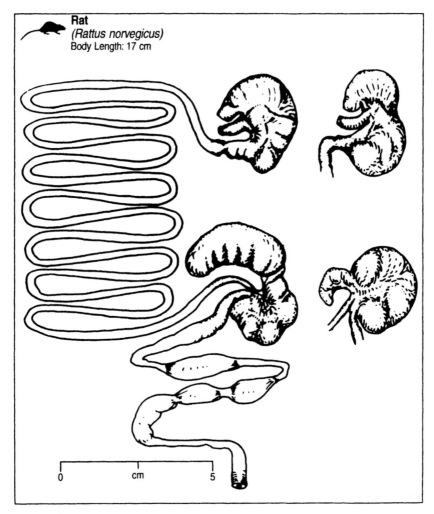

Fig. 4.1. Gastrointestinal tract of rats fed at 12-hour intervals for two weeks prior to sacrifice. The specimen on the left was obtained 4 hours after feeding. The stomach and cecum to the right were obtained immediately after feeding. Note the effect of time after feeding on the volume and shape of these two organs. The digestive tract was immediately dissected free of its attachments, arranged in the pattern indicated and drawn to scale, with special attention to the structural characteristics of the stomach and large intestine. Body length represents the distance from mouth to anus. (From Stevens 1977.)

(1986) for orders and families and Nowak and Paradiso (1983) for genera and species. This approach underlines the point that animals that are closely related can show extreme variations in gastrointestinal anatomy, and, conversely, species with no immediate genetic relationship can demonstrate marked similarities.

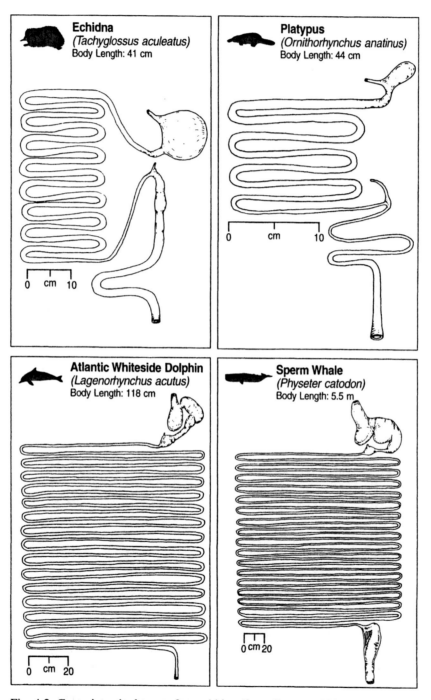

Fig. 4.2. Gastrointestinal tract of an echidna (from Stevens, 1980), the platypus, the Atlantic whiteside dolphin, and the sperm whale.

Monotremata

The monotremes are listed as a single order under subclass Prototheria, the egg-laying mammals. This subclass and order are now represented by only the echidnas (spiny anteaters) and the platypus, which are restricted to Australia and New Guinea. The gastrointestinal tracts of an echidna and platypus are illustrated in Fig. 4.2. Echidnas are terrestrial anteaters. Their stomach is lined completely with stratified squamous epithelium, and its contents are maintained at a pH of 6.2 to 6.5 after a meal of termites (Krause 1970; Harrop and Hume 1980). The pylorus is muscular and somewhat elongated. The intestinal tract was said by Flower (1872) to show no evidence of a sphincter or marked distinction between a small and large bowel, other than the presence of a small vermiform cecum lined with lymph glands. The platypus is a semiaquatic animal that feeds on insects, mollusks, and worms. Its stomach also is lined with stratified squamous epithelium plus Brunner's glands near its junction with the intestine. As in the echidna, the small and large intestine are not separated by a sphincter or constriction, but a somewhat longer vermiform cecum is present (Fig. 4.2).

The general pattern of the monotreme intestine is similar to that of the reptile and demonstrates only two major differences from that of birds (Mitchell 1905). Most birds have a specialized duodenal loop, not seen in monotremes or most other mammals, and a straight, short segment of hindgut. The hindgut of monotremes and most other mammals, extends a considerable distance in a series of arches, loops, or spirals.

Pholidota

The Pholidota include eight species of anteaters that inhabit parts of Africa and Asia, have horny scales, short legs, and a long body and tail similar to that of rep-

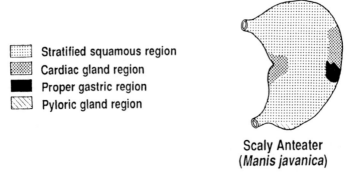

Stratified squamous region
Cardiac gland region
Proper gastric region
Pyloric gland region

Scaly Anteater
(*Manis javanica*)

Fig. 4.3. Distribution of gastric epithelium in the scaly anteater. (From Bensley 1902–03.)

tiles. The pangolins, or scaly anteaters, in this order were originally classified as edentates, because of the many characteristics shared with the edentate anteaters. The stomach of *Manis javanica* (scaly pangolin) is small, round, and almost entirely lined with stratified squamous epithelium (Fig. 4.3). The glandular stomach appears to be entirely represented by two small regions of cardiac glands on the lesser and greater curvature and a glandular mass on the greater curvature. The latter consists of proper gastric gland follicles whose ducts empty into the stomach via a single orifice. The stomach has thick walls and often contains stones, indicating that it may also serve a triturating function. The intestinal canal showed no division between the midgut and hindgut, other than a gradual enlargement, and there was no cecum (Mitchell 1905).

Tubulidentata

This order is represented by one living species, the aardvark, which feeds on ants and termites. It retains its milk teeth as an adult, and the termites and ants are generally chewed before they are swallowed (Eisenberg 1981). Aardvarks also feed on the fruits of *Cucumis humifructus*, but the seeds apparently pass through the digestive tract unharmed. Flower's (1872) examination of an aardvark specimen indicated that the hindgut had a prominent ileocecal valve and a large cecum (18 cm long and 8 cm in diameter at its apex). The colon was approximately one fourth the length of the small intestine and sacculated over its proximal segment.

Cetacea

The cetaceans, which include seventy-eight species of dolphins, porpoises, and whales, are referred to as either toothed (suborder Odontoceti) or baleen (suborder Mysticeti) whales. The various species can be subdivided according to diet into the following three groups: (1) Sarcophagi, which include the killer whale or grampus, feed on other mammals such as seals; (2) Tenthophagi, which include the sperm whale, narwhal, and beluga, feed primarily on cephalopods; (3) Ichthyophagi, which include the common porpoise, most dolphins, and finwhales, feed on fish; and (4) Pteropodophagi, which include the right and bowhead whales, feed primarily on pteropods and small crustaceans.

Toothed whales have a mouth designed for rapidly ingesting and swallowing their prey, but baleen whales lack true teeth and feed by ingesting crustaceans or other animals in the water and trapping them against the sieve formed by ridges of palatal mucous membrane–the whalebone or baleen–as water leaves the mouth. The whale then closes its mouth and raises the tongue, like a piston, to express the remaining water before swallowing its prey. Finwhales have a very large mouth

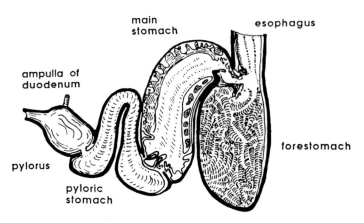

Fig. 4.4. Sagittal section of a dolphin stomach. (From Pernkopf 1937.)

and feed on fish, but the Greenland whale, which only eats pteropods and crustacea, has a gigantic mouth that is as large as its entire thoracic and abdominal cavity.

The gastrointestinal tracts of two species of toothed whales are illustrated in Fig. 4.2. The esophagus of cetaceans opens into a large, multicompartmental stomach (Fig. 4.4). Slijper (1962) stated that the first compartment in most cetaceans was lined with a stratified squamous (noncornified) epithelium, and the second compartment was lined with proper gastric mucosa. The remaining compartments were lined with pyloric mucosa. According to Ridgway (1972), the first stomach compartment of toothed whales was lined with stratified squamous epithelium, and the second compartment was lined with simple tubular glands that contain (proper gastric mucosa) mucus, chief, and oxyntic cells. The subsequent pyloric compartments were lined only with mucous-type cells. The proximal duodenum was dilated into an additional compartment that received the common bile duct. In some whales, the presence of stones in the compartment lined with stratified squamous epithelium suggested that it may function primarily as a gizzard. However, there also is evidence of microbial fermentation in the stomach of some whales (see Chapter 8).

There is considerable variation in the number of gastric compartments among cetacean families. The compartment lined with stratified squamous epithelium is absent in beaked whales and members of several other families. The number of main compartments can range from one to three, with more than one pyloric compartment in three families (Rice and Wolman 1990). The connection between the main and pyloric chambers ranges from a constricted and long to a wide and short channel. In the pygmy sperm whale *Kogia breviceps*, the channel was 12 cm long but only 1.5 cm in diameter when expanded (Rice and Wolman 1990). The function of such severe constriction is unknown, but it is likely to result in very long retention of food in the main compartment.

Among the toothed whales, there was considerable variation in the ratio of intestinal length to body length; bottlenose whales, 5.5/1; beaked whales, 6/1; beluga whales, 10/1; narwhal, 11/1; various species of dolphin, 7 to 14/1 (Slijper 1962). Baleen whales tended to show lower ratios (finwhale, 4/1; little piked and humpback whales, 5.5/1). These ratios did not necessarily correlate with either the diet or size of the species.

The hindgut cannot be distinguished by any changes in diameter in many species, but a short, distinct hindgut is seen in others (Fig. 4.2). Ridgway (1972) found that a cecum was absent from all toothed whales, with the exception of the Gangetic dolphin *Platanista gangetica*, but the hindgut was distinguished in the Pacific white-striped dolphin by an absence of villi in the terminal 30 cm of intestine. The intestine of baleen whales was said to differ in that it has a short, conical cecum. However, a specimen of a young toothed sperm whale (Fig. 4.2) showed a short, enlarged hindgut and some evidence of a cecum. Flower (1872) concluded that the gastrointestinal tract, olfactory organs, and rudimentary hind limbs suggest a close relationship between whales and terrestrial mammals.

Macroscelidea

The Macroscelidea (elephant shrews), which are indigenous to Africa, are insectivorous but may be more closely related to lagomorphs than to the order Insectivora (Eisenberg 1981). The elongated snout, which extends well beyond the nasal bones, is used in probing for arthropods among leaf litter and loose soil. All species of elephant shrews have a cylindrical cecum and a large colon, folded twice upon itself (Fig. 4.5). However, Woodall (1987) noted that species from more mesic habitats had a shorter colon and smaller cecum than those from more arid regions.

Insectivora

This order contains six families and 406 species of principally insectivorous animals. Although the gastrointestinal tract varies among different species, the hindgut is simple in structure and lacks a cecum. The hedgehog *Erinaceus europaeus*, which is actually omnivorous, provides a good example (Fig. 4.5). Its stomach is simple, and the pylorus has a relatively thick layer of circular muscle. There is no external distinction between small and large intestine, although the intestine narrows in an area that would correspond to the ileum of other mammals. The gastrointestinal tract of the common English mole, family Talpidae, (Fig. 4.5), is similar to that of the hedgehog except that the stomach is very large

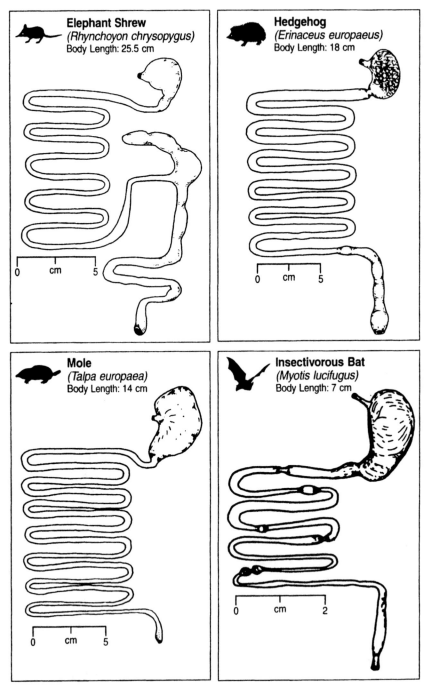

Fig. 4.5. Gastrointestinal tracts of a hedgehog (from Clemens 1980), a mole, an insectivorous bat (from Stevens 1980), and an elephant shrew.

with a well-developed fundus. The intestine of animals in this genus appears to differ from that of all other mammals in that the rectum passes ventral to the pelvis, rather than through it.

According to Flower (1872), the intestine of tenrecs, small, prolific animals that are indigenous to Madagascar and live primarily on worms, also showed no indication of any division into small and large bowel other than a slight enlargement of the terminal straight segment. Members of the shrew family (Soricidae) have a stomach that is rounder than that of other Insectivora, and its cardiac inlet and pyloric outlet are close to one another. The intestine is extremely short, only three to four times the length of the shrew's body.

Scandentia

This order is represented by a single family (Tupaiidae) of tree shrews, which resemble small, long-snouted squirrels. Present-day forms are confined to eastern India and Southeast Asia. Tree shrews feed on a mixture of fruits and insects, and, according to Flower (1872), they have a small cecum and short colon.

Chiroptera

This large order of 853 species is divided into two suborders, the Microchiroptera (seventeen families) and the Megachiroptera (three families of large fruit bats, or flying foxes, from Africa, Asia, and Australia). The digestive system of microchiropteran bats has attracted considerable interest because of their wide range of diets. Flower (1872), McMillan and Churchill (1947), Brown (1962), and Rouk and Glass (1970), described variations in the morphology and histology of the stomach of insectivorous, frugivorous, and sanguivorous (blood-feeding) species. Variations in compartmentalization and the epithelial lining of the microchiropteran bat stomach are shown in Fig. 4.6. Insectivorous bats tend to have a relatively simple globular stomach, but the stomach of some nectivorous and frugivorous species is both large and complex. The stomach of *Desmodus rotundus* and *D. rufus*, South American blood-feeding species, is unique among mammals, consisting of a long, blind, sacculated tube. This convoluted stomach is approximately twice the length of the animal. According to Rouk and Glass (1970), the stomach of *D. rotundus* contains acinar and tubuloacinar glands unlike those of other mammals. The studies of Ito and Winchester (1960, 1963) provide descriptions of mucosal structure.

The intestinal tracts of an insectivorous and a vampire bat are illustrated in Figs. 4.5 and 4.7. The hindgut is short and difficult to distinguish from the midgut. A cecum is present in only a few microchiropterans such as *Rhinopoma harwickie*

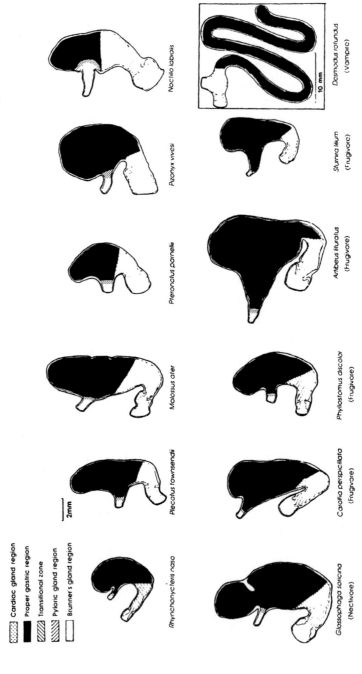

Fig. 4.6. Distribution of gastric epithelium in various species of insectivorous (*top row*), nectivorous, frugivorous, and vampire bats (*bottom row*). (From Forman 1972.)

Cardiac gland region
Proper gastric region
Transitional zone
Pyloric gland region
Brunner's gland region

Rhynchonycteris naso

Plecotus townsendii

Molossus ater

Pteronotus parnellii

Pizonyx vivesi

Noctilio labialis

Glossophaga soricina
(Nectivore)

Carollia perspicillata
(Frugivore)

Phyllostomus discolor
(Frugivore)

Artibeus lituratus
(Frugivore)

Sturnira lilium
(Frugivore)

Desmodus rotundus
(Vampire)

2mm

10 mm

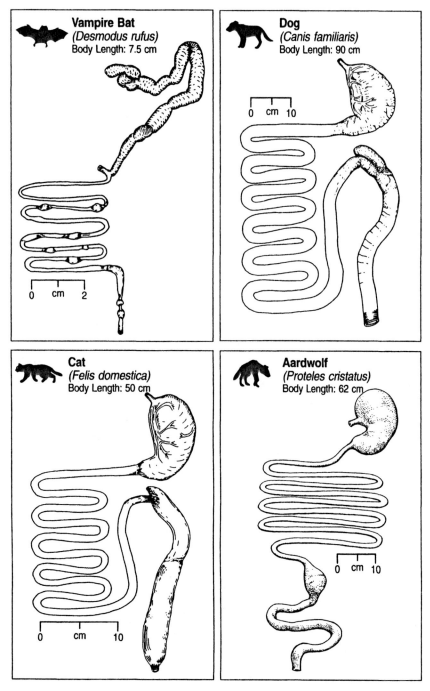

Fig. 4.7. Gastrointestinal tracts of a vampire bat (from Stevens 1980), dog (from Stevens 1977), cat, and aardwolf.

and *Megaderma spasma*. In most species of bats there is no external indication to distinguish between the small and large intestine, other than a slight increase in diameter and mucosal changes from villiform to longitudinal folds.

The gastrointestinal tract of megachiropterans had received less attention. The stomachs of two Australian species of *Pteropus* had a pronounced proximal pouch (Tedman and Hall 1985) similar to some frugivorous microchiropterans, but their intestine was much longer than that of microchiropterans, irrespective of the diet of the latter species.

Carnivora

The gastrointestinal tract of the 284 species in this order is characterized by a relatively simple stomach and short intestinal tract. As indicated by their name, most of these animals are flesh eaters. The following comprise the terrestrial Carnivora: families Canidae (dogs, wolves, foxes, jackals), Felidae (cats), Viverridae (civets, genets, mongooses), Hyaenidae (hyenas, aardwolf), Procyonidae (raccoons, pandas, ring-tail cats), Ursidae (bears), and Mustelidae (mink, skunks, badgers, weasels, otters, wolverines). The otters are semiaquatic, and the sea otter *Enhydra lutris* is almost completely aquatic. The fully aquatic Carnivora, which consist of the earless seals (Phocidae), eared seals (Otariidae), and walruses (Odobenidae), include the largest species, ranging from 90 to 3,600 kg in body weight.

The gastrointestinal tracts of the domesticated dog and cat are relatively short, and the hindgut contains a small cecum and an unsacculated colon (Fig. 4.7). The cecum consists of a coiled appendage located just distal to the ileocecal valve of the dog. Domestic cats have a gastrointestinal tract resembling that of dogs except that the cecum is not as coiled. Some species of Felidae were said to have a sacculated segment of proximal colon (Mitchell 1905). The gastrointestinal tracts of Hyaenidae and Viverridae were reported to be quite similar to that of the domestic cat, although the cecum was very rudimentary in one viverrine species and absent in another (Mitchell 1905). The gastrointestinal tract of the aardwolf (Fig. 4.7), which feeds almost exclusively on one genus of termite (Anderson, Richardson, and Woodall 1992), appears similar to those of the dog and cat, except for the cecum, which consists of a bilateral distension of the proximal hindgut.

The stomachs of Procyonidae, Ursidae, and Mustelidae are simple, and the distal segment of their intestine is marked only by a sudden change in the mucosa, with no cecum. Fig. 4.8 includes the gastrointestinal tract of a mink, raccoon, and black bear. The intestine of the omnivorous raccoon is longer than that of the dog and cat, approximately 2.7 times the body length, and the hindgut is shorter, with neither a cecum nor a distinct ileocolonic valve. The hindgut of the giant panda

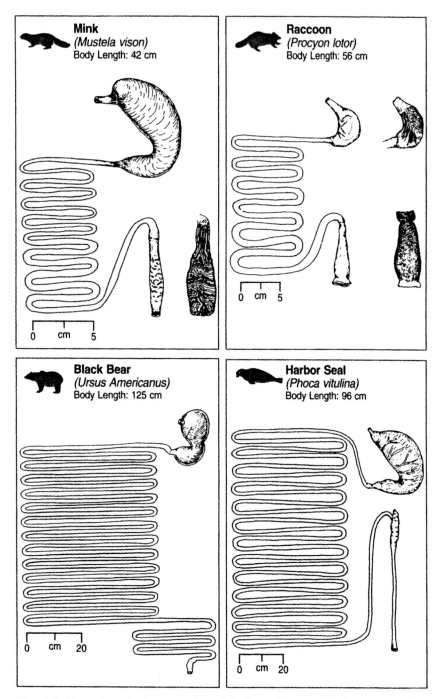

Fig. 4.8. Gastrointestinal tracts of a mink (from Stevens 1977), raccoon (from Argenzio and Stevens 1984), black bear, and harbor seal.

also appears to be relatively simple (Dierenfeld et al. 1982). The intestinal tract of the omnivorous black bear showed a larger stomach than that of the dog or raccoon, and the intestine was approximately ten times the length of the body, with no distinction between the midgut and hindgut, similar to that of the raccoon. The stomach of the mink was similar in size to that of the bear, and the intestine was approximately three times the body length (like that of the raccoon), with a hindgut that was short and simple like that of the raccoon and bear.

Aquatic carnivores (suborder Pinnipedia), feed entirely on fish or marine animals. The common seal *Phoca vitulina* has an elongated tubular stomach, sharply bent at the beginning of the pyloric antrum (Fig. 4.8). Its small intestine is quite long. The cecum is small, and the colon is approximately as long as the body. Other members of this group appear to demonstrate a similar digestive tract anatomy (Ridgway 1972).

Marsupialia

As previously noted, the marsupials and the eutherian (placental) mammals have undergone separate development from a similar origin. Because the 242 species of marsupials have a limited ecological distribution, they are listed as a single order under Infraclass Metatheria. However, as Flower (1872) pointed out, the various families of Marsupialia could be considered as equivalents to the orders of Eutheria. Their digestive system shows as great a variation.

The gastrointestinal tracts of marsupials and their relationship to diet, nutrition, and ecology are discussed by Hume (1982). The carnivores can be divided into two groups. The first group consists of two American families: the Caenolestidae (shrew-opossums) and the Didelphidae (opossums), which include some omnivores. The second group consists of four Australian families: the Dasyuridae, Myrmecobiidae, Notoryctidae, and Thylacinidae. The Dasyuridae contain forty-nine species, ranging in size from the tiny shrew-sized planigales to the 10-kg Tasmanian devil. The others each contain only one species respectively: the insectivorous numbat *Myrmecobius fasciatus*, the marsupial mole *Notoryctes typhlops*, and the Tasmanian tiger *Thylacinus cynocephalus*, which is probably extinct.

The gastrointestinal tracts of two species of Dasyuridae, the brush-tailed phascogale and tiger quoll, are illustrated in Fig. 4.9. They consist of a simple stomach and relatively short intestine with no external distinction between midgut and hindgut. The gastrointestinal tracts of other marsupial carnivores also have a relatively short intestine and, with the exception of the opossums, lack a cecum.

Marsupial omnivores can be divided into three groups. The first includes the family Peramelidae (bandicoots and bilbies). The second consists of some members of Didelphidae. The third group includes Australian arboreal species such as

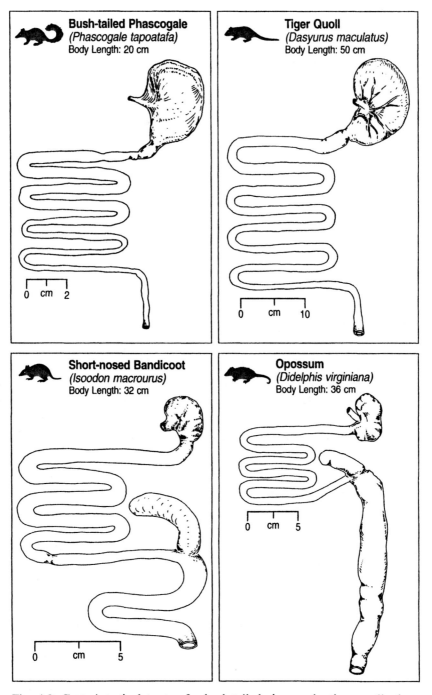

Fig. 4.9. Gastrointestinal tracts of a bush-tailed phascogale, tiger quoll, short-nosed bandicoot, and opossum.

the striped, Leadbeater's, eastern and mountain pygmy possums, and the yellow-bellied and sugar gliders. Fig. 4.9 also includes the gastrointestinal tracts of the short-nosed bandicoot and an American opossum. These animals also have a relatively short intestinal tract, but the hindgut is enlarged and includes a well-developed cecum.

The marsupial herbivores can be divided into nonmacropods and macropods. The nonmacropods include all members of families Vombatidae (wombats), Phalangeridae (cuscuses, brushtail possums, scaly-tailed possum), Phascolarctidae (koala), and Pseudocheiridae (greater glider and ringtail possums). The gastrointestinal tracts of the common wombat, koala, and greater glider are illustrated in Fig. 4.10. The stomachs of the koala and wombat contain a mass of proper gastric glandular mucosa near the gastroesophageal junction. The hindgut of each of these animals is well developed and serves as the major site for microbial fermentation. However, this consists of a voluminous colon in the wombat, a large and voluminous cecum in the greater glider, and a large voluminous cecum and well-developed colon in the koala.

The macropod marsupials have a complex voluminous stomach, which serves as the major site of microbial fermentation. These animals are contained in two families, the Potoroidae (rat-kangaroos), which are small (1–3 kg), and the Macropodidae (kangaroos and wallabies), which vary in size from 1-kg hare-wallabies to 70-kg kangaroos. The gastrointestinal tract of a kangaroo is illustrated in Fig. 4.10. The stomach is a basically tubular organ, which is drawn into haustrations by three bands of longitudinal muscle (taeniae). The forestomach is divided into the sacciform region (the blind sac orad from the opening of the esophagus) and the tubiform region (between the gastroesophageal junction and hindstomach). The rat-kangaroo forestomach is dominated by the sacciform region; the kangaroo forestomach is dominated by the tubiform region; and the wallaby forestomach tends to be intermediate between these two extremes. The hindstomach is lined with proper gastric and pyloric gland mucosa.

The distribution of stratified squamous epithelium and glandular mucosa in three macropod species is illustrated in Fig. 4.11. The luminal surface has a groove (the gastric sulcus) that runs along the lesser curvature of the tubiform forestomach. This groove, which may serve for the rapid passage of milk through the forestomach of suckling young, is found in many mammals that have a forestomach.

Edentata

The edentates (anteaters, armadillos, sloths) are believed to be a relatively primitive group of New World placental mammals. Although their name indicates an

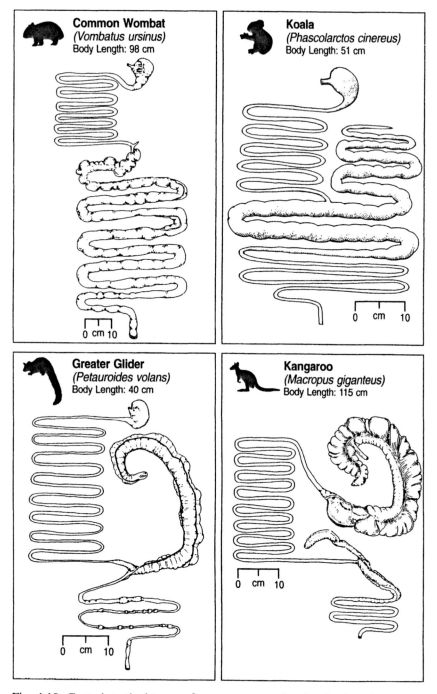

Fig. 4.10. Gastrointestinal tracts of a common wombat, koala (from Harrop and Hume 1980), greater glider, and kangaroo (from Stevens 1977).

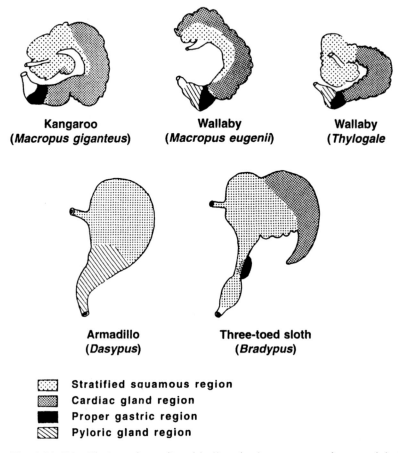

Kangaroo
(*Macropus giganteus*)

Wallaby
(*Macropus eugenii*)

Wallaby
(*Thylogale*

Armadillo
(*Dasypus*)

Three-toed sloth
(*Bradypus*)

Stratified squamous region
Cardiac gland region
Proper gastric region
Pyloric gland region

Fig. 4.11. Distribution of gastric epithelium in three macropod marsupials (from Langer, Dellow, and Hume 1980), an armadillo (from Bensley 1902–03), and a three-toed sloth (from Bensley 1902–03).

absence of dentition, all but the anteaters have teeth in one form or another. This order, which includes three families and thirty-one species, represents two extreme examples of alimentary tract specialization to diet, the Myrmecophagidae, which live on ants and termites, and the sloths (Bradypodidae), which are arboreal folivores. The anteaters are generally characterized by their long head, long tongue, extremely large submaxillary glands, and a mouth that is toothless (although the inner surface of the cheek may contain ridges or spines, which could aid in mastication). The stomach is relatively simple, consisting of an expanded area containing proper gastric glands and a smaller pyloric area lined with a thicker epithelium and heavy muscular walls. According to Flower (1872), the pylorus of some species, such as the South American great anteater *Myrmecophaga tri-*

dactyla, is lined with stratified squamous epithelium, giving it the characteristics of the gizzard of birds. The pylorus of the tamandua anteater and the silky, or two-toed, anteater *Cyclopes didactylus* also is thickened, but not to gizzard-like proportions.

The small intestine of the great anteater was reported to be seven times the body length, and the large intestine was equivalent to the length of the body (Flower 1872). Transition between the two was indicated by an abrupt enlargement of diameter, but there was no valve, and only a poorly defined cecal pouch was noted. The small intestine of the tamandua anteater was approximately twenty times the length of the colon. A valve and a short, round cecum opened into a colon, which was approximately twice the diameter of the ileum and enveloped in a thick muscular coat. The intestine of the two-toed anteater was distinguished by the presence of paired, foliated ceca that opened to the colon via very small apertures.

Although the digestive tract of armadillos (family Dasypodidae) (Fig. 4.12) shows many similarities to the anteaters, armadillos feed on a wider variety of animal food as well as some plant material. The stomach appears to contain only stratified squamous epithelium and pyloric glandular mucosa (Fig. 4.11). The common long-nosed armadillo *Dasypus novemcinctus* has an intestine similar to that of the great anteater, with a short canal and no cecum. But the intestine of the Brazilian lesser long-nosed armadillo *Dasypus septemcinctus* was reported to have a valve and a pair of short, round ceca (Flower 1872).

The sloths are indigenous to South American jungles and divided into two genera: *Bradypus*, the three-toed sloth, and *Choloepus*, which has only two toes on its front feet. They are arboreal folivores that differ from the anteaters and armadillo in having a short, round face, short tongue, and normal-sized submaxillary and slightly larger parotid salivary glands (Montgomery and Sunquist 1978).

The stomach of the sloths is as complex as that of any species, and the two genera differ from each other only in that the stomach of *Bradypus* is more complex than that of *Choloepus*. Fig. 4.12 illustrates the gastrointestinal tract of *Bradypus tridactylus*, and details of the stomach are illustrated in Fig. 4.13. The stomach consists of three major divisions. The first is a large compartment, partially divided into anterior and posterior sacs. The anterior sac has a prolonged cecal appendage. The esophagus enters the posterior sac, which is also subdivided into an upper and lower pouch. The second compartment is very small. A groove runs across its lesser curvature from the cardia to the third major compartment, in a manner similar to that seen in the kangaroo stomach. The third compartment is tubular. Stratified squamous epithelium is distributed over much of the first compartment, the entire second compartment, and most of the third compartment (Fig. 4.11). The first compartment also contains a large island of cardiac glandular mucosa, similar to that described in the scaly anteater, and the third compartment

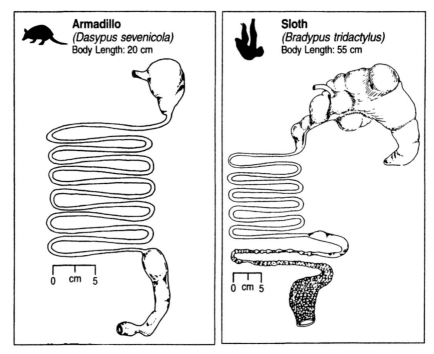

Fig. 4.12. Gastrointestinal tracts of an armadillo and sloth. (From Stevens 1980.)

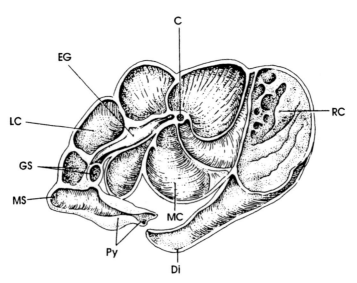

Fig. 4.13. Stomach of *Bradypus tridactylus*. Cardia (C); esophageal (ventricular) groove (EG); left cecum (LC); middle cecum (MC); right cecum (RC); diverticulum (Di); muscular stomach (MS); pyloris (Py); gastric sulcus (GS). (After Pernkopf, from Grassé 1955.)

contains an oval patch of proper gastric glands on its greater curvature. The stomach of *Choloepus* was said to differ primarily in the fact that the anterior sac of the first compartment was shorter and less complicated in its arrangement of internal septa (Flower 1872).

The hindgut of the sloth is enlarged in its proximal segment. Mitchell (1905) described a pair of small ceca near the ileocecal junction of *B. infascatus*. The specimen of *B. tridactylus* shown in Fig. 4.12 shows no evidence of these ceca. It does show, however, distention of the rectum with fecal pellets. As protection from predators, these animals descend from the trees to defecate and urinate only about once a week.

Rodentia

The rodents represent a group of animals that have been very successful, as measured by their ecological distribution and the number of families (30), species (1,700), and individuals. Rodents are generally herbivores, although some, such as the rat, are omnivores, and some others are carnivores (Landry 1970). They are of special interest to physiologists for both their success and the widespread use of some species as experimental animals.

Rodents show a wide range of species variation with respect to the structure of their stomach and hindgut. The gastrointestinal tracts of a rat, guinea pig, hamster, woodchuck, chinchilla, beaver, brush-tailed porcupine, capybara, and vole are illustrated in Figs. 4.1, 4.14, and 4.15. The stomachs of guinea pigs, woodchucks, and capybara (a large, South American rodent) are relatively simple in structure. The same is true of a number of other species, including squirrels and beavers, although the beaver stomach has a partition, or fold, attached to its greater curvature and a glandular mass (the cardiogastric gland) near the gastroesophageal junction. However, the stomach of many herbivorous rodents, such as the hamster and vole, is compartmentalized to varying degrees (Vorontsov 1962, Carleton 1981, Perrin and Kokkin 1986), and the proximal portion (forestomach) of many species, including the rat, is lined with stratified squamous epithelium (Fig. 4.16). The forestomach is papillated in some African species (Perrin and Kokkin 1986).

The hindgut of rodents demonstrates a range of complexity, often to a degree inversely proportional to that seen in the stomach of a given species (Lange and Starland, 1978). The cecum is usually large in relation to the rest of the digestive tract. As noted in Fig. 4.1, the cecum of the Norway rat can be quite voluminous. The relative capacity of the jerboa and woodchuck cecum is less than that of rats, but guinea pigs, beavers, capybara, and voles have a greater cecal capacity. The cecal contents of guinea pigs are said to equal 5 percent to 10 percent and those of the capybara equal 12 percent of the body weight. The cecum of many rodents

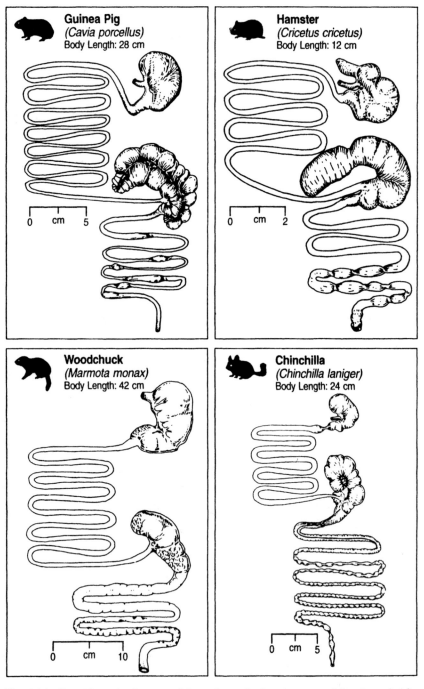

Fig. 4.14. Gastrointestinal tracts of the guinea pig, hamster, woodchuck, and chinchilla.

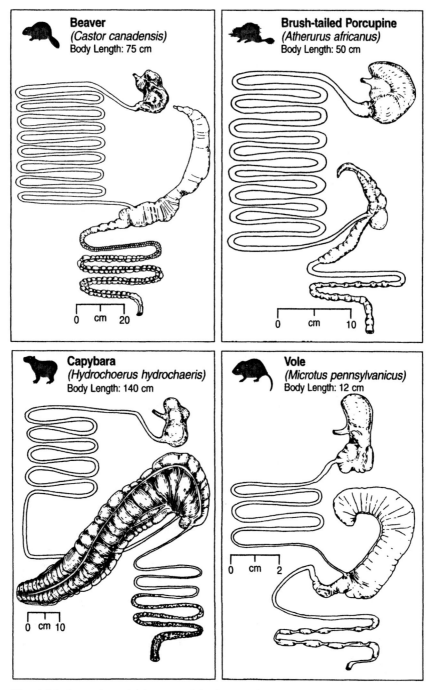

Fig. 4.15. Gastrointestinal tracts of the beaver, brush-tailed porcupine, capybara, and vole.

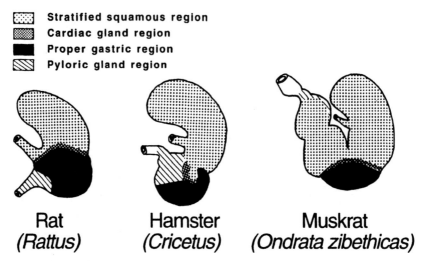

Fig. 4.16. Distribution of gastric epithelium in three species of rodents. (From Bensley 1902–03.)

is haustrated with taeniae. The initial segment of colon also is haustrated in a number of species such as the chinchilla (Fig. 4.14). The cecum of the prehensile-tailed porcupine *Coendou prehensilis* terminates in what appears to be a long vermiform appendix, like that of the rabbit (Fig. 4.29). The proximal colon of the vole and lemming forms a tight spiral (see Figs. 6.7 and 6.8). Behmann (1973) provided a well-illustrated comparative review on the functional anatomy of the rodent cecum and colon.

Primates

This order of 230 species includes the lemurs and their relatives, monkeys, apes, and humans. The order Primates is divided into two suborders: Strepshirhini (lower primates) and Haplorhini (higher primates). The lower primates include the following seven extant families: Lemuridae (true lemurs), Cheirogaleidae (dwarf lemurs), Megaladapidae (sportive lemurs), Indriidae (indri, and sifakas), Daubentonidae (aye-aye), Loridae (pottos, and lorises), and Galagonidae (bush babies). Lemurs are variously insectivorous, omnivorous, or herbivorous-frugivorous. The lorises feed on insects, fruit, and small vertebrates, and the aye-aye and dwarf lemurs are principally insectivorous. The remainder of the strepshirhines eat a mixed diet, which may include fruit, leaves, and exudates.

The gastrointestinal tract of the bush baby is illustrated in Fig. 4.17. Both Flower (1872) and Mitchell (1905) described the gastrointestinal tracts of a number of other strepshirhines. The cecum on the ruffed lemur *Varecia variegata* was longer

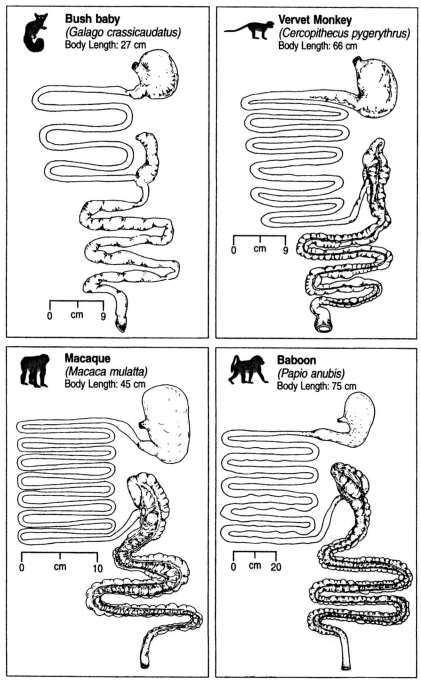

Fig. 4.17. Gastrointestinal tracts of a bush baby and vervet monkey (from Clemens 1980), a macaque and baboon.

than that of the bush baby and well developed. The small intestine of the *Indri* was considerably longer than that of the bush baby, and its cecum was twice the body length and slightly haustrated. The intestinal tracts of two species of lorises were reported to be shorter than that of the bush babies, but the cecum was longer, slightly haustrated, and contracted at its apex into a vermiform appendix. The digestive tracts of two other African species of lorises (*Perodicticus potto* and *Arctocebus calabarensis*) and an inhabitant of Sri Lanka (*Loris tardigradus*) were similar except for a much shorter cecum. The aye-aye *Daubentonia madagascarensis* has teeth that resemble those of rodents to such a degree that it was once so classified. Its colon was much like that of most species of this order.

The higher primates can be subdivided into the following six families: Tarsiidae (tarsiers), Cercopithecidae (Old World monkeys), Cebidae (capuchins and other New World monkeys), Callitrichidae (marmosets and tamarins), Hylobatidae (gibbons), and Hominidae (great apes and humans). The tarsiers inhabit islands of the Indo-Malayan Archipelago, feeding on insects and small vertebrates. The colon of one species was found to be quite short, only about one fifth as long as the small intestine. The cecum was approximately one half the length of the colon and arranged in a spiral. The Cercopithecidae include many genera. Some of these are African monkeys, while others, such as the *Macaca* (rhesus and pig-tailed monkeys), are almost entirely Asian species. They tend to be herbivores, and many are folivores. However, baboons can develop the behavioral pattern of predatory carnivores in the wild state (Harding and Strum 1976). Most members of this family have cheek pouches (which aid in food storage), a simple stomach, a relatively short small intestine, and a haustrated cecum. Fig. 4.17 shows the gastrointestinal tract of a vervet monkey, a macaque, and a baboon. Three taeniae draw the colon of these animals into marked haustra. The anatomy of many of the species in this family has been described by Hill and Rewell (1948), Hill (1966a,b), and Chivers and Hladik (1980).

The genera *Presbytis* (Asian langur) and *Colobus* (African colobus) monkeys are of particular interest because of their complex stomach. The stomach wall contains two bands of longitudinal muscle (taenia), and the normal tonus of these muscles results in a series of haustra, similar to those of the kangaroo stomach. The stomach of *Presbytis entellus* is shown in Fig. 4.18. The early study of Owen (1835) concluded that the stomach was subdivided into cardiac, middle, and pyloric compartments, normally separated by relatively small, constricted orifices. There was no evidence of rumination in captive animals, and he concluded that the large, complex stomach allowed rapid ingestion and storage of food by these timid animals, since they lacked the advantage of the cheek pouches seen in related monkeys. Hill (1952) reported that the small intestine of *Presbytis* was about eight times the length of the body. The large intestine was twice the length of the

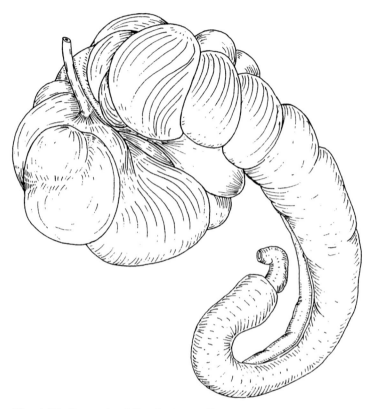

Fig. 4.18. Stomach of *Presbytis entellus*. (From Owen 1835.)

body, and the cecum was 25 percent of the body length. As demonstrated in Fig. 4.19, the stomach of *Colobus abyssinicus* appears similar to that of the langur. However, its small and large intestines are shorter, and the cecum is less developed.

The Cebidae, which include all New World monkeys, tend to be omnivores, although many species feed on fruit and leaves, and the night monkey *Aotus trivirgatus* is said to feed on insects and bats (Vaughan 1986). The gastrointestinal tracts of three species are illustrated in Fig. 4.19, and those of a spider monkey (*Ateles* sp.), capuchin monkeys (*Cebus* spp.), howler monkeys (*Alouatta* spp.), and two species of the family Callithricihdae (*Saguinas midas* and *Callithrix penicillata*), were described by Flower (1872) and Mitchell (1905). These animals showed considerable variation in the relative length of the small intestine, cecum, and colon. The cecum was not haustrated in many of these species, but the colon was haustrated as a result of one or more taeniae in most of these animals. The proximal colon of the spider and capuchin monkeys showed an expansion similar to that illustrated in the night and woolly monkeys.

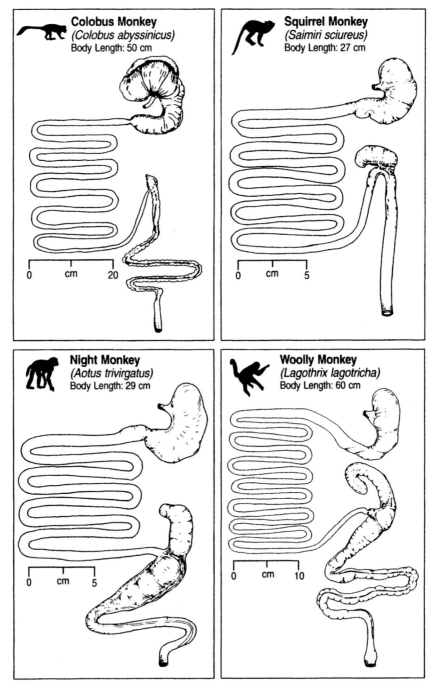

Fig. 4.19. Gastrointestinal tracts of a colobus monkey (from Stevens 1983), squirrel monkey, night monkey, and woolly monkey.

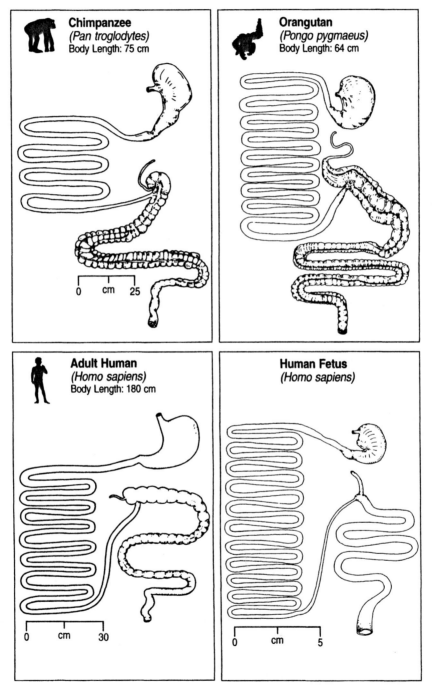

Fig. 4.20. Gastrointestinal tracts of a chimpanzee, orangutan, an adult human (from Wrong, Edmonds, and Chadwick 1981), and human fetus.

Family Hominidae is composed of the great apes and humans. The gorilla, chimpanzee, and orangutan are generally considered to be strict herbivores, although there is evidence that chimpanzees may hunt and eat termites and smaller monkeys (van Lawick-Goodall 1968). The gastrointestinal tract of a chimpanzee is illustrated in Fig. 4.20. Its colon is haustrated by the presence of three taeniae, which extend over most of its length. They also extend over the cecum, which terminates in a vermiform appendix. The gastrointestinal tract of the gorilla is similar, except that the small intestine is relatively long, and the hindgut is more voluminous (Mitchell 1905; Raven 1950). The cecum had a similar vermiform appendage, and three taeniae passed over the cecum and first 10 percent of the colon.

Orangutans and gibbons are arboreal folivores, although gibbons feed occasionally on insects and small vertebrates. The small intestine and colon of the orangutan are longer than those of chimpanzees, and its proximal colon is expanded (Fig. 4.20). A specimen of the gastrointestinal tract of a gibbon (*Hylobates* sp.) showed that it was similar to that of other apes, except for a shorter colon (Flower 1872).

The human gastrointestinal tract is well described elsewhere, and only a few major features of the large intestine will be highlighted for comparison. The cecum appears in the fetal gut of humans as a conical diverticulum (Fig. 4.20). The intestine then lengthens, but its terminal segment remains narrow in caliber. In the sixth month, faint longitudinal bands of muscle can be seen running from the apex of the cecum over the colon. At this stage, cecal growth is arrested and this, plus the shortening of its lateral longitudinal muscle bands and continued growth of the colon, results in the development of the adult form (Fig. 4.20). Therefore, a bulge of colon forms what is really a cecum that does not correspond either to the apex of the fetal cecum or the cecum of most other species. However, the vermiform appendix remains, and the colon is haustrated throughout almost its entire length, like that of most apes and many monkeys.

Suzuki et al. (1991) described species variations in the distribution of gastric epithelium. A distribution similar to that of the human stomach was seen in the macaco lemur; thick-tailed bush baby; night monkey; dusky titi; squirrel monkey; tufted capuchin; common and silvery marmosets, saddle-backed, white-lipped, and golden lion tamarins; Goeldi's monkey, and chimpanzee (Fig. 4.21, *A*). However, a relatively large area of cardiac mucosa was found in the stomach of many other monkeys (Fig. 4.21, *B*), and the stomachs of the black-and-white colobus, Francois' and dusky lutongs and proboscis monkey contained both an expanded area of cardiac mucosa and small region of stratified squamous epithelium (Fig. 4.21, *C*). Hill (1952) concluded that the *Colobus, Procolobus, Semnopithecus, Trachypithecus,* and *Presbytis* stomachs were quite similar, and the stratified squa-

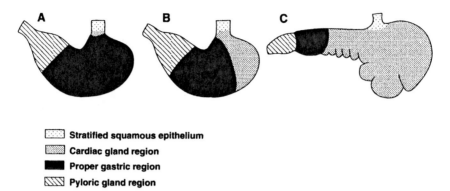

Stratified squamous epithelium
Cardiac gland region
Proper gastric region
Pyloric gland region

Fig. 4.21. Distribution of gastric epithelium in primates. **A**, Macaco lemur; thick-tailed bush baby; night monkey; dusky titi; squirrel monkey; tufted capuchin; common and silvery marmosets; saddle-backed, white-lipped, and golden lion tamarins; Goeldi's monkey, and chimpanzee. **B**, Humboldt's woolly monkey; diana monkey; Anubis, Hamadryas, and Guinea baboons; white-collared mangabey; Rhesus monkey; crab-eating monkey; Japanese monkey, and moor monkey. **C**, Black and white colobus; Francois' and dusky lutongs; and proboscis monkey. (From Suzuki et al. 1991.)

mous epithelium, evidenced by its "... firm, horny texture, opaque white appearance and rolled margin," continued into the first compartment to form a "cardiac shield" around the gastroesophageal junction. Kuhn (1964), Peng et al. (1983), and Ayer (1948) reached similar conclusions for *Procolobus*, *Rhinopithecus*, and *Semnopithecus* respectively.

Chivers and Hladik (1980) reviewed an extensive series of studies relating measurement of various parameters of the digestive tract to the diet of faunivorous, frugivorous, and folivorous primates. The faunivores had a simple stomach and hindgut and a long small intestine. Folivores had a larger stomach and/or an enlarged cecum and colon. Weight, volume, and surface area ratios between the stomach plus hindgut and the small intestine varied in a similar manner.

Artiodactyla

This order contains the ungulates (hoofed mammals) with an even number of toes. Although some swine are omnivores, with a simple stomach, most artiodactyls are herbivores and have a stomach that is both voluminous and highly compartmentalized. Because of the domestication of many members of this order for the production of food and fiber, their digestive systems have received extensive study, and much of our knowledge of microbial digestion in herbivores is derived from these animals.

There has been much disagreement among taxonomists over whether the 187

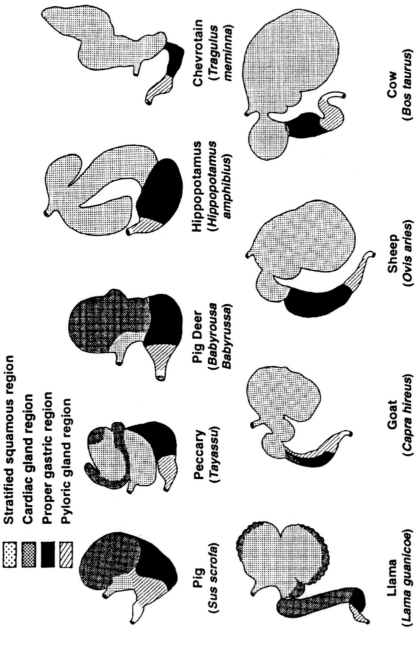

Fig. 4.22. Distribution of gastric epithelium in various species of Artiodactyla. (From Bensley 1902–03; Langer 1974; Moir 1965; Sisson 1975; Vallenas et al. 1971a.)

Stratified squamous region

Cardiac gland region

Proper gastric region

Pyloric gland region

Pig
(*Sus scrofa*)

Peccary
(*Tayassu*)

Pig Deer
(*Babyrousa Babyrussa*)

Hippopotamus
(*Hippopotamus amphibius*)

Chevrotain
(*Tragulus meminna*)

Llama
(*Lama guanicoe*)

Goat
(*Capra hireus*)

Sheep
(*Ovis aries*)

Cow
(*Bos taurus*)

species within this order should be classified according to their ability to ruminate (regurgitate, remasticate, and reswallow their food), and the number of stomach compartments, or other characteristics. The present discussion follows the
classification of Simpson (1945), which subdivides the order into suborders
Suiformes, Tylopoda, and Ruminantia. The Suiformes include the families Suidae
(swine), Tayassuidae (javelinas, or peccaries), and Hippopotamidae (hippos). The
Tylopoda consists of the family Camelidae (New and Old World camels). The Ruminantia consists of the family Tragulidae (chevrotains, or mouse deer) and the
more advanced families Cervidae, Giraffidae, Antilocapridae, and Bovidae. Both
the Tylopoda and the Ruminantia ruminate. Therefore, the latter are sometimes
referred to as *the advanced ruminants*, or Pecora.

Fig. 4.22 illustrates variations, compartmentalization, and the distribution of the
different types of epithelium in the stomachs of artiodactyls. The Suidae include
the domestic pig, warthog, and babyrusa. The gastrointestinal tract of a domestic
pig is shown in Fig. 4.23. Its stomach is relatively simple, with a small conical
pouch projecting along the esophagus. The mucosal surface of the lesser curvature is marked by a longitudinal groove, the gastric sulcus.

The cranial half of the pig's stomach is lined with a small area of nonglandular, stratified squamous epithelium, and a large area of cardiac glandular mucosa
(Fig. 4.22). The caudal portion contains regions of proper gastric and pyloric glandular mucosa, which are approximately equal in size. The intestine of the domestic pig is relatively long, and the colon forms about one fourth of this. The
cecum is conical and haustrated. The colon is haustrated throughout much of its
length and arranged in a spiraling coil, which first diminishes and then, upon re-
versing its direction, increases the diameter of its spiral. A spiral colon, which
also was noted in tarsiers and lemmings, is characteristic of most of the Artiodactyla.

The stomach of the African warthog *Phacochoerus* appeared to have no cardiac pouch and less of a nonglandular area, and the African bush pig *Potamochoerus* had a stomach similar to that of the domestic pig (Flower 1872). However, the stomachs of the Malayan pig deer, babyrousa, and collared peccary
Tayassu tajacu are subdivided into a number of compartments and lined with
larger regions of nonglandular stratified epithelium (Fig. 4.22). A gastric sulcus
with elevated lips also passes from the gastroesophageal junction into the pylorus
of these animals (Langer 1988).

The hippopotamus has one of the most complex stomachs of all artiodactyl
species (Fig. 4.24). The cranial portion, which is the most complex in its compartmentalization, is lined with nonglandular stratified epithelium (Fig. 4.22). The
caudal compartment contains the proper gastric glands and a terminal region of
pyloric glandular mucosa. There appears to be no evidence of cardiac glandular

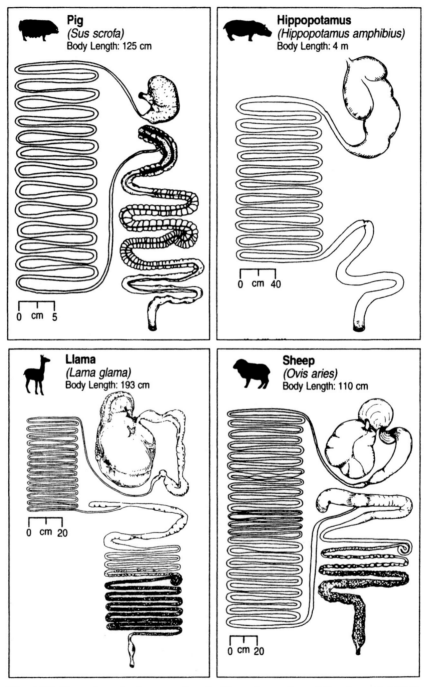

Fig. 4.23. Gastrointestinal tracts of a pig (from Argenzio and Southworth 1974), sheep (from Stevens 1977), hippopotamus, and llama. The stomach of the hippopotamus is less than its normal size because of protracted illness and anorexia before its death.

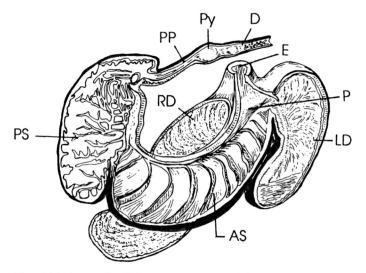

Fig. 4.24. Stomach of *Hippopotamus amphibius*. Esophagus (*E*), pillar (*P*), anterior stomach (*AS*), right diverticulum (*RD*), left diverticulum (*LD*), posterior stomach (*PS*), pars pylorica (*PP*), pylorus (*Py*), duodenum (*D*). (From Frechkop 1955.)

mucosa. The large intestine differs from other artiodactyls in that it is short and includes no cecum (Fig. 4.23). The ileocolic junction is marked principally by the cessation of villi, the beginning of longitudinal folds of mucosa, and a large patch of lymph follicles covered by a reticulated mucous membrane. The colon comprises only one tenth of the total intestinal length, and it is arranged in transverse folds rather than spirals. Therefore, the hippopotamus represents two extremes of the Artiodactyla gut, one of the most complex stomachs, and the simplest and shortest hindgut.

All other members of order Artiodactyla ruminate. The only living representatives of the Camelidae are the camels and their South American relatives–the llama, alpaca, guanaco, and vicuña. New World camelids have a complex, three-compartment stomach. The first compartment is partially subdivided into a cranial and caudal sac by a transverse muscular pillar crossing its ventral surface (Figs. 4.25 and 4.26). The first and second compartments communicate via a relatively large opening. Both compartments are lined with stratified squamous epithelium except for their ventral portions, which contain a large number of recessed glandular pouches. A strong muscular sphincter separates the second compartment from the elongated third compartment, which is entirely lined with glandular epithelium. The glandular sacs of the first two compartments and the mucosal surface of the proximal four-fifths of the third compartment are lined with a cardiac glandular mucosa (Cummings, Munnell, and Vallenas 1972). Proper

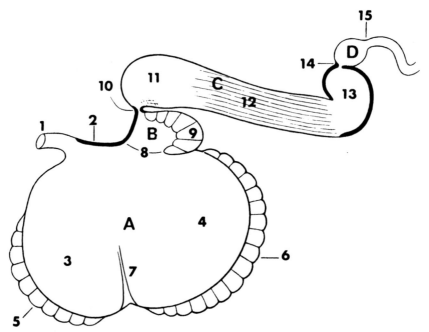

Fig. 4.25. Schematic representation of the llama and guanaco stomach. **A**, First compartment; **B**, second compartment; **C**, third compartment; **D**, duodenal ampulla; *1*, esophagus; *2*, ventricular groove; *3*, cranial sac; *4*, caudal sac; *5*, sacculated area in cranial sac; *6*, sacculated area in caudal sac; *7*, transverse pillar; *8*, entrance to second compartment; *9*, glandular cells of second compartment; *10*, tubular passage to third compartment; *11*, initial one fifth of third compartment (note reticulated area on lesser curvature); *12*, middle three fifths of third compartment (lines represent mucosal pleats); *13*, terminal one fifth of third compartment; *14*, pylorus; *15*, duodenum. (From Vallenas, Cummings, and Munnell 1971.)

gastric and pyloric glands are confined to the terminal one fifth of the third compartment. A gastric sulcus, or ventricular groove, joins the esophagus to the third compartment. The camel and dromedary appear to have a stomach similar to that of the New World species in both its compartmentalization and distribution of epithelium.

Fig. 4.23 includes the llama gastrointestinal tract. The intestinal tract is relatively long in comparison with other Artiodactyla and almost equally divided between small and large intestine. The diameter of the cecum and the proximal colon are enlarged in comparison with the remainder of the intestine, but they do not appear to be extensively haustrated. The intestinal tract of the dromedary *Camelus dromedarius* appears to differ only in having a much longer cecum and a shorter, more coiled spiral colon (Mitchell 1905). The camelids have the longest hindgut of all artiodactyls.

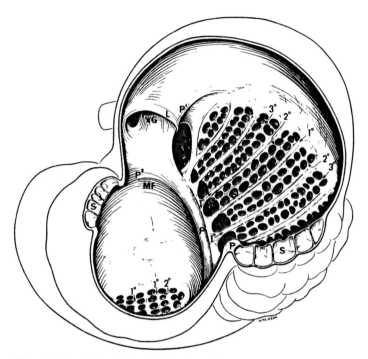

Fig. 4.26. Left lateral longitudinal section of llama stomach. A portion of the left wall of the first compartment has been removed to allow visualization of the mucosal surface and glandular pouches. *E*, Entrance to second compartment; *L*, lip of ventricular groove; *MF*, mucosal folds in cranial sac; *P*, transverse pillar; P^1, caudal limb of pillar; P^2, cranial limb of pillar; *S*, glandular saccules; *VG*, ventricular groove, 1°, primary crests; 2°, secondary crests; 3° tertiary crests. (From Vallenas, Cummings, and Munnell 1971.)

The Ruminantia include the chevrotains and a large number of domesticated and wild species of ruminants such as cattle, sheep, goats, deer, antelope, elk, buffalo, yak, and giraffe. The stomach of chevrotains, small (2.3–4.6 kg), deerlike animals found in Southern Asia and West Africa, has a large, nonglandular blind sac folded into a sigmoid pouch, or reticulorumen (Fig. 4.27). The esophagus enters a partially separated, cranial diverticulum called the *reticulum* because of the reticular pattern of its epithelial lining. The reticulorumen is lined with nonglandular, stratified squamous epithelium (Fig. 4.22). In the Asian genus *Tragulus* its outlet passes directly into a glandular compartment containing regions of proper gastric and pyloric glandular mucosa, but no evidence of cardiac glandular mucosa (Langer 1974; Agungpriyono et al. 1992). Individuals of the African genus *Hyemoschus* were reported to have an intervening, rudimentary omasum containing one or two leaves (Moir 1968). The glandular stomach is homologous to the glandular abomasum of more advanced ruminants. The intestinal tract of a num-

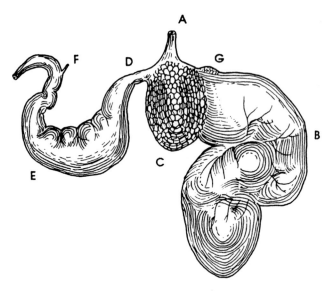

Fig. 4.27. Stomach of the African water chevrotain, *Hyemoschus aquaticus*. Letters refer to the esophagus (*A*), rumen (*B*), reticulum (*C*), reticuloabomasal junction (*D*), abomasum (*E*), entrance of the common bile duct (*F*), and spleen (*G*). (From Flower 1872.)

ber of Indian species of *Tragulus* revealed a relatively long hindgut, which included a cecum and a spiral colon (Mitchell 1905).

The highly compartmentalized stomach of more advanced ruminants consists of a forestomach (reticulum, rumen, and omasum) and a glandular stomach (abomasum). Fig. 4.28 provides a diagrammatic representation of the bovine stomach. The reticulorumen of domestic cattle is partially subdivided by a transverse fold and muscular pillars into a cranial reticulum and the dorsal and ventral sacs of the rumen. The gastroesophageal opening is connected by a ventricular groove, with muscular lips, to the reticulo-omasal orifice, which is located close to the gastroesophageal junction. The reticulo-omasal orifice contains a sphincter, which separates the reticulum from the omasum. The omasum is a globular-shaped organ. Most of the space in this organ is filled with long, thin, closely-spaced plates of tissue, called *omasal leaves*. The leaves, which project from its dorsal and dorsolateral curvature gave it the German term *Psalter* or *Bible*. The lesser curvature of the omasum contains a small area free of leaves, called the *omasal canal*, and a continuation of the ventricular groove. The omasum joins the third major compartment, the abomasum, via a relatively large omaso–abomasal orifice.

Both the reticulorumen and omasum are lined with nonglandular stratified squamous epithelium, but the abomasum contains proper gastric and pyloric glan-

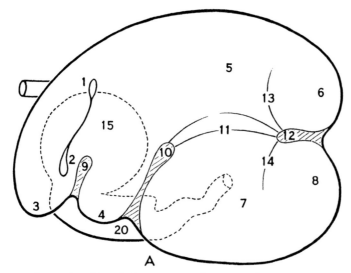

Longitudinal Section of Reticulorumen

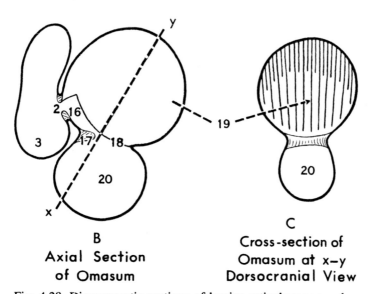

Fig. 4.28. Diagrammatic sections of bovine reticulorumen and omasum. Structures and compartments of major importance are numbered as follows: *1*, Cardia; *2*, reticulo-omasal orifice; *3*, reticulum; *4*, cranial sac of rumen; *5*, dorsal sac of rumen; *6*, caudodorsal blind sac; *7*, ventral sac of rumen; *8*, caudoventral blind sac; *9*, ruminoreticular fold; *10*, cranial pillar; *11*, right longitudinal pillar; *12*, caudal pillar; *13*, dorsal coronary pillar; *14*, ventral coronary pillar; *15*, omasum; *16*, omasal canal; *17*, omasal pillar; *18*, omaso-abomasal orifice; *19*, omasal lamina (leaf); *20*, abomasum. (From Sellers and Stevens 1966.)

dular mucosa (Fig. 4.22). Stomachs of sheep and goats are similar to that of cattle. Bensley found no cardiac mucosa in the abomasum of sheep, but Langer (1974) described a region of cardiac mucosa on the greater curvature and pyloric mucosa along the length of the lesser curvature of the goat abomasum. The relative size of the omasum varies in different species of ruminants. It is largest in domestic cattle, where the surface area of its leaves can constitute one third of the entire forestomach.

Hofmann (1968, 1973, 1983) examined the relationship between the structural characteristics of the forestomach and the diet in a wide range of East African ruminants. He found that these animals could be categorized in the following ways: (1) as concentrate selectors, which browsed on the more highly nutritious, succulent portions of the plant; (2) as bulk and roughage eaters that were relatively nonselective in their browsing and grazing; and (3) as an intermediate group that varied in their feeding habits according to area or availability of food. The forestomach of concentrate selectors such as Guenther's dik-dik (Fig. 4.29), which weighs 3.5–5 kg, has a relatively small reticulorumen and omasum in comparison with the intermediate feeders and, particularly, the nonselective bulk and roughage eaters.

The gastrointestinal tract of a sheep is shown in Fig. 4.23. The small intestine is very long in relation to the hindgut, when compared with the llama or the pig. The cecum and proximal colon appear more voluminous than those of the llama but show a similar lack of haustration. According to Sack and Ballantyne (1965), the cecum of a musk ox calf was approximately one third the length of the animal's body.

In summary, the stomach of individual artiodactyl species varies from a simple, nonvoluminous, noncompartmentalized organ to an extremely voluminous, highly compartmentalized organ. Although there are major species variations in the distribution of the different types of gastric epithelium, the stomach of all species contains an area of nonglandular stratified squamous epithelium emanating from the gastroesophageal junction. This varies in its relative size and, in some species, contains islands of cardiac mucosa. In others, it simply joins the zone of cardiac or proper gastric glandular mucosa. A second common characteristic is the presence of a ventricular groove or gastric sulcus along the mucosal surface of the lesser curvature of the stomach. In young ruminants, the muscular lips of this groove close at the time of suckling, forming a tube that shunts milk past the forestomach and directly into the abomasum.

The intestinal tract of artiodactyls shows wide variation in the relative length of the small versus the large intestine, as well as large intestinal capacity and degree of haustration. There tends to be an inverse relationship between the relative

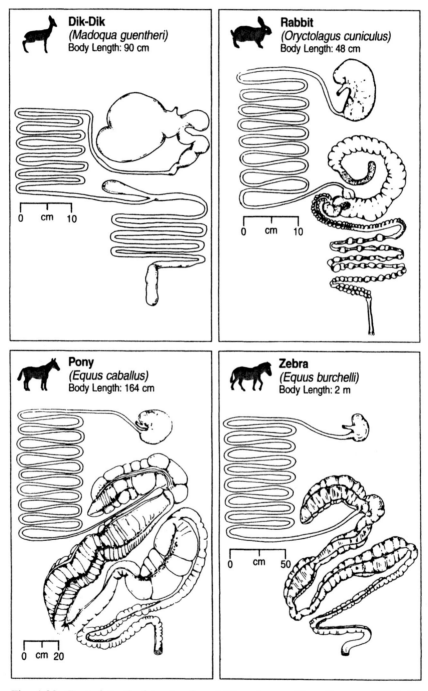

Fig. 4.29. Gastrointestinal tracts of a rabbit and pony (from Stevens 1977) dik-dik and a zebra.

complexity and capacity of the stomach versus the large intestine. However, the length and complexity of the large intestine also correlates with the need for water conservation, as evidenced by differences between the hippopotamus and camelid species.

Dermoptera

The two species of colugo (*Cynocephalus* spp.) are often referred to as *flying lemurs*, even though they glide rather than fly and are not really lemurs. Their diet consists of leaves, buds, flowers, and fruit. Wharton (1950) reported that the intestinal tract of *C. volans* was nine times the body length, and the cecum, which was approximately four times the body length, was divided into two compartments. Flower (1872) stated that the colon was relatively large in diameter and haustrated in its more proximal segment, as a result of three taenia.

Lagomorpha

The rabbits, hares, and pika, which constitute this order, are herbivores, presently represented by two families and fifty-eight species. They are similar to rodents in several respects and were once included in the same order. The gastrointestinal tract of a rabbit is shown in Fig. 4.29. The stomach is simple, but relatively long with a well-developed oral pouch. The cardiac glandular region projects only a short distance into the stomach. Proper gastric glands line a large portion of the stomach, including the pouch, and the terminal one fourth is lined with pyloric glandular mucosa. Therefore, these animals, which are strict herbivores, have a stomach lined entirely with glandular epithelium and, largely, with proper gastric glands. The small intestine is dilated immediately above the ileocolic valve in a structure called the *sacculus rotundus*, which has lymph follicles in its wall. Another group of these follicles partially encircle the colon distal to the valve.

The rabbit cecum is very large in relation to the rest of the gut. It forms a spiral that occupies much of the abdominal cavity and has a capacity approximately ten times that of the stomach. Its mucous membrane includes a projecting fold that spirals around the inside of the cecum as many as twenty-five times over its length, greatly increasing the internal surface area (Snipes 1978). The cecum terminates in a vermiform appendix with a small lumen and thick walls containing lymph follicles. The colon begins as a thin-walled sac, but very soon it becomes bilaterally haustrated by the presence of three taeniae. Two of these disappear after a short distance so that the haustrations continue on only one side of the colon. The third taenia also disappears approximately one third the distance to the anus (Snipes et al. 1982).

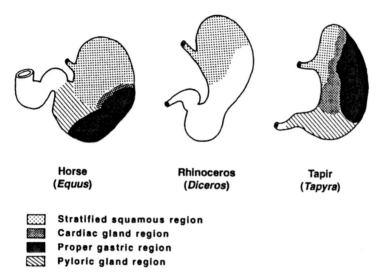

Horse
(*Equus*)

Rhinoceros
(*Diceros*)

Tapir
(*Tapyra*)

Stratified squamous region
Cardiac gland region
Proper gastric region
Pyloric gland region

Fig. 4.30. Distribution of gastric epithelium in Perissodactyla. (From Bensley 1902–03.)

Perissodactyla

This group of ungulates with an odd number of toes is represented by the families of equine, rhinoceros, and tapir species. The equids are bulk-feeding grazers that spend up to eighteen hours a day feeding. The rhinos, which are presently represented by five species, include both grazers and browsers. The tapir family includes four extant species, most of which are browsers that inhabit tropical forests.

Fig. 4.29 shows the gastrointestinal tract of the pony. The stomach is simple. The cranial half of the equine stomach is lined with nonglandular stratified squamous epithelium (Fig. 4.30). The caudal half is progressively lined with a narrow segment of cardiac mucosa and wider areas of proper gastric and pyloric glandular mucosa. Stomachs of the tapir and rhinoceros appear more elongated than that of the horse, and the tapir stomach contains a much smaller area of nonglandular epithelium.

The large intestine of horses is very complex and extremely voluminous. Although the cecum is relatively large and haustrated, the colon is much more voluminous and is both haustrated and further subdivided into compartments. The cecum opens into an expanded segment of colon (ventral colon). The next segment (dorsal colon) is, at first, narrow in diameter and then expanded once more. The terminal third of the colon has a lesser diameter and is referred to as the *small colon*. The large intestines of the zebra (Fig. 4.29), rhinoceros (Fig. 4.31), and tapir (Mitchell 1905) show a somewhat similar construction.

Proboscidea

This order contains only two surviving species, which inhabit Asia (*Elephas max-imus*) and Africa (*Loxodonta africana*). These are the largest living land animals. Their diet consists of trees, shrubs, grasses, and aquatic plants. The gastrointestinal tract from an African elephant is illustrated in Fig. 4.31. The stomach is simple and relatively narrow, with a cone-shaped cul-de-sac near its gastroesophageal junction. The cecum is conical and haustrated, and the colon is large in diameter and haustrated over a major portion of its length. Mitchell (1905) gave a similar description of the intestine from a young African elephant and noted that the ileum passed some distance into the wall of the hindgut, which suggested to him the presence of "primitively" paired ceca.

Sirenia

The Sirenia is a group of aquatic herbivores, which are believed to be related to both the ungulates and the whales. It includes two genera: the dugongs, *Dugong*, which inhabit a wide range of southern ocean and the manatees, *Trichechus*, of the African and American tropics. They feed on grasses and algae that normally inhabit coastlines, bays, and river estuaries. A third, very large (up to 7.5 m in length) member of this order, Steller's sea cow *Hydrodamalis gigas*, was exterminated within thirty years of its discovery in 1768 (Husar 1975).

A dugong gastrointestinal tract is shown in Fig. 4.31. Kenchington (1972) described the stomach of *Dugong dugong*, including a histological examination of gastric mucosa. The stomach was partially divided into two compartments by a ridge near its center (Fig. 4.32). A glandular mass near the gastroesophageal junction consisted of cardiac glandular mucosa. The remainder of the first compartment appeared to be lined with proper gastric mucosa. The second compartment, which he referred to as the *pyloric region*, was lined with a mucosa containing deep gastric pits with prominent gland cells and underlying submucosal aggregations of Brunner's glands. A sphincter separated the second compartment from a third compartment, which appeared to be a duodenal bulb, with a pair of diverticula and no sphincter between it and the remainder of the small intestine. The small intestine entered a very muscular cecum and a colon that was extremely long, but thin-walled, with no haustra.

The stomach of the manatee resembles that of the dugong, with a glandular mass in the first compartment and a pair of appendages in the second compartment (Flower 1872; Mitchell 1905). However, the cecum differed from that of most other mammals, in that the ileum enters a very expanded section of unsacculated proximal colon from which a pair of short, conical ceca project back along each side of the ileum. The ceca are not muscular as in the dugong. The expanded

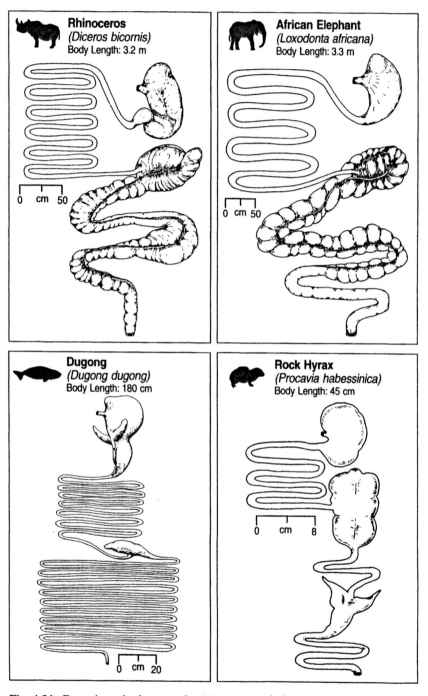

Fig. 4.31. Gastrointestinal tracts of a rhinoceros and elephant (from Clemens and Maloiy 1982), a hyrax (from Clemens 1977) a dugong.

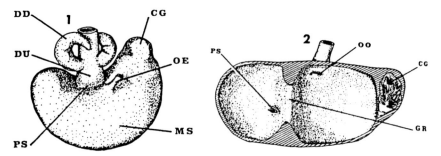

Fig. 4.32. Stomach of a dugong. The drawing on the left shows the cardiac gland area (*CG*), esophagus (*OE*), pyloric sphincter area (*PS*), duodenum (*DU*), duodenal diverticulum (*DD*), and main sac (*MS*). The drawing on the right gives a sectional view showing the esophageal entrance (*OO*), cardiac gland (*CG*), gastric ridge (*GR*), and pyloric sphincter (*PS*). (From Kenchington 1972.)

section then narrows to form a very long and coiled segment. Snipes (1984) provided a detailed description of the anatomy of the cecum from the West Indian manatee *T. manatus*.

Hyracoidea

The hyraxes are rodentlike herbivores. They are presently represented by three genera and eleven species indigenous to Africa and the Middle East and are believed to be related to the elephants and sirenians. The gastrointestinal tract of a hyrax is illustrated in Fig. 4.31. The large stomach contains a cranial outpocketing, or cul-de-sac, lined with nonglandular epithelium, and the first segment of intestine opens into a cecum, which is sacculated and quite capacious, the contents contributing 5.2 percent of body weight (Rübsamen, Hume, and Engelhardt 1982). The intestine leaving the cecum is histologically similar to hindgut, that is, it does not have villi (Rahm 1980) and decreases to a diameter approximately equal to that of the small intestine. However, it then opens into a wide segment containing paired ceca, or "colonic appendages" (Rübsamen et al. 1982). This serial arrangement of ceca is unique to these animals. The remainder of the hindgut gradually diminishes in diameter again.

Summary

The gastrointestinal tracts of some mammals are similar to those of various reptilian species. The simplest gastrointestinal tract is seen in some Insectivora, Carnivora, and some carnivorous marsupial species. It consists of a simple stomach lined with cardiac, proper gastric and pyloric glandular mucosa, and a relatively short intestinal tract, which may not contain either a valve or cecum to indicate separation of the midgut and hindgut. The hindgut in these species is neither larger than the midgut in its diameter nor is it haustrated. In some of these animals, this may represent a primitive condition. However, lack of a clear distinction between midgut and hindgut in species belonging to other orders, such as the cetaceans, probably represents a regressive characteristic.

The stomach and hindgut of many mammals show some major structural innovations as compared with lower vertebrates. The stomach of a number of mammalian species is highly developed into a capacious, compartmentalized organ. This can be seen in individual species belonging to the orders Chiroptera, Artiodactyla, Sirenia, Cetacea, Rodentia, Edentata, Marsupialia, and Primates. It provides for increased storage capacity and, in all but the bats, a site for microbial fermentation. The stomach of most of these species contains a gastric sulcus connecting the esophagus to the region of proper glandular mucosa. Black and Sharkey (1970) concluded that this was an obligatory adaptation to prevent dilution and fermentation of milk in suckling young. With the exception of the bats and sirenians, expansion of the stomach is associated with an increase in the areas lined with stratified squamous epithelium and often those lined with cardiac glandular mucosa. However, stratified squamous epithelium is found in the simple stomachs of the monotremes, scaly anteaters, armadillos, the domestic pig, and the Perissodactyla as well.

The hindgut shows a similar range in relative capacity and complexity. The hindgut of most carnivorous mammals is simple, and it lacks a cecum or even a valvular separation in some species. The midgut and hindgut of most mammals can be distinguished from one another by the presence of a valve or sphincter, and a cecum, which is paired in a few species. The cecum and colon show considerable species variation in their capacity and degree of haustration. The cecum is haustrated in many species, and this often extends to the proximal colon. In a few animals such as the horse, pig, macaque, baboon, ape, and human, these haustrations extend over the entire length of the colon. The cecum is particularly well-developed in the lagomorphs, many rodents, and some marsupials. The greatest capacity and complexity of the colon is seen in the Perissodactyla and Proboscidea. The physiological significance of the dimensions, capacity, haustration, compartmentalization, and other structural characteristics of the gastrointestinal tract is discussed in subsequent chapters.

5

Motor activity

The motor or muscular activities of the digestive tract are responsible for the mixing of food with secretions and for the transit of food and digesta through the alimentary tract at a rate optimal for digestion and absorption. As noted in Chapter 2, the esophagus and gastrointestinal tract of vertebrates are generally invested with an inner layer of circular muscle and an outer layer of longitudinal muscle. An additional oblique layer, or sling, of muscle may be present in the cranial wall of the stomach. The circular muscle is thicker in some sections, forming sphincters or valves at various points along the digestive tract. The longitudinal layer is actually spiroform in some regions, such as the esophagus, and it consists of only two or more longitudinal bands of muscle, or taeniae, distributed at intervals around the stomach, cecum, or colon of some species.

The variety of mechanisms that have been adopted for the ingestion and mastication of food were mentioned in earlier chapters. The esophageal muscle of some species is striated over varying lengths, but the remainder of the tract contains only smooth muscle. The smooth muscle of the gut contracts more slowly and shows a remarkable ability to adjust its tonus to accommodate marked changes in volume, with little or no increase in intraluminal pressure. This is most evident in the ability of the stomach muscle to relax as it fills during a meal. The following discussion concentrates on the motor activity of the digestive tract and its contributions to the mixing and transit of digesta.

Early investigators divided the motor activities of the digestive tract into different types of stationary mixing contractions and the propulsive contractions of peristalsis. The latter were defined as a moving ring of muscular activity, which consisted of a wave of relaxation followed by a wave of contraction that propelled digesta aborally along the gut. Similar propulsion in the orad direction was labeled antiperistalsis. Christensen (1971) stated that the esophagus appears to be the only segment of the mammalian tract that regularly demonstrated these moving rings of contraction and, in humans, even deglutition could be accomplished in about 10 percent of the young and a majority of elderly individuals by a simultaneous contraction of the entire esophagus. He concluded that the motor activity can be most readily described as simply mixing and propulsive contractions. Nevertheless, the terms *peristalsis* and *antiperistalsis* continue to be used to describe most types of propulsive and retropropulsive movement of digesta.

Esophagus

Deglutition

Deglutition, or swallowing, involves a complicated series of events that have been described in detail for dogs and humans (Davenport 1982; Goyal and Paterson 1989). Deglutition is initiated by the voluntary action of food being pushed by the tongue upward and backward against the palate and into the pharynx. This stimulates receptors that initiate nerve impulses to the deglutition center in the brain stem, resulting in reflex contraction of pharyngeal muscles. The soft palate is pulled upward to close off the posterior nares. The palatopharyngeal folds pull medially to form a slit, through which food must pass. The vocal cords are pulled together, and the epiglottis is moved backward over the glottis, the opening of the larynx, preventing entry of food into the trachea. At the same time, the upper esophageal sphincter, which includes the first 3–4 cm of the human esophagus, relaxes, and contraction of the superior constrictor muscle of the pharynx forces food into the esophagus. This initiates a peristaltic wave of contraction over the pharynx and esophagus (Fig. 5-1, *A*). The entire pharyngeal stage of deglutition interrupts respiration for 1–2 seconds, during which the respiratory center is inhibited by impulses from the deglutition center.

The circular muscle at the terminal 2–5 cm of the human esophagus serves as a lower esophageal or gastroesophageal sphincter. As peristaltic waves approach the terminal esophagus, they are preceded by a wave of reflex relaxation. This is associated with relaxation of the gastroesophageal sphincter and receptive relaxation of stomach muscle. Peristaltic waves originating in the pharynx pass over the length of the human esophagus in approximately 8–10 seconds. If food remains in the esophagus, distension can generate secondary waves of peristalsis at any point along its length.

Much of the research conducted on the esophagus has centered on the mechanisms that provide for a competent gastroesophageal valve or sphincter. This is partly accomplished by constriction of the muscle at or near the gastroesophageal junction. However, it may be aided by the following: (1) flaps or rosettes formed by gastric mucosal folds; (2) constrictions enforced by the diaphragmatic crura; (3) the stopcock effect of an acute angle between the terminal esophagus and cranial stomach; and/or (4) a sling formed by the gastric oblique muscle. Therefore, all of these factors need to be considered for comparisons among species.

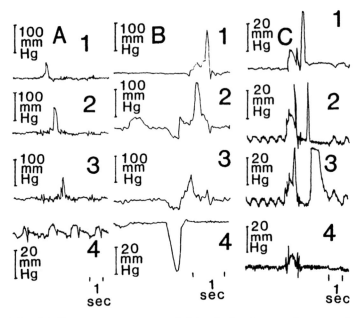

Fig. 5.1. Pressure events recorded in the bovine esophagus and pleural cavity during deglutition, regurgitation, and eructation. Numbers on the traces in these three series represent (1) cervical esophagus, 70 cm cranial to cardia, (2) thoracic esophagus, 40 cm cranial to cardia, (3) thoracic esophagus, 10 cm cranial to cardia, and (4) pleural cavity. Note peristaltic deglutition wave **(A)** with no associated change in respiratory cycle, compared with regurgitation **(B)**, which shows the marked drop of pressure in the pleural cavity and thoracic esophagus followed by an antiperistaltic wave of esophageal contraction. Eructation **(C)** is recorded in the esophagus as a rise in pressure of low amplitude and approximately one second duration, followed by a high pressure wave of short duration. The first wave of pressure, which is recorded simultaneously in the pleural cavity, represents the effect of abdominal press. The second, more rapid component, is an antiperistaltic wave of contraction that is followed, on this trace, by a peristaltic wave of deglutition. (From Sellers and Stevens 1960.)

Botha (1958, 1962) investigated the comparative anatomy of the gastro-esophageal junction in a dogfish, sturgeon, frog, tortoise, chicken, bat, mole, hedgehog, four species of Carnivora (dog, fox, ferret, and cat), four Rodentia (rat, mouse, guinea pig, and hamster), three Artiodactyla (pig, sheep, and ox), a rabbit, a horse, and a rhesus monkey. Functional studies were conducted on all species except the mole, hedgehog, bat, and fox. The circular muscle was thickened in the distal esophagus of the sturgeon but not the dogfish. He concluded that folds of gastric mucosa could act as a valve to prevent gastroesophageal reflux in both species of fish. The frog and tortoise stomachs showed a well-marked cardiac angle and large folds of gastric mucosa near the cardia, but no constriction or thickening of muscle was seen near their gastroesophageal junction. No sphincter, mucosal folds, or cardiac angle that might prevent reflux from the proventriculus and ventriculus (gizzard) were seen in the chicken. He suggested that fibers of gizzard muscle, passing over the terminal proventricular wall, may help serve this purpose.

Each of the mammals showed a functionally competent occlusion of the terminal esophagus. This was represented by a well-defined sphincter in the bat, pig, horse, rabbit, and rodents. It consisted of smooth muscle in the bat, pig, horse, and guinea pig; both smooth and striated muscle in the rabbit; and striated muscle in the rat, mouse, hamster, and mole. The terminal esophagus of the ferret, cat, dog, and monkey showed a similar constriction at its junction with the stomach but no anatomical sphincter or thickening of the circular muscle. All species demonstrated varying amounts of gastric mucosal folding and a cardiac sling of oblique muscle.

Under anesthesia, all mammals demonstrated primary peristalsis, and many demonstrated a secondary peristalsis as well. Peristaltic waves passed most rapidly over the section of esophagus containing striated muscle. The smaller the species, the more rapid the rate of passage. Esophageal peristalsis in the rabbit, pig, and cat also was rapid until it reached the terminal few cm, which underwent a contraction that was slower and delayed. The delayed contractions in the distal esophagus of these species could be attributed to slower contraction of smooth muscle. However, a similar delay in esophageal emptying was witnessed in the rat, mouse, hamster, guinea pig, ferret, and dog, whose esophagus contain only striated muscle. In the monkey, the only species examined that had an extensive section of esophageal smooth muscle, the peristaltic wave slowed down at the level of the aortic arch, continuing at a slower pace until it reached the terminal esophagus. The latter then contracted, on its own, forcing its contents into the stomach. Contraction of the longitudinal muscle shortened the esophagus during peristalsis, pulling the cardia forward. This was slight in some species, but marked in others, sometimes pulling the gastroesophageal junction into the thoracic cavity.

Reflux could be induced in all of these animals by either a sufficient increase in the intragastric pressure or manipulation of the gastroesophageal junction or cardia. The effect of sectioning the diaphragmatic crura varied. In some species, it appeared to have no effect on the competence of the gastroesophageal junction, but it appeared to have some effect in the ferret and dog, and reflux was much easier to induce in the cat after the crura were cut.

Regurgitation

Most mammals are capable of emesis, or vomiting, a reflex act that protects against absorption of toxic substances. The rat and horse are exceptions. The rat has highly developed senses of taste and smell, which aid in its selection of food. The horse appears to be less fortunate and, when a horse does vomit, its stomach is often found to be ruptured. This may be due to the highly muscular lower esophageal sphincter, or cardiac sling. Distension or irritation of the gastrointestinal tract initiates a nervous reflex via an emetic center in the brain. This inhibits the esophageal sphincters, increases the tonus of stomach muscles, and shortens the longitudinal fibers of the esophagus. These events, supplemented by constriction of the abdominal muscles (abdominal press), force gastric contents into the esophagus and mouth. There is no evidence of esophageal antiperistalsis during emesis in humans, and Botha (1958, 1962) observed no antiperistalsis in any of the species he studied.

Many raptors, such as the hawks and owls, are capable of controlling their regurgitation of digesta. This is a normal, physiological function that is associated with esophageal antiperistalsis and expels the relatively indigestible skin and skeletal parts of their prey (Duke et al. 1975; Kostuch and Duke 1975). Among mammals, the ruminating artiodactyls (camelids and ruminants) also demonstrate controlled regurgitation as part of the act of rumination, which consists of the regurgitation, remastication, reinsalivation, and redeglutition of food.

Grazing cattle spend approximately one third of their day ruminating. This is an involuntary act, which can be initiated by the presence of coarse, fibrous material in the forestomach and inhibited by any stimulus that tends to distress the animal. It allows these animals to ingest in haste and masticate at leisure and in greater safety. It also allows for thorough mastication of plant material that has already been softened and partly digested in the fermentation fluid (Hume and Warner 1980). Gordon (1968) concluded that the ability to ruminate in a recumbent position also conserves energy, which he estimated to be equivalent to 10 percent of the daily energy intake of sheep.

Rumination was extensively studied by French physiologists in the middle 1800s and by Dutch and German physiologists in the 1920s. Results of these and more recent studies have been reviewed by Bell (1958), Gordon (1968), and Stevens and Sellers (1968). The complete cycle of rumination is closely integrated with cyclic contractions of the ruminant forestomach and is discussed, along with that subject, later in this chapter. We will concentrate here only on the regurgitation and deglutition phases that directly involve the esophagus.

The regurgitation phase of rumination begins with inspiration against a closed glottis, which results in a strong, negative pressure in the thoracic segment of the esophagus (Fig. 5.1, *B*). This is accompanied by relaxation of the gastroesophageal sphincter, resulting in the aspiration of digesta from the rumen antrum into this segment of the esophagus. Digesta are then swept to the mouth by an antiperistaltic wave of esophageal contraction. The regurgitation phase of rumination differs from emesis in both the presence of esophageal antiperistalsis and the absence of abdominal press. Reflux of abomasal contents into the reticulorumen, in association with abdominal press, also can occur with blockage of the small intestine (Hammond et al. 1964), but trajectory emesis, with abdominal press and ejection of rumen contents from the mouth, occurs only rarely with raticulitis or ingestion of certain plant toxins.

Studies of regurgitation by kangaroos reviewed by Hume (1982) indicate that although these animals may regurgitate food at irregular intervals, the bolus is usually swallowed without further mastication. Addition of grain to a hay diet increased the frequency of regurgitation, suggesting that this served as a stimulus to salivary secretion rather than as an aid to mastication (Dellow 1982). Prolonged periods of rhythmic jaw movements, which were observed in the absence of feeding or regurgitation, also may stimulate salivary secretion.

Eructation

Eructation, the release of gas from the stomach via the esophagus, is a phenomenon that appears to be common to most species. However, it has been most extensively studied in ruminants because of its critical importance to the survival of these animals (Sellers and Stevens 1966; Dougherty 1968). Microbial fermentation in the forestomach of cattle can produce as much as a liter of gas per minute. Although much of this is CO_2, which is readily absorbable, approximately half of the gas is the less absorbable CH_4. Furthermore, at the height of fermentation, neither gas can be absorbed from the forestomach at a rate sufficient to prevent dilation. Therefore, ruminants are very subject to tympany, or bloat–a potentially life-threatening condition seen in animals on certain types of legume pasture or in animals that are allowed to rapidly ingest food containing readily fermentable carbohydrates.

Eructation, like rumination, is intimately tied to the cyclic contractions of the forestomach and, except for its esophageal components it occurs at a time in the reticuloruminal cycle when the cardia is exposed to gas. Contraction of the dorsal sac of the rumen and abdominal press increase the pressure of gas at the ruminal antrum. Relaxation of the gastroesophageal sphincter allows the gas to be forced into the esophagus and an antiperistaltic wave of esophageal contraction passes it to the mouth (Fig. 5.1, *C*). Heywood and Wood (1985) observed similar antiperistaltic contractions during simultaneous radiographic and electromyographical studies of eructation in sheep. The presence of fluid at the cardia (the area around the entry of the esophagus) of sheep appears to inhibit eructation. However, even when digesta were completely removed from the rumen of cattle, eructation only occurred at the time of dorsal sac contraction (Stevens and Sellers 1960).

Retrograde propulsion of gas by antiperistaltic waves of esophageal contraction were believed to be confined to ruminants. However, Heywood and Wood (1988) observed similar antiperistaltic contractions in simultaneous radiographic and electromyographic studies of naturally occurring eructation in the dog. They concluded that this may be present in other mammals with an esophagus composed completely of striated muscle, but the antiperistalsis associated with regurgitation in some species of birds indicates that this is not confined to striated muscle.

Although the ruminant esophagus shows a number of unusual functional characteristics, it represents a good model for studies of the nervous control of esophageal function. For example, Doty (1968) referred to the studies by Roman in sheep as supplying some of the most definitive information available on feedback alterations of the course and vigor of esophageal contraction.

Stomach

The motor activity of animals with a simple stomach has been most extensively studied in the dog, cat, and human (Kelly 1981; Malagelada and Azpiroz 1989). The stomach of these animals can be divided into two distinct regions of motor activity. The proximal one third undergoes receptive relaxation upon receipt of a bolus and can distend to a large size for food storage, with little increase in intragastric pressure. As the gastric phase of digestion progresses, depolarization of the membranes of the circular muscle cells in this region results in either rapid stationary contractions or a sustained tonic contraction of the muscle. The stationary contractions aid in the mixing of gastric contents. The sustained contractions press fluid from the proximal segment of the stomach and gradually move the more solid gastric contents into the distal segment. However, the muscle in this region shows neither rhythmic changes in the electrical potential of their cell membranes nor peristaltic waves of contraction.

The myoelectric activity of the distal region of the stomach is quite different from that of the proximal segment. The electrical potentials of its muscle cell membranes undergo slow, rhythmic cycles of partial depolarization, referred to as the *slow waves* or *basal electrical rhythm*. These are initiated by pacemaker muscle cells on the greater curvature of the stomach and are transmitted to a ring of circular muscle around the entire stomach. They then travel at a fixed rate for varying distances in an aboral direction toward and, sometimes over, the pylorus and its sphincter. Although the slow waves of depolarization do not initiate contraction, they sensitize the muscle cells to contract at the frequency or multiples of the frequency of slow wave production. When these muscle cells are further depolarized by extrinsic stimuli, the slow waves are associated with spike potentials, which result in waves of peristaltic muscular contraction. Contractions of distal stomach muscle mix gastric chyme with gastric secretions, propel the gastric chyme to the pylorus, and pass fluid and small particles of food into the duodenum. Both the presence and amplitude of these contractions are modulated by neurohumoral stimuli, which are discussed in Chapter 10.

Fluoroscopic studies of the tammar wallaby *Macropus eugenii* stomach showed two types of haustral contraction (Richardson and Creed 1981). One of these consisted of waves of contraction that spread in an aboral direction from one haustrum to the next. The other was an intermittent narrowing of haustra. A basal electrical rhythm of 5–6 cycles per minute was recorded along the entire stomach of the tammar wallaby and the quokka *Setonix brachyurus* (Richardson and Wyburn 1983, 1988), and haustral contractions were observed to pass aborally at the same frequency as the slow waves. Spike potentials were superimposed on the slow waves of the quokka stomach but were completely absent from the haustrated region of the tammar stomach.

The motor activities of the ruminant forestomach are very different from those of the simple stomach, in that they are cyclic, almost continuous, and under extrinsic nervous control. The complex, cyclic contractions of the bovine forestomach were first described by the classical studies of Schalk and Amadon (1921, 1928) and Wester (1926). More recent information has been reviewed by Sellers and Stevens (1966), Titchen (1968), and Phillipson (1977). The ruminant forestomach consists of the reticulorumen and omasum (see Fig. 4.28). The reticulum, which is the most cranial compartment, is partially separated from the rumen by a ruminoreticular fold. The cranial and dorsal sacs of the rumen are contiguous but separated from the ventral sac of the rumen by a ring of muscular pillars.

Cyclic contractions of the bovine forestomach occur in continuous succession at a rate of 0.9 ± 0.1 per minute during rest periods. This increases during feeding and decreases slightly during rumination. Ruckebusch (1975) has shown that these cycles are modified, but not entirely arrested, during sleep. A reticuloruminal cycle of muscular contraction begins with a double (cattle) or biphasic (sheep) contraction of the reticulum, which is immediately followed by sequential contraction of the cranial, dorsal, and ventral sacs of the rumen. The pillars also contract during this sequence with the cranial pillar contracting first and just before contraction of the dorsal sac. This primary contraction of the compartments and pillars of the rumen may or may not be followed by a secondary contraction of the rumen, which involves most of these same structures, but it is not preceded by the reticular contractions. There is disagreement over whether the primary and secondary contractions follow the same progression through the rumen. Ruckebusch and Tomov (1973) concluded that the secondary contractions of the sheep rumen may originate in the ventral blind sac of the rumen.

Fig. 5.2 shows a series of reticuloruminal cycles in cattle, consisting of double contractions of the reticulum followed by primary and, on alternate cycles, secondary contraction of the rumen. The effects of these contractions on the transit of digesta through the rumen are illustrated in Fig. 5.3. Cyclic contraction of the reticulorumen circulate and macerate ingested food and mix it with rumen microbes. Reticuloruminal contents consist of a ventral layer of fluid and small particulate digesta, a floating layer of coarse, fibrous plant material, and a pocket of gas that occupies the most dorsal aspect of the rumen. Fibrous plant material is carried by reticular and cranial sac contractions back into the dorsal sac and then forced ventrally by contractions of the dorsal sac. Subsequent contraction of the ventral sac of the rumen forces its contents upward and, if the cranial pillar is completely relaxed, some of the digesta are returned to the reticulum. However, Reid and Cornwall (1959) and Reid (1963) found that when a primary wave of ruminal contraction is followed by a secondary wave, the cranial pillar remains partly contracted. This blocks the return of digesta to the reticulum, resulting in a mixing of ruminal dorsal and ventral sac contents and exposure of the gastro-esophageal junction to gas for its eructation.

Particulate food of high specific gravity, such as poorly masticated kernels of corn, tend to sink rapidly into the cranial sac of the rumen immediately after ingestion. The next cyclic contraction of the cranial sac passes these forward into the reticulum, rather than over the cranial pillar into the more posterior compartments of the rumen.

The reticulo-omasal sphincter and omasal canal of cattle also undergo cycles of primary and secondary contractions with each reticuloruminal cycle at approximately the same time that the dorsal sac of the rumen undergoes contractions (Fig. 5.2). Palpation indicated closure of the reticulo-omasal sphincter with each contraction of the omasal canal, and the omasal canal showed a negative pressure at the height of reticular contractions and immediately following each contraction of the omasal canal. The omasal body also underwent strong prolonged contractions, but their timing was independent of those of the reticulorumen and omasal canal (Fig. 5.2), and their amplitude was inhibited by distention of the abomasum with a balloon. Although it once was assumed that digesta were pumped into the omasum by reticular contractions, the location of the reticulo-omasal orifice in a pocket formed by the ventral lips of the reticular groove made this difficult to explain.

Stevens, Sellers, and Spurrell (1960) examined the relationship between reticulo-ruminal-omasal motility and the flow of digesta between the reticulum and omasum, which was recorded with a Pitot tube arrangement of catheters in the omasal canal. Pitot tube measurements showed a rapid, brief flow of digesta at the height of the second reticular contraction and a slower but more prolonged

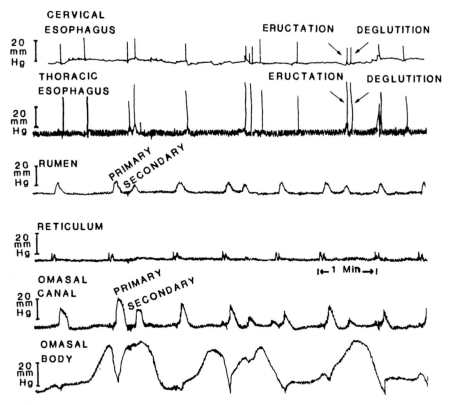

Fig. 5.2. A six-minute recording of reticuloruminal cycles and associated events in the esophagus and omasum. Note interruption in omasal body pressure waves during primary contraction of dorsal rumen and omasal canal. Upper traces show eructation as a biphasic wave occurring during secondary rumen contractions and the frequent waves of deglutition associated with the swallowing of saliva. (From Sellers and Stevens 1966.)

CYCLE WITH ONLY PRIMARY
CONTRACTION OF RUMEN

CYCLE WITH PRIMARY AND SECONDARY
CONTRACTIONS

Fig. 5.3. Movement of digesta during primary and secondary contractions of the reticuloruminal cycle.

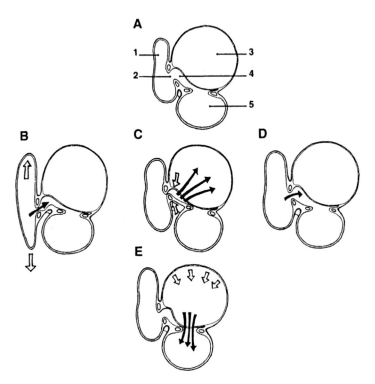

Fig. 5.4. Movement of digesta through the bovine omasum. (Movement of digesta is shown by *closed arrows*, and movement of the forestomach walls is indicated by *open arrows*.) Diagrammatic axial section (see Fig. 4.28) shows the cranial reticulum (*1*), reticulo-omasal orifice (*2*), omasal leaf portion of omasal body (*3*), omasal canal (*4*) and cranial abomasum (*5*). **A,** All structures are relaxed during much of the cyclic contraction of the forestomach. **B,** During the second reticular contraction, the reticulo-omasal orifice and omasal canal are pulled ventrally, producing a negative pressure in the canal and a closing and then opening of the orifice, which results in aspiration of digesta from the base of the reticulum. **C,** Primary contraction of the rumen is associated with a primary contraction of the reticulo-omasal orifice and omasal canal, which forces fluid and small digesta particles between the leaves of the omasal body and into the abomasum. This is followed by relaxation of these structures (**D**), and aspiration of digesta into the canal. This is repeated if the forestomach undergoes a secondary contraction. **E,** At intervals that vary and are unrelated to the cyclic contractions of the forestomach, a wave of contraction passes over the omasal body, releasing its contents into the abomasum.

flow of digesta during relaxation of the omasal canal, after its primary or secondary contraction. Reflux of digesta from the omasal canal back into the reticulum also was recorded when the omasal canal and the reticulo-omasal orifice were relaxed at the time of an omasal body contraction. These results suggested the following hypothesis that the bovine omasum acts as a two-phase filter pump in the manner illustrated in Fig. 5.4. The reticulo-omasal orifice is open during most of the cyclic contraction of the reticulum, rumen, and omasal canal. Reticular contractions appear to result in its passive closure and a negative pressure in the omasal canal. However, the stronger, second contraction of the reticulum reopens the orifice and results in a brief, but strong, aspiration of digesta from the base of the reticulum. This would account for the observation that whole kernels of corn can pass out of the reticulum and through the remainder of the digestive tract. Primary and secondary contractions of the reticulo-omasal orifice and omasal canal force digesta between the leaves of the omasal body and, in small amounts, into the abomasum. The subsequent relaxation of the omasal canal results in a negative pressure that aspirates digesta from the reticulum. This process would serve to fill the omasal body with fluid and small particles of digesta and retain larger particles for periodic reflux into the reticulum.

Transfer of digesta from the omasum into the abomasum is completed by strong and prolonged waves of contraction that pass over the omasal body (Fig. 5.2). These contractions are not related to the cyclic contractions of the reticulum, rumen, reticulo-omasal sphincter, or omasal canal. Palpation of the omaso-abomasal orifice indicated that digesta passed from the omasum into the abomasum in small dribbles during contractions of the omasal canal but in large quantities during contractions of the omasal body.

Omasal body contractions were stimulated by its distention, indicating that its contractions were controlled by the degree of fill. However, the amplitude of contraction was inhibited by abomasal distention, indicating that the omaso-abomasal orifice was subject to feed-back inhibition as well.

More recent studies of omasal function in cattle and sheep reviewed by Ehrlein (1980) and Mathison et al. (1994), appear to confirm many of the above conclusions. Opening of the reticulo-omasal orifice at the height of the second contraction was confirmed by a number of studies. Although there is disagreement over the periods of reticulo-omasal sphincter relaxation and evidence of differences between sheep and cattle, most studies indicate that it is relaxed during much of cyclic contraction (Deswysen 1987; Bueno and Ruckebusch 1974). Continuous distension of the reticulo-omasal orifice with a canula resulted in more than a twofold increase in reticulo-omasal flow (Bueno 1972). Backflow of digesta into the reticulum was confirmed by Ehrlein and Lebzien (cited by Deswysen 1987), and McBride, Milligan and Turner (1984). Omasal outflow was doubled by its diversion from the abomasum (Phillipson and Ash 1965). There is good evidence that the rate of digesta flow through the omasum increases during feeding and rumination, when the rumen and omasal canal are undergoing their most rapid rates of contraction. Digesta transit increases with an increase in either the rate or amplitude of reticular contractions, but this could be due to similar increases in the

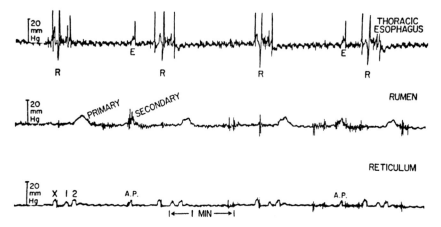

Fig. 5.5. Three complete cycles of reticuloruminal contraction during rumination. The double reticular contraction (*1* and *2*) and a primary and secondary ruminal contraction are labeled for the first cycle. The extra reticular contraction associated with regurgitation (*X*) is also labeled. Esophageal pressure changes during regurgitation (*R*) consist of a small positive-pressure wave, followed by the negative deflection caused by inspiration against a closed glottis and then the large positive deflection caused by antiperistaltic esophageal contraction (see Fig. 5.1, *B*). Regurgitation is usually soon followed by deglutition (unlabeled) of the fluid expressed from the bolus at the beginning of mastication. A wave of deglutition, carrying the bolus to the rumen, also is seen on the esophageal trace at the end of each cycle of rumination and just before the next regurgitation. Note the close integration of the rumination and reticuloruminal cycles. Eructation (*E*) is labeled, and the associated increase in pressure resulting from abdominal press can be seen on the reticular trace (*A.P.*) and superimposed on the secondary wave of ruminal contraction. (From Stevens and Sellers 1968.)

rumen and omasal canal, and Mathison et al. (1994) concluded that reticular contractions were not the principal factor governing outflow from the reticulorumen.

As mentioned earlier, rumination and eructation are closely integrated with the cyclic contractions of the reticulorumen. Fig. 5.5 shows a series of reticuloruminal cycles during a period of rumination. At the time of regurgitation, the reticulum undergoes an extra contraction, just before the normal two, which floods the cardia with additional fluid. Digesta are then aspirated into the esophagus and transferred by antiperistalsis into the mouth. Excess fluid is expressed from the bolus and immediately swallowed, but mastication continues until the remainder of the bolus is swallowed, just before the next cycle of rumination.

The entire series of events associated with cyclic contractions of the forestomach, rumination, and eructation are closely integrated under central nervous system control. For example, distension of the distal esophagus of cattle in the absence of rumination initiated secondary peristalsis rather than antiperistalsis

(Sellers and Titchen 1959), and dental abnormalities that disturb the mastication patterns of sheep can result in a change in forestomach motility (Ruckebusch, Fargeas, and Dumas 1970). The bolus of food is never swallowed until a few seconds before the initiation of another cyclic contraction of the forestomach. Eructation occurs only at a given phase of the reticuloruminal cycle (Fig. 5.2). It is seen only during contraction of the dorsal sac of the rumen, more commonly on the secondary contraction in cattle, when relaxation of the reticulum and contraction of the rumen lowers the level of digesta in the rumen and exposes the cardia to rumen gas. However, it is confined to this period of the cycle even after removal of digesta and continuous exposure of the cardia to gas.

Vagotomy eliminates the cyclic contractions, as well as rumination and eructation. Sheep subjected to total vagotomy and sustained for two to five months by intra-abomasal infusion of nutrients, regained motility of the forestomach, but it was no longer synchronized into the cycles seen in the vagus-intact animal (Gregory 1982). Although pentagastrin inhibited forestomach motility in sheep (Carr, McLeay, and Titchen 1970) and cattle (Ruckebusch 1971), recent evidence suggests that this acts through gastric centers in the medulla (Grovum and Chapman 1982). The control of forestomach motility, rumination, and eructation in ruminants is one of the most complex examples of central brain stem control and integration of a biological function.

The forestomach compartments of the llama also undergo cyclic, sequential contractions (Ehrlein and Engelhardt 1971; Vallenas and Stevens 1971a). Figs. 4.25 and 4.26 illustrate the llama stomach with its three major compartments and its glandular pouches. Vallenas and Stevens (1971a) found that the cycle began with a contraction of the second compartment (Fig. 5.6). This was followed by six or seven contractions of the caudal and then the cranial sacs of the first compartment. Each contraction was associated with eversion of its glandular pouches (Fig. 5.7). During rumination, regurgitation of the bolus was associated with extra contractions of the cranial sac of the first compartment. Eructation occurred at the height of caudal sac contraction. Cyclic contractions of the llama forestomach occurred at a rate of 0.6 ± 0.1/min, which is similar to the rate of reticuloruminal cycles in cattle, and the rate was similarly increased by feeding and decreased during rumination. Ehrlein and Engelhardt (1971) recorded more frequent contractions of the second compartment, but a more complex cycle of forestomach motility was described by Heller, Gregory, and Engelhardt (1984). The latter study indicated that the motility of the first two compartments of the llama stomach was under vagal nerve control in a manner similar to that noted in cattle and sheep.

Rumination has been described in both infant and adult humans (Brown 1968; Einhorn 1977; Menking et al. 1969; Feldman and Fordtran 1978; Fleischer 1979). Brown noted that it was mentioned by Aristotle and Galen and was well docu-

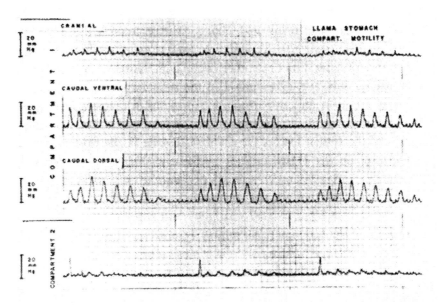

Fig. 5.6. Pressure events recorded simultaneously from compartments 1 and 2 of the llama stomach. Note that cycles begin with a contraction of compartment 2. This is followed by a series of contractions of the caudal and cranial sacs of compartment 1. (From Vallenas and Stevens 1971a.)

mented as early as 1618. It appears to be an involuntary, painless act, usually seen in patients with signs of emotional distress. Monographic and cineradiographic studies have indicated normal esophageal peristalsis, and normal activity and competence of the gastroesophageal sphincter in the absence of rumination. There was no evidence that regurgitation was associated with antiperistaltic contractions of the esophagus, as seen in ruminants.

In contrast to the forestomach, the abomasum of ruminants appears to contract in a manner similar to the stomach of humans and dogs (Ehrlein 1970). Although distension of the abomasum inhibits contractions of the omasal body, abomasal contractions are not coordinated with those of the reticulum, rumen, or omasal canal and are not permanently affected by vagotomy (Duncan 1953).

Midgut

The motor activity of the small intestine is more varied and complex than that of the simple stomach (Weisbrodt 1981; Davenport 1982; Mathias and Sninsky 1985; Fiorenza, Yee, and Zfass 1987). The major functions of the small intestinal muscle are to mix its contents and propel digesta from the stomach to the hindgut, or large intestine. Although mixing is aided by movement of mucosal folds and villi,

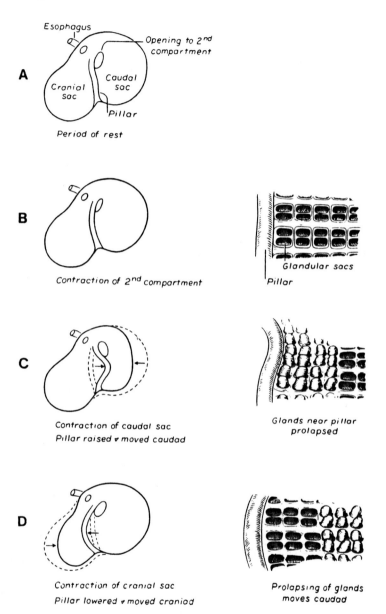

Fig. 5.7. Basic stages in cyclic contraction of the first compartment of the llama and guanaco stomach. Drawings **A–D** indicate contractions of pillar and sacs. In **B,** contraction of the second compartment can be observed only as a decrease in size of the orifice separating compartments 1 and 2. Cyclic eversion of caudal sac glandular pouches is indicated by drawings (*on the right*) of this area during three stages of contraction. (From Vallenas and Stevens 1971a.)

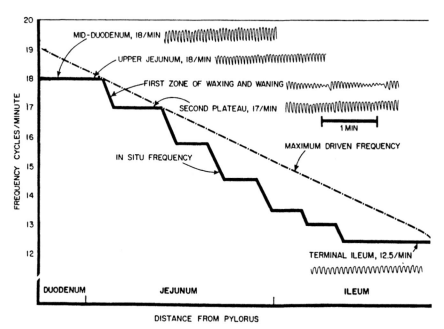

Fig. 5.8. Frequency of the basic electrical rhythm of the small intestine of the anesthetized cat. The stepwise line shows the observed frequency, 18/minute in the duodenum and upper jejunum and decreasing to 12.5/minute in the terminal ileum. Segments of the electrical record from which the frequency was measured are shown. (Modified from Davenport 1982.)

which is believed to result from contraction of the muscularis mucosa, it is largely dependent on rhythmic segmental contractions of the circular muscle layer. These segmental contractions are the most common type of motor activity, and their maximal frequency is determined by the slow waves or basal electrical rhythm (BER) of the muscle membrane potential, in a manner similar to that described for the simple stomach. The frequency of the slow waves decreases in a stepwise fashion between the duodenum and the terminal ileum of most species (Fig. 5.8), and this was believed to aid in the aboral propulsion of digesta. However, no gradient in slow wave frequency was found along the small intestine of the guinea pig (Calligan, Costa, and Furness 1985) or the tammar wallaby and quokka (Richardson and Wyburn 1988). As in the stomach, the frequency of the basal electrical rhythm is intrinsic to the muscle cells, but the number of contractions associated with these cycles is modulated by neuroendocrine control.

Most of the propulsion of chyme through the small intestine is accomplished either by the contractions just described, which move it short distances, or by peristaltic rushes, which are accompanied by migrating myoelectrical complexes (MMC), and may travel over the entire length. In many species, feeding increases the frequency of segmental contractions, and peristaltic rushes are largely confined to interdigestive periods. The three phases of the MMC are illustrated in Fig. 5.9. Phase I is a quiescent period, when slow waves are not accompanied by action potentials. The slow waves are accompanied by action potentials intermittently in phase II and continuously in phase III. In carnivores and humans, feeding results in prolongation of phase II. However, in herbivores, such as the ruminants, horse, and macropod marsupials (Richardson and Wyburn 1983), which are normally more continuous feeders, the MMC pattern is uninterrupted by feeding. The effect of feeding on the MMC of the omnivorous pig appeared to vary between that of an herbivore, when pigs were fed an herbivorous diet, and that of a carnivore, when pigs were fed two or three high-energy meals a day.

Myoelectrical activity has been recorded in the small intestine of a wide range of mammals (Ruckebusch 1981). The MMC was found to pass over the upper small intestine of the dog at a rate of 6–12 cm/min, slowing down to a rate of 1–3 cm/min in the distal half of the small intestine. Six or eight sequences of MMC were measured in the small intestine of humans during an overnight fast, with an average of about 10 per day. However, both the number of complexes per day and their velocity of propagation vary with species (Fig. 5.10). Both appear to increase with the length of the intestine. In carnivores, the MMC were limited to interdigestive periods, but large, continuous-feeding herbivores averaged 15–25 MMC, distributed throughout the day. Roche and Ruckebusch (1978) recorded myoelectrical activity in the oral as well as in the aboral direction in the small intestine of the chicken.

A valve, or sphincter, separates the midgut and hindgut of most adult amphibians, reptiles, birds, and mammals. This serves to retain digesta in the midgut for effective digestion by endogenous enzymes and for the absorption of end products. It also allows retrograde propulsion of hindgut contents without microbial overgrowth of the midgut contents. Philips (1988) reviewed studies of the ileocolonic junction and concluded that the passage of digesta may be controlled by a mechanical valve, muscular sphincter, external ligaments, or a combination of these barriers.

Hindgut

The motor activity of the hindgut, or large intestine, is more complex than that of the midgut and much less understood (Christensen 1994). Although a number of early investigators concluded that the presence of a cecum of any length requires a mechanism for its filling, Cannon (1902) appears to have made the earliest radiographic observations of the retropulsion of contents from the proximal colon into the cecum of the cat. He also noted that the contents of more distal segments of the colon were separated into "globular masses" by segmental contractions, which could be seen to move slowly in an aboral direction. These observations were followed by a comparative study of large intestinal motility and digesta transit by Elliott and Barclay-Smith (1904) on the anesthetized hedgehog, ferret, dog, cat, rat, guinea pig, and rabbit. The first four species were selected for the relative simplicity of their hindgut. The rat was chosen to represent an omnivore, and the last two species were selected to represent herbivores, with the rabbit as the extreme in cecal length and capacity. Colonic antiperistalsis was observed in all species except the dog and resulted in a reflux of colonic digesta into the cecum of animals other than the hedgehog and ferret, which lack a cecum.

From these studies, Elliott and Barclay-Smith (1904) concluded that the colon of the rabbit could be functionally divided into the following three segments: (1) a proximal segment with fluid contents similar to that of the cecum and a motor activity dominated by antiperistaltic contractions; (2) an intermediate segment of colon in which fecal nodules first appeared and peristaltic contraction predominated; and (3) a distal colon with definite fecal nodules and a motility similar to that of the intermediate colon–except that stimulation of its sacral parasympathetic nerve supply resulted in its evacuation by a massive simultaneous contraction. These functional segments of the colon were less distinct in the omnivorous rat and least distinct in the carnivores. The entire colon of the dog appeared to be equivalent to the distal segment of the rabbit colon.

It is interesting to note that Elliott and Barclay-Smith (1904) concluded their discussion with the observation that the proximal colon of mammalian herbivores tends to be characterized by marked sacculation or haustration, a condition seen throughout the human large intestine. They cited a study of 1,000 Egyptian mummies that indicated that their cecum was considerably larger than that of present-day humans. Therefore, in spite of reservations about deducing function from structure, these investigators concluded that "There can be little doubt that the human colon is rather of the herbivorous than carnivorous type."

It is surprising that the above study attracted relatively little attention until recent times, because it would seem to provide a reasonable working hypothesis for

Fig. 5.9. Phases of migrating myoelectrical complexes (MMC) during interdigestive (fasting) and feeding periods in carnivores and humans. **A** shows slow waves with no spike potentials during phase I, intermittent spike potentials during phase II, and consistent spike potentials during phase III. Lower diagrams **B** show the effects of interdigestive periods (B) and feeding (C) on the MCC, which originate in the stomach and lower esophageal sphincter (LES) and pass through the small intestine. **C,** Feeding interrupts the cycle and increases the duration of phase II. (From Hendrix 1987.)

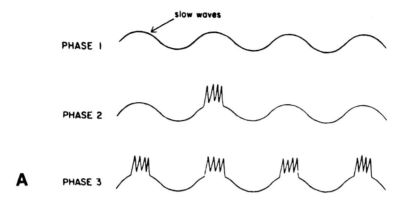

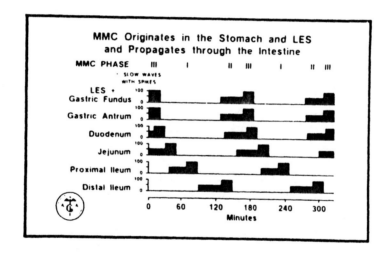

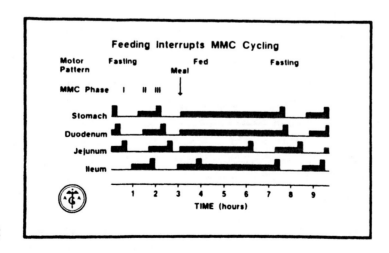

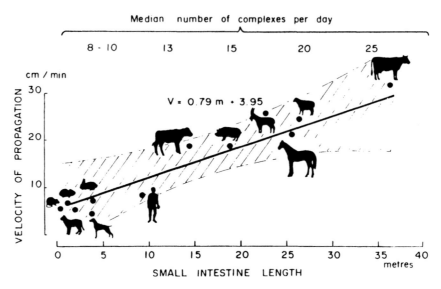

Fig. 5.10. Relationship between the velocity of propagation of myoelectric complexes and the length of the small intestine. The median daily number of jejunal complexes is high in ruminants and low in carnivores because of the obliterating effect of feeding in the latter species. The hatched area shows the 95 percent confidence limits. (From Ruckebusch 1981.)

functional studies of the mammalian hindgut. This may be explained by the inherent distrust of experiments conducted on anesthetized animals and the fact that, until quite recently, the hindgut was subjected to little study. It is difficult to study in conscious, intact animals, because its motility patterns tangle catheters and its anatomical disposition almost defies radiographic examination. It also is difficult to effectively fistulate in many sites. As noted below, even its electromyographic responses show a marked difference from those of the esophagus, stomach, or small intestine.

Ruckebusch and Fioramonti (1976) found a pacemaker area in the rabbit colon at a site near Elliott and Barclay-Smith's (1904) point of separation between the proximal and intermediate colon. Hukuhara and Neya (1968) described a pacemaker area in the colon of both the rat and guinea pig that was located some distance from the ileocecal junction. This area appeared to initiate antiperistaltic waves in the direction of the cecum and peristaltic waves in the opposite direction. Hukuhara and Neya also observed occasional, strong peristaltic contractions that moved from the cecum down the colon, passing over the waves coming from the opposite direction. The inevitable collision of the antiperistaltic and peristaltic waves appeared to result in the passage of some of the digesta aborally.

The human proximal colon demonstrates retropulsion and prolonged retention of its contents, while the distal transverse, the descending, and the sigmoid colon tend to be divided into uniform segments by haustra, contracting rings that may be stationary or move slowly in an aboral direction (Christensen 1989). Relatively long segments of the transverse, descending, and sigmoid colon undergo simultaneous contraction, resulting in a massive aboral movement of their contents. These contractions provide rapid movement of colonic digesta toward the rectum, where they stimulate the defecation reflex. Colonic contractions can be stimulated by a gastrocolic reflex or the release of gastrin, which accounts for the relationship between feeding and defecation.

Electromyographical studies of the cat colon showed a basal electrical rhythm consisting of slow waves similar to those found in the small intestine (Christensen, Caprilli, and Lund 1969; Christensen, Anuras, and Hauser 1974). They originated in the circular muscle, and their frequency increased from the ileal-cecal-colonic junction to the midcolon and tended to remain constant over the remaining length of the distal colon. This gradient in the frequency of the slow waves was consistent with the spread of these waves toward the cecum from pacemaker tissue located in the midcolon (Fig. 5.11). In addition to this, another independent electrical phenomenon was noted. It consisted of migrating spike bursts (MSB), that were usually directed aborally from a point near the hepatic flexure of the colon. The MSB were associated with strong contractions of circular muscle and, presumably, mass movement of digesta.

Sarna et al. (1980) found three distinct segments of electrical control activity in the human colon, with the middle segment having the dominant frequency. A pacemaker area also has been described at the pelvic flexure of the equine colon, and it was concluded that this may be responsible for retropulsion of digesta in the ventral colon and cecum of these animals (Sellers et al. 1982). However, in animals that have a cloaca, the pacemaker for antiperistaltic contractions appears to be located at the terminal hindgut. Antiperistaltic contractions were observed to pass over the entire hindgut of the tortoise *Geoclemys reevsii* (Hukuhara, Naitoh, and Kameyama 1975), and there have been numerous demonstrations that urine is refluxed from the cloaca through the colon and into the ceca of chickens (Browne 1922; Koike and McFarland 1966; Akester et al. 1967; Skadhauge 1968, 1973), Japanese quail (Akester et al., 1967; Fenna and Boag 1974), road runners *Geococcyx californianus* (Ohmart, McFarland, and Morgan 1970), and turkeys (Dzuik 1971). Electromyographical studies of the hindgut of the opossum, a marsupial that has retained the cloaca, showed a progressive increase in the frequency of slow waves from the ileocecal junction to the rectum (Anuras and Christensen 1975).

Therefore, the pacemaker that initiates antiperistaltic waves of contraction may have originated in the cloaca of lower vertebrates, suggesting that the motor ac-

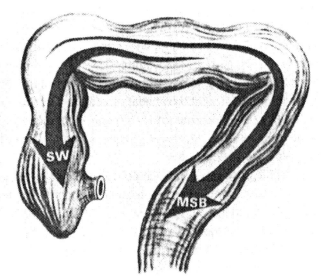

Fig. 5.11. A proposed scheme relating digesta flow in the cat colon to electrical slow waves (*SW*) and migrating spike bursts (*MSB*). Electrical slow waves are oriented in such a way that they appear to spread toward the cecum, away from a pacemaker area about midway along the colon. Since slow waves appear to pace rhythmic contractions, such contractions should tend to produce flow with a polarity in the same direction (*arrow*, *SW*). This polarity of slow wave spread is probably not fixed. The migrating spike bursts begin at a variable position in the proximal colon and migrate toward the rectum. As contractions accompany migrating spike bursts, they should tend to produce flow with a polarity in the same direction (*arrow*, *MSB*). The migrating spike burst also has the capacity for reversal of direction. (From Christensen, Anuras, and Hauser 1974.)

tivity of their entire hindgut is equivalent to that described by Elliott and Barclay-Smith (1904) for the proximal segment of the rabbit colon. However, with the loss of the cloaca and the lengthening of the hindgut in most mammals, the pacemaker appears to have migrated to a more proximal position in the colon.

A gastrocolic reflex stimulation of colonic motility as a result of gastric distension has been noted in the dog (Tansy, Kendall, and Murphy 1972) and horse (Sellers, Lowe, and Brondum 1979), and a ruminocecal reflex has been described in sheep (Fioramonti and Ruckebusch 1979). A gastrocecal reflex also has been described in the rabbit, horse (Ruckebusch and Vigroux 1974), and pig (Fioramonti and Bueno 1977).

Summary

The sequential and episodic events of digestion and absorption are highly dependent on the motor activities of the digestive tract. Food is rapidly transported by the esophagus from the mouth to a stomach in most vertebrates, where it is retained for the processes of gastric digestion. As food particles are reduced in size, they are released into the midgut. This reduction in particle size is aided by regurgitation of less digestible components of prey in raptors and by regurgitation and remastication of plant fiber in ruminants. Retention of particulate digesta in the stomach varies from a few hours, in most species, to extremely long periods in carnivores that ingest their prey in one piece, and in herbivores with a complex voluminous stomach.

Transit of digesta through the midgut can vary with the prevalence of different types of motor activity in a species. Digesta transit is delayed once more in the hindgut of most vertebrates, allowing for resorption of electrolytes and water secreted into the upper digestive tract. Retention of digesta in the hindgut of reptiles and birds is aided by antiperistaltic contractions, which originate at or near the cloaca and reflux urinary excretions and digesta the length of the colon. An intervening sphincter or valve prevents the further reflux of hindgut contents into the midgut, where they could interfere with its digestive and absorptive functions. The advent of a cecum or ceca provides an additional site for digesta retention and also requires a mechanism for reflux filling.

Separation of the exits from the urinary and digestive tracts in most mammals removed the need for the hindgut to serve as a site for the recovery of urinary electrolytes. This was associated with a lengthening of the colon and displacement of the pacemaker, which initiates retropulsion of digesta, to a more proximal segment of the colon. Delay in digesta transit increases the opportunity for the multiplication of indigenous microorganisms, which aid in the conservation of water and nitrogen, and provide most herbivores with a major fraction of their nutritional requirements (see Chapters 8 and 9). Therefore, specialization of the cecum and proximal colon for retention of fluid and small particles, and specialization of the distal colon for recovery of electrolytes and water would be advantageous to terrestrial vertebrates, especially the herbivorous species. Adaptations of the hindgut for these purpose are discussed in following chapters.

6

Digesta transit and retention

Measurement of the rate at which digesta move through the digestive tract provides a link between physiological studies on gut motility and the nutrition of the animal (Warner 1981). Obviously there is a need to match the rate of digesta transit with the rates of feeding, digestion, and absorption to maximize the conversion of ingested food to assimilable products in minima of time and gut volume. An entire process is optimal only when each stage functions optimally with respect to the preceding stage (Penry and Jumars 1987).

Rate of digesta transit can be assessed with marker substances that are not normally secreted, digested, or absorbed by the gut; that are not toxic to the gut microbiota; and that are easily assayed in digesta and feces. Because fluid and particulate digesta move through all but the simplest digestive tracts at different rates, it is important to mark both phases of digesta. Studies based on only one phase, or on the passage of a nonspecific marker such as chromic oxide, which does not associate exclusively with the fluid or particulate phases, are of limited usefulness.

Although the rate of passage of digesta has been measured in many species, useful species comparisons are often limited by differences in experimental procedures and design. For example, rate of passage can vary markedly between diets and feeding regimen (feeding intervals), and the rate at which particulate digesta move through the digestive tract of different species is dependent on particle size and specific gravity, as well as on the structural and functional characteristics of the gut. Materials used to mark ingesta also can affect estimates of digesta rate of passage.

The advantages and limitations of various substances as markers have been reviewed by Faichney (1975), Warner (1981), Clemens (1982), and Van Soest (1994). Chromium ethylenediaminetetraacetic acid (Cr-EDTA), Co-EDTA, and polyethylene glycol 4000 (PEG) all generally satisfy the criteria for a fluid marker in that they associate almost exclusively with the fluid phase, are inert, are not normally

118

absorbed from the gut, and are readily assayed in digesta or feces after oral administration. Small amounts (1–5 percent) of PEG and EDTA complexes can be absorbed and excreted in the urine of some species (Van Soest, Sniffen, and Allen 1988). PEG also suffers from the limitation that it preferentially complexes with tannins in the presence of proteins (Jones and Mangan 1977), and so it should not be used with animals feeding on tanniniferous browse species. As demonstrated by Foley and Hume (1987a), this property of PEG can be used experimentally to reverse the detrimental effects that tannins have on protein availability in the gut of animals feeding on high-tannin plant material.

Particulate markers present a greater problem. For example, chromic oxide (Cr_2O_3) has been widely used to estimate digesta rate of passage in domestic mammalian species. The disadvantage is that it does not move specifically with either the fluid or the particulate phases of digesta and therefore cannot be recommended in studies of digesta flow kinetics, although fecal concentration ratios can give reasonable estimates of total-tract digestibility if the Cr_2O_3 is administered more-or-less continuously (Faichney 1972). Plastic particles are indigestible and retain their dimensions and specific gravity if they escape mastication or trituration in a gizzard. They can be cut to match the size and dimensions of specific food particles, but they do not necessarily match the specific gravity of normal digesta particles. Feed particles stained with a dye or labeled with isotopes better represent the true characteristics of particulate ingesta. However, digestion of these particles can change their dimensions during transit and release the marker to the fluid phase (in the case of dyes) or to migrate to other, progressively smaller particles (in the case of ruthenium-phenanthroline [Ru-P]).

The most suitable particle marker for measuring the effect of a particle's size on its rate of transit appears to be chromium fixed to food particles by a mordant that renders the particles indigestible, even though the mordanting procedure increases the specific gravity of the marked particles (Lechner-Doll, Kaske, and Engelhardt 1991). The combination of Co-EDTA for fluid and Cr-mordanted particles is now the most widely used system for measuring the rate of digesta passage across a range of vertebrate species (Udén et al. 1982; Foley and Hume 1987b; Sakaguchi et al. 1987; Hume, Morgan, and Kenagy 1993).

Numerous terms are used in the literature to describe rate of passage of indigestible markers through the gut (Van Soest 1994). Not all measure the same thing. The two used in this book, because they are perhaps the most useful for comparison across a wide range of digestive systems, are "transit time" and "mean retention time". Transit time is time from administration of the marker to its first appearance in the feces (Warner 1981, Van Soest 1994). Mean retention time (or mean time as Van Soest (1994) calls it) is the time the average particle or fluid marker molecule takes to be excreted after a pulse dose of the markers, and is the

Table 6.1. *Gut transit times in reptiles at known body temperatures*

Species Temp (°C)		Transit timea	Reference
Turtles		*Hours*	
Pseudemys scripta	25	61 ± 3	Parmenter (1981)
Chrysemys picta	25	59 ± 4	"
Chelydra serpentina	25	31 ± 5	"
Sternotherus odoratus	25	49	"
Stemotherus minor	25	57	"
Snakes		*Days[b]*	
Diadophis punctatus	35	1.5	Skoczylas (1978)
	33	2–3	"
	25	3–5	"
	18–20	7–8	"
	15	14	"
Lampropeltis getulus	23	7	"
Nerodia sipedon	21	3.5	"
Natrix natrix	30	4	"
	25	6.7	"
Thamnophis sauritus	21–26	4–5	"
Vipera berus	21	7	"
Lizard			
Uta stansburiana	32	1.2	Waldschmidt, Jones,
	22	4.6	and Porter (1986)

[a] Gut transit time taken to be the time of first appearance of the marker in the feces. (Warner 1981.)
[b] Effect of body temperature on gut transit time illustrated by data linked by brackets.

integrated average of the distribution of marker concentration in the feces after dosing. It is the best single measure of digesta passage through the entire digestive tract (Warner 1981). Mean retention time (MRT) is calculated as $\sum_{i=1}^{n} M_i T_i$ divided by the total marker administered ($\sum_{i=1}^{n} M_i$), where M_i is the amount of marker in the ith defecation at time T_i after dosing, and n is the total number of defecations (Faichney 1975, Van Soest 1994). The general terms "passage" and "transit" are used interchangeably to refer to the flow of digesta through the gut (Van Soest 1994).

Warner (1981) provided a comprehensive review of studies of digesta transit through the gut of sixty-one species of mammals and four species of birds, determined with a variety of procedures and markers. Many estimates are based on the use of only a single marker and are therefore of only limited usefulness. Compared with mammals, there is a paucity of information on rates of passage in fish, amphibians, reptiles, and birds. Tables 6.1 and 6.2 show transit times measured in sixteen species of reptiles; Table 6.3 shows several measures of digesta pas-

Table 6.2. *Mean retention time (hours) of digesta markers in the tract of four reptiles*

Species	Liquid marker	Particulate markers		
		2 mm long	5 mm long	10 mm long
Caiman crocodilus	41	162	162	162
Chrysemys picta belli	35	56	57	60
Geochelone carbonaria	<48	170	185	363
Iguana iguana	<48	207	221	386

Values represent two trials on four specimens of each species except *C. crocodilus* (three specimens). Liquid marker was polyethylene glycol or $BaSO_4$. Particulate markers were cut from 2.2 mm diameter polyethylene tubing. (From Guard 1980.)

Table 6.3. *Mean retention times of markers in the digestive tract of birds*

Species	Body mass (g)	Mean retention time (hours)		Reference
		Fluid	Particles	
Rufous hummingbird (*Selasphorus rufus*)	3.2	0.8	—	Karasov et al. (1986)
Cedar waxwing (*Bombycilla cedrorum*)	35	0.7	—	Martinez del Rio, Karasov, and Levey (1989)
European starling (*Sturnus vulgaris*)	71	0.9	—	Karasov and Levey (1990)
American robin (*Turdus migratorius*)	79	0.8	—	Karasov and Levey (1990)
Rock ptarmigan (*Lagopus mutus*)	460	9.9	1.9	Gasaway, Holleman, and White (1975)
Ring-necked pheasant (*Phasianus colchicus*)	1,400	—	5.0	Duke, Petrides, and Ringer (1968)
Sooty albatross (*Phaebetria fusca*)	2,460	6.3	13.5–17.3	Jackson and Place (1990)
Rockhopper penguin (*Eudyptes chrysocome*)	2,520	3.8	14.0–20.2	Jackson (1992)
Goose (*Anser anser*)	5,700	5.7	—	Clemens, Stevens, and Southworth (1975)
Emu (*Dromaius novaehollandiae*)	38,000	3.9	4.7	Herd and Dawson (1984)
Ostrich (*Struthio camelus*)	45,800	—	48.0	Mackie (1987)

Table 6.4. *Mean retention times of fluid and particle markers in the digestive tract of mammalian foregut fermenters*

Species	Body mass (kg)	Diet	Mean retention time (hours)		Reference
			Fluid	Particles	
Long-nosed potoroo (*Potorus tridactylus*)	0.97	Pelleted concentrates plus oat hulls	20	20	Wallis (1994)
Rufous rat-kangaroo (*Aepyprymnus rufescens*)	2.96	"	31	34	
Rufous hare-wallaby (*Lagorchestes hirsutus*)	1.19	Commercial pellets	23	38	Bridie, Hume, and Hill (1994)
Eastern grey kangaroo (*Macropus giganteus*)	20.8	Chopped alfalfa hay	14	30	Dellow (1982)
Hill kangaroo (wallaroo) (*M. robustus*)	21.3	Pelleted straw plus concentrates	19	30	Freudenberger and Hume (1992)
Goat (*Capra hircus*)	34.1	"	35	39	"
Sheep (*Ovis aries*)	49.5	Pelleted alfalfa hay plus oats	35	50	Faichney and White (1988)
Llama (*Llama guanacoe*)	110–160	Hay plus concentrates	36	52–60	Heller, Cercasov, and Engelhardt (1986)
Cow	220	Grass hay	20	55	Udén et al.
(*Bos taurus*)	610	"	20	66	(1982)

sage in eleven species of birds; and Tables 6.4, 6.5, and 6.6 show MRTs measured in thirty species of mammals in which the passage of both phases of digesta has been measured with specific markers. In some cases, 50 percent excretion times are recorded. While not the same as MRT, the two measures are close enough to allow for broad comparisons among animal taxons. Warner (1981) found that within a species the MRT could vary not only with diet and frequency of feeding but also with ambient temperature, pregnancy, exercise, and age. He also found a

Table 6.5. *Mean retention times of fluid and particle markers in the digestive tract of mammalian colon fermenters*

Species	Body mass (kg)	Diet	Mean rentention time (hours)		Reference
			Fluid	Particles	
Horse	210	Chopped alfalfa hay	22	27	Orton, Hume, and Leng (1985)
(*Equus caballus*)	132	Timothy hay	18	23	Udén et al. (1982)
Pig	176	Hay/grain	39	48	Warner (1981)[a]
(*Sus scrofa*)					
Wombat	28	Pelleted hay	36	62	Barboza (1993)
(*Vombatus ursinus*)					
Wombat	25	"	30	52	
(*Lasiorhinus latifrons*)					

[a] Calculated from data of Clemens, Stevens and Southworth (1975a).

substantial coefficient of variation both between and within animals, even under highly controlled conditions.

The rate of passage through the gut of ectothermic vertebrates is highly temperature dependent. This is well illustrated in fish in tables compiled by Fänge and Grove (1979). For example, in the cyprinid species *Rutilus rutilus*, total emptying time for meals from the digestive tract was seven hours at 25°C, but thirty-one hours at 6°C. In another cyprinid, *Leuciscus baicalensis*, gut emptying time increased from 15 hours at 25°C to 130 hours at 0.5°C. The effect of temperature on digesta passage is represented by a Q_{10} of about 2.6, and gastric emptying times approximated about 50 percent of total tract emptying times in the species surveyed by Fänge and Grove (1979).

The strong influence of temperature on passage times notwithstanding, Horn (1989) summarized available information on rates of passage of food particles through the fish digestive tract. The data are not extensive, and the few data available on herbivorous species are mainly from freshwater forms; interest in marine herbivores is comparatively recent. Horn concluded that herbivorous fish passed food through their gut faster than carnivorous species. Total gut emptying times for carnivores ranged from 10 to 158 hours, but times for herbivores were mostly less than 10 hours. The difference between animal and plant diets is seen within an omnivorous species, the hemiramphid *Hyporhamphus melanochir*, which feeds on seagrass during the daytime but switches to crustaceans at night when they

Table 6.6. *Mean retention times of fluid and particle markers in three types of mammalian cecum fermenters*

Species[a]	Body mass (kg)	Diet	Mean retention time (hours)		Reference[b]
			Fluid	Particles	
Type A					
Rat	0.1–0.3	50% alfalfa meal	12	13	1
(*Rattus norvegicus*)	"	90% purified, 10% bran	23	27	2
"	"	"	29	33	2
"	"	90% purified, 10% pectin	35	35	2
"	"	Pelleted hay, grain	20	22	3
Bandicoot	0.80	Mealworms	24	11	16
(*Perameles nasuta*)	0.80	Sweet potato tubers	33	27	16
Elephant shrew (*Elephantulus edwardii*)	0.05	High-protein cereal	3.2	3.4	12
Type B					
Guinea pig	1.00	50% alfalfa	23	31	4
(*Cavia porcellus*)	0.50	meal cubes	20	17	5
"	0.41	"	19	21	6
"	0.38	"	16	16	1
"	0.62	Commercial pellets	13	13	7
Nutria (*Myocastor coypus*)	4.4	"	44	45	4
Mara (*Dolichotis patagonum*)	7.4	"	27	27	6
Degu (*Octodon degus*)	0.18	"	20	16	5
Marmot (*Marmota caligata*)	2.31		25	29	8
Ground squirrel (*Spermophilus columbianus*)	0.66	"	23	22	8
Chipmunk (*Eutamias amoenus*)	0.06	"	13	14	8
Brushtail possum	2.53	Eucalypt foliage	51	49	9
(*Trichosurus vulpecula*)	2.00	Semipurified	64	71	10
"	2.50	"	36	33	11

Table 6.6. (*cont.*)

Species[a]	Body mass (kg)	Diet	Mean retention time (hours)		Reference[b]
			Fluid	Particles	
Hamster (*Mesocricetus auratus*)	0.11	50% alfalfa meal	?	9	1
Vole (*Microtus townsendii*)	0.06	Commercial rodent chow	15	13	8
Leaf-eared mouse (*Phyllotis darwini*)	0.08	50% alfalfa meal	9.1	8.8	5
Type C					
Rabbit	1.76	44% alfalfa meal	80	17	13
(*Oryctolagus cuniculus*)	—	Commercial pellets	39	27	3
	2.78	44% alfalfa meal	49	38	11
	3.10	20% alfalfa meal	197	91	11
Ringtail possum	0.75	"	210	112	11
(*Pseudocheirus peregrinus*)	0.62	Eucalypt foliage	63	37	14
Greater glider (*Petauroides volans*)	1.14	"	50	23	9
Koala (*Phascolarctos cinereus*)	6.55	"	200	130	15

[a] Three types of cecum fermenters: Type A are omnivores with nonhaustrated cecum and little or no selective retention of fluid marker. Type B are herbivores with larger, haustrated cecum; most show no selective digesta rentention, but some show selective retention of fluid marker. Type C are herbivores/folivores with a large cecum and marked selective digesta retention (MRT of fluid marker generally at least 50 percent greater than that of particle marker) that use high-fiber grasses, *Eucalyptus* foliage, or tree bark. (Modified from Hume and Sakaguchi 1991.)

[b] References:

1. Sakaguchi, Itoh, Uchida and Horigome (1987)
2. Luick and Penner (1991)
3. Clemens and Stevens (1980)
4. Sakaguchi and Nabata (1992)
5. Sakaguchi and Ohmura (1992)
6. Sakaguchi, Nippashi and Endoh (1992)
7. Sakaguchi, Heller, Becker and Engelhardt (1986)
8. Hume, Morgan and Kenagy (1993)
9. Foley and Hume (1987c)
10. Wellard and Hume (1981)
11. Sakaguchi and Hume (1990)
12. Woodall and Currie (1989)
13. Sakaguchi, Kaizu and Nakamichi (1992)
14. Chilcott and Hume (1985)
15. Cork and Warner (1983)
16. Moyle, Hume, and Hill (1995)

emerge. The seagrass passed through the short tubular gut in 4.4 hours, twice as fast as the crustaceans did (8.3 hours). It seemed to Horn that continuous feeding and rapid passage of food resulted in a more favorable benefit/cost ratio for herbivores (even though digestibility is low) than the more carnivorelike attributes of intermittent feeding, longer transit times, and higher digestibilities. However, this conclusion appears to be most applicable to fish with short, simple intestines. Longer transit times have been found among marine herbivores. For example, transit time in the cold temperate marine species *Cebidichthys violaceus* exceeded fifty hours (Horn 1989), and Rimmer and Wiebe (1987) reported a transit time of twenty-one hours in *Kyphosus sydneyanus*, a warm temperate species. A question that is raised by these data is whether, among fish, some marine herbivores are using a different strategy from that used by freshwater herbivores. It may be that microbial fermentation is more important in these particular marine herbivores, as suggested by Horn's (1989) description of four types of alimentary canals in marine herbivorous fish (see Fig. 3.10).

Transit times in reptiles are characteristically long, several days, in line with their low metabolic rates (only 10 percent of those of mammals–see Chapter 1) and consequently low food intakes. The influence of temperature on food passage in reptiles is similar to that seen in fish, and is illustrated in two snakes and a lizard in Table 6.1. Guard (1980) measured rates of passage of fluid and particulate markers in two carnivorous and two herbivorous reptiles (Table 6.2). The fluid marker (PEG) passed more quickly than particles of three different sizes (cut from 2.2 mm-diameter polyethylene tubing) through the gut of all four species. Radiological examination of the carnivorous caiman and terrapin showed the stomach to be the major site of particle retention. However, the cecum and proximal colon were major sites of marker retention in the herbivorous lizard and tortoise. The proximal colon of herbivorous lizards features both circular and semilunar valves (see Fig. 3.12); the number of these valves appears to increase with the degree of herbivory of the species (Iverson 1980). The valves retard the aboral movement of particles more than they do fluid.

One of the shortest MRTs (or 50 percent excretion time) recorded in a vertebrate is less than one hour in hummingbirds (Table 6.3). This is related to their small body mass and exceptionally high metabolic rate (Nagy 1987). The longest MRT reported in birds is forty-eight hours for a particulate marker in the ostrich (Mackie 1987). The longest fluid marker MRT recorded is 17.9 hours in the hoatzin, the only foregut-fermenting bird so far described (Grajal and Parra, In press). The longest fluid marker MRT recorded in a hindgut-fermenting bird is 9.9 hours in the rock ptarmigan (Gasaway, Holleman, and White 1975). The MRT of a particle marker was only 1.9 hours in the same birds, strongly indicating that fluid is selectively retained in the large ceca of the ptarmigan.

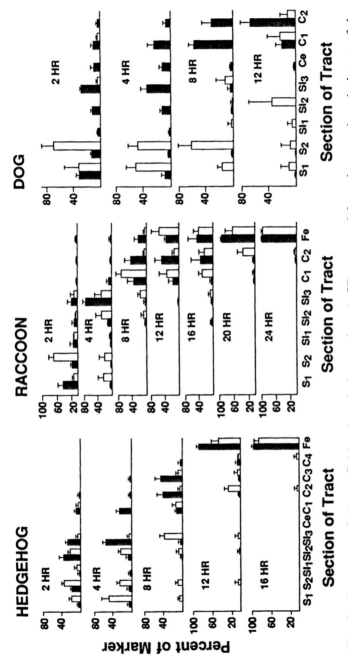

Fig. 6.1. Percentage of digesta fluid and particulate markers (±SE) recovered from the gastrointestinal tract of the hedgehog, raccoon, and dog at various times following oral administration at the time of feeding. Hedgehogs and raccoons were fed a commercially prepared, pelleted, low-concentrate, high-fiber diet. Dogs were fed a meat diet. Animals were fed at twelve-hour intervals for at least two weeks before the study. Fluid markers consisted of PEG or ^{51}Cr-EDTA. Plastic marker consisted of polyethylene tubing with an outside diameter of 2 mm, cut into lengths of 2 mm. Animals were sacrificed in groups of three at the times designated following the meal, and sections of the gut were immediately separated by ligatures for recovery of markers. Solid bars = fluid marker; open bars = particulate marker; S_1, S_2 = stomach; SI_1, SI_2, SI_3 = small intestine; Ce = cecum; C_1, C_2, C_3, C_4 = colon; Fe = feces. (Modified from Banta et al. 1979; Clemens and Stevens 1979; Clemens 1980.)

127

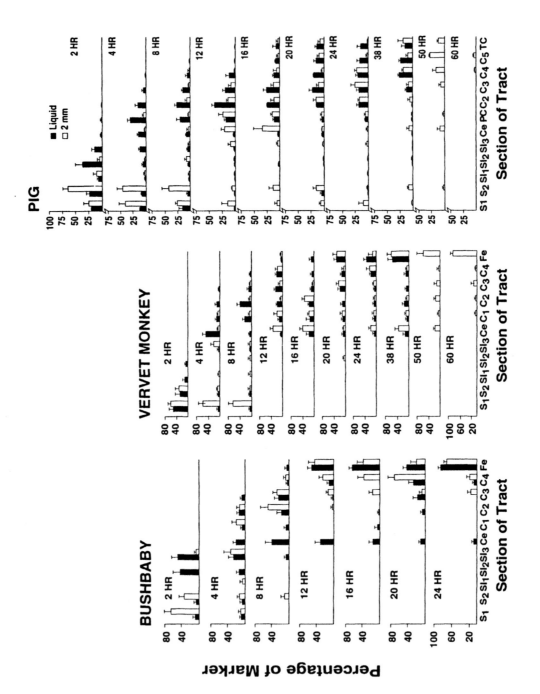

128

The widest range in MRTs is found in mammals (see Tables 6.4, 6.5, and 6.6). In some small carnivores such as shrews and mink, MRT is as short as it is in small birds. The longest MRTs are found in sirenians and arboreal folivores, notably the koala and the sloths. Lanyon, Marsh and Sanson (1995) recently reported MRTs of inert beads in the dugong *Dugong dugon* of 146 to 166 hours, similar to earlier estimates in the West Indian manatee *Trichechus manatus* of 120 to 168 hours (Lomolino and Ewel 1984). Both sirenians were reported to be coprophagic, which may account at least in part for these long MRTs. Neither the koala nor the sloths are coprophagic. Yet in the two-toed sloth Honigmann (1936) recorded a minimum transit time of 96 hours and a maximum time of 1128 hours, while Montgomery and Sunquist (1978) recorded MRTs of 3 mm diameter glass beads of 200 to 300 hours. The time for 5 percent excretion was 60 hours and for 95 percent excretion it was 50 days! Much of the delay was due to retention of feces in the rectum, for up to 7 or 8 days, but the major site of retention was the forestomach (see Fig. 4.12).

Figs. 6.1, 6.2, and 6.3 present the results of a complete series of studies on the transit of fluid (Cr-EDTA or PEG) and particle markers (polyethylene tubing, 2 mm in diameter and 2 or 5 mm in length) through the digestive tract of the hedgehog, raccoon, bush baby, vervet monkey, dog, pig, rabbit, and rock hyrax. Most of the fluid marker had left the stomach within four hours of administration in all species except the hyrax. Much of the fluid marker had reached the hindgut of the rabbit within two hours, the vervet monkey within four hours, and the remaining species within eight hours. The greatest variation in retention times in these simple-stomached animals was in the hindgut. Most of the fluid marker had been excreted in the feces of the hedgehog, raccoon, and bush baby and 60 percent are excreted by the dog within twenty-four hours. However, after thirty-eight hours the vervet monkey, hyrax, and pig had excreted only 50 percent, 40 percent, and 20 percent respectively of the fluid marker. The rabbit study was terminated after twenty-four hours, but at that time, the excretion rate was similar to that of the hyrax. The cecum was the principal site of fluid marker retention in the hyrax and rabbit, whereas the colon was the main retention site in the other species.

Fig. 6.2. Percentage of digesta fluid and particulate markers (±SE) recovered from the gastrointestinal tract of the bush baby, vervet monkey, and pig. The bush babies and vervet monkeys were fed a commercially-prepared primate diet, the pigs were fed the same diet as the hedgehogs and raccoons in the studies described in Fig. 6.1. The dietary regimen and recovery of markers were the same as those described for Fig. 6.1, except that longer particles (2 × 5 mm) were administered to the bush baby and vervet monkey. PC = proximal colon; TC = terminal colon. Other symbols as in Fig. 6.1. (Modified from Clemens, Stevens, and Southworth 1975a; Clemens 1980.)

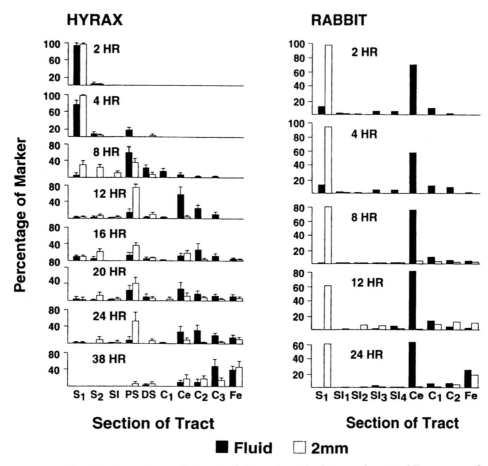

Fig. 6.3. Percentage of digesta fluid and particulate markers (±SE) recovered from the gastrointestinal tract of the rabbit and hyrax. Rabbits were fed a commercially prepared, pelleted rabbit diet, and the hyrax were preconditioned to a diet of alfalfa (lucerne) leaves. Other conditions of the experiment were identical to those described in Figs. 6.1. Symbols as in Fig. 6.1. (Modified from Pickard and Stevens 1972; Clemens 1977.)

Particles were retained in the stomach of all species for a longer time than the fluid marker. The role of the stomach in the retention of particulate digesta is illustrated clearly in Fig. 6.4, which shows the results of intragastric administration of markers to the pony. Although 75 percent of the fluid had left the stomach by 0.5 hours, 75 percent of the large and small particles remained in the stomach 1.5 hours after administration.

Particles also were retained longer in the hindgut of all species, except the dog and rabbit. The proportion of particles excreted in the feces within the first twenty-four hours was 100 percent in the hedgehog, 90 percent in the raccoon, 80 per-

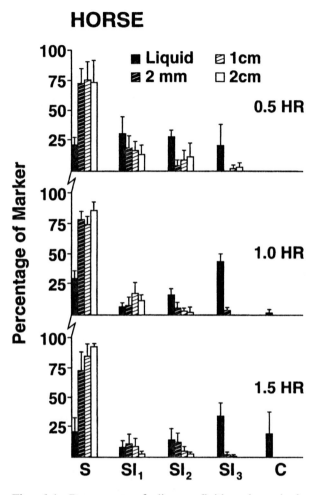

Fig. 6.4. Percentage of digesta fluid and particulate markers (±SE) recovered from the stomach, small intestine, and cecum during the first 1.5 hours
after feeding and the administration of markers to ponies. Animals were fed a
commercially-prepared, pelleted, hay-grain diet. Plastic markers consisted of
polyethylene tubing with an outside diameter of 2 mm, cut into lengths of 2 mm,
1 cm or 2 cm. Other conditions were identical to those described in the previous
three figures. Symbols as in Fig. 6.1. (From Argenzio, et al. 1974.)

cent in the dog, 20 percent in the rabbit, 15 to 20 percent in the vervet monkey
and pig, and 10 percent in the hyrax.

Although fluid digesta often move through the gastrointestinal tract faster than
particles, this is not always the case, as illustrated in Figs. 6.1, 6.2, and 6.3 above,
and by Gasaway, Holleman, and White (1975) with the rock ptarmigan, and so it
is appropriate that the phenomena of differential flow and selective retention of
different components of digesta be discussed in some detail.

Retention of particles

In all vertebrates studied to date, fluid appears to leave the stomach more rapidly than particulate matter of any size (Mayer 1994), as illustrated in the pony (Fig. 6.4). The forestomach of sheep and cattle have received a great deal of study with respect to their mechanism for the passage of fluid and particulate digesta (Hydén 1961; Balch and Crampling 1965; Grovum and Williams 1973; Faichney 1975; Van Soest, Sniffen, and Allen 1988; Kaske, Hatipoglu, and Engelhardt 1992). Because of the ventral location of the reticulo-omasal orifice and the filtering arrangement of the omasum, particles cannot escape the reticulorumen until they are reduced to a given size and have gained a specific gravity that is neither too light nor too heavy to allow their release from the reticulum (Kaske et al. 1992). King and Moore (1957) found that plastic particles with a volume of 20–30 mm^3 and a specific gravity of 1.2 had the highest rate of transport out of the reticulorumen of cattle. This maximizes retention of large particles in the reticulorumen and thereby maximizes the digestibility of plant cell walls (i.e., fiber). On unchopped roughage diets of high-fiber content, the rate of breakdown of the food particles can be so slow that the reticulorumen becomes distended with a mass of indigestible fiber and food intake is inhibited (Van Soest 1994). That is perhaps why ruminants are not competitive with large hindgut fermenters on high-fiber diets such as savanna grasslands; passage of large particles through tubiform fermentation chambers such as the proximal colon is not constrained in the way that it is through the ruminant forestomach (Hume and Sakaguchi 1991).

Fluid and particles pass more quickly through the forestomach of kangaroos than through that of ruminants, and with a greater degree of separation in the transit of fluid versus particles (Fig. 6.5). This is due to the tubular nature of digesta flow through the "colonlike" kangaroo stomach, with only limited mixing of contents along the axis of the forestomach. Dellow (1982) observed radiographically that although there was effective local mixing of contrast medium and digesta, contrast medium did not mix with the entire contents of the forestomach. Instead, marked digesta were transported slowly along the length of the forestomach so that after six hours, in the frequently fed animal, newly ingested food was not marked with previously administered contrast medium. Contractions of the forestomach wall propel digesta aborally in such a way that fluid is expressed through a matrix of particulate material. This leads to selective retention of particles. The longer the forestomach, the greater the separation of fluid and particles and the greater the difference in whole-tract MRTs between the slower particles and the faster fluid.

The other principal site of digesta particle retention is the hindgut. Argenzio, Lowe, Pickard, and Stevens (1974) administered fluid or particle markers of three different sizes into the stomach, cecum, or dorsal colon of ponies and used com-

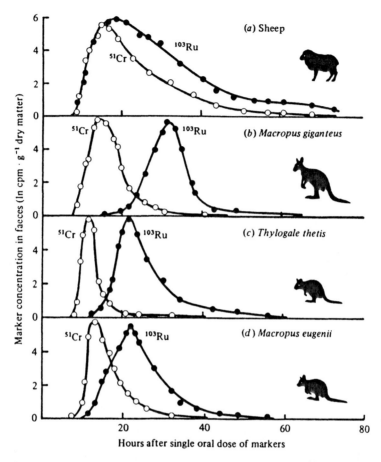

Fig. 6.5. The pattern of appearance of the fluid marker ^{51}Cr-EDTA and the particulate marker ^{103}Ru-phenanthroline in the feces after a single oral dose in sheep, eastern grey kangaroo (*Macropus giganteus*), red-necked pademelon (*Thylogale thetis*), and tammar wallaby (*M. eugenii*), which were fed chopped lucerne hay ad libitum. Note the faster elimination of both markers in the macropod marsupials, and the greater separation of the two markers in the macropods than in the sheep. (From Hume 1982, after Dellow 1982).

partmental analysis to generate equations for calculating marker flow between large intestinal segments. Curves describing changes in marker concentration over time for the cecum, ventral colon, dorsal colon, and feces are shown in Fig. 6.6. Fluid markers passed rapidly from the stomach to the cecum and from the cecum to the proximal (large) colon but were retained there for prolonged periods. Junctions between the cecum and ventral colon, the ventral and dorsal large colon, and the dorsal and small (distal) colon appeared to operate as major barriers to digesta flow, even though there is no evidence of sphincters at any of these sites. There was no retrograde flow of either fluid or particle markers from the dorsal colon

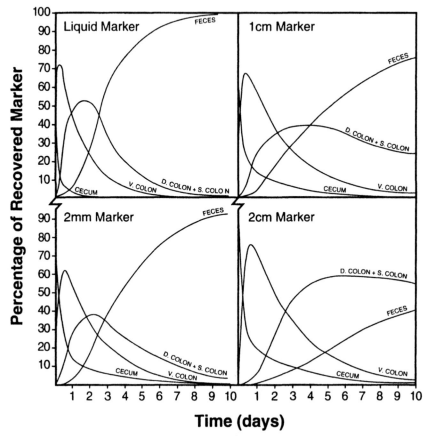

Fig. 6.6. Retention of fluid and particulate markers by the hindgut of ponies fed a pelleted hay-grain diet. Fluid markers were PEG or ^{51}Cr-EDTA. Particulate markers were polyethylene tubing with an outside diameter of 2 mm, cut into lengths of 2 mm, 1 cm or 2 cm. Curves for the percentage of markers present in the cecum, ventral colon, and dorsal plus small colon were derived from experiments in which the markers were administered directly into the cecum. Fecal appearance curves were derived from a separate experiment in which markers were administered orally and collected over the next ten days. Average standard error for these experiments was ±18 percent. (From Argenzio et al. 1974.)

into the ventral colon, nor from the ventral colon into the cecum. The rate of particle marker movement decreased with increasing particle size in all segments (Fig. 6.6). After ten days, all of the orally administered fluid marker but only 40 percent of the largest particles had been excreted in the feces. The principal site of retention of particles was in the dorsal plus small colon, while fluid and the smallest particles moved through the ventral and dorsal colon at similar rates (Fig. 6.6). Thus the cecum, ventral colon, dorsal colon, and probably the small colon represent four distinct mixing pools of digesta.

This study also demonstrated that the fluid marker was retained longer in the ileum on a high-fiber diet than on a commercial pelleted hay-grain diet of lower fiber content. This may account for the higher bacterial counts and short chain fatty acid (SCFA) concentrations in the ileum of horses fed grass (Mackie and Wilkins 1988) than those seen by Argenzio, Southworth, and Stevens (1974) in ponies on a high-concentrate pelleted diet (see Chapter 8).

Udén et al. (1982) measured passage of fluid (with Co-EDTA) and Cr-mordanted particles through the entire digestive tract of horses and ponies fed long and chopped (but not pelleted) timothy (*Phleum pratense*) hay respectively. Both the horses and ponies excreted the fluid and particle markers more rapidly than the ponies in the study of Argenzio et al. (1974), and there was a much smaller difference in relative rates of excretion of fluid and particles, with only a five hour difference in MRTs of the two markers (Table 6.5; Fig. 6.6). Orton, Hume and Leng (1985) also found only a small difference (three to five hours) in the MRTs of the fluid marker Cr-EDTA and the particle marker Ru-P in horses fed a long hay–concentrate mix ad libitum. These results suggest that the greater bulk consumption associated with unpelleted hay diets stimulates gut motility and increases the rate of passage of digesta, especially particulate digesta, through the equine hindgut compared with pelleted diets of much smaller particle size.

A similar effect of bulk was seen in dogs fed either a meat or a cereal diet (Banta et al. 1979); those on the cereal (high-fiber) diet had a much more rapid transit of both fluid and particulate markers.

Retention of fluid

Selective retention of fluid digesta has been demonstrated in the cecum in a wide range of small mammals and a number of birds. Studies of coprophagy (ingestion of feces) in the rabbit provide some insight into the mechanisms by which the hindgut can selectively retain fluid digesta. In 1882, Morot, a French veterinarian, reported that the domestic rabbit excreted hard, dry fecal pellets during the day, and soft, moist pellets at night. Only the soft feces, which tend to attach to the hairs around the anus, were ingested. Southern (1940, 1942) later demonstrated that wild rabbits produce their soft feces during the day, after returning to their burrow from early morning foraging, suggesting that the diurnal variation in fecal composition was related to feeding patterns. Morot had concluded that the hard feces represented material that had passed through the digestive tract twice, but Madsen (1939) thought that the difference between hard and soft feces was due to a difference in the rhythmic activity of the intestine.

Eden (1940) then reported that the soft feces contained about 30 percent protein and 15 percent crude fiber, as opposed to 10 percent protein and 30 percent fiber in the hard fecal pellets. From this finding, plus analysis of ash content and the relative protein levels of soft feces and cecal contents, he concluded that the soft feces had a composition quite similar to that of cecal digesta. This suggested that the differences in fecal composition might be explained by direct passage of some digesta from the ileum to the colon to form the dry pellets, and a periodic, massive evacuation of cecal contents to form the softer version. Thacker and Brandt (1955) reached a similar conclusion, postulating that the soft feces were produced by contraction of the spiral musculature of the cecum and rapid passage through the colon.

Subsequent studies (Sperber 1968; Pickard and Stevens 1972; Björnhag 1972) showed that the fluid and small particles were separated from larger particles of digesta in the proximal colon of the rabbit and that this was accomplished by mechanical expression rather than by absorption of fluid from the proximal colon. The fluid and small particles, which include microorganisms, were retained in the proximal colon and cecum by cyclic retropulsive digesta flow, and the larger particles were evacuated more rapidly as hard feces. Retropulsive digesta flow is assisted by net secretion of water into the proximal colon (Björnhag 1987). Clauss (1983) demonstrated that the diurnal cycle is under the influence of aldosterone. This sets up what can be envisaged as an internal water cycle (Björnhag 1987) that operates during hard feces formation but is switched off during soft feces formation (Clauss 1983).

Björnhag (1981) observed that the retrograde transit of digesta occurred along the wall of the haustra, leaving the bulk of the lumen free for aboral transit of gas and other contents, and that this process was periodically interrupted by the formation and transit of soft feces. Fioramonti and Ruckebusch (1976) noted a decrease in the motility of the cecum and proximal colon and an increase in the motility of the distal colon during production of soft pellets and that these passed through the colon 1.5–2.5 times faster than hard pellets. A pacemaker was found in the fusus coli, at the junction of the proximal and distal colon (Ruckebusch and Fioramonti 1976). Both the electrical (Ruckebusch and Hörnicke 1977) and motor (Ehrlein, Reich, and Schwinger 1983) activity of the haustrated proximal colon increased during the formation of hard feces. Formation of soft feces was accompanied by a decrease in motility of the proximal colon and cecal base and increased motility of the colon distal to the pacemaker area. Subsequent studies (Pairet, Bouyssou, and Ruckebusch 1986) showed that the formation of soft feces was accompanied by a 30 percent reduction in the myoelectrical activity of the distal colon and indicated that endogenous prostaglandins play a role in soft feces production, probably under the control of aldosterone (Clauss 1983).

As already indicated, the ability to selectively retain fluid or small particles in the cecum and proximal colon is not limited to rabbits. Björnhag (1987) reviewed

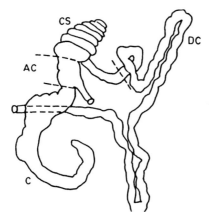

Fig. 6.7. Cecum and colon of Scandinavian lemming. *AC*, ampulla coli; *C*, cecum; *CS*, colonic spiral; *DC*, distal colon. (From Sperber, Björnhag, and Ridderstrale 1983.)

the extent to which the phenomenon occurred in other lagomorphs (hare, pika), rodents such as guinea pigs, chinchillas, rats, voles and lemmings, and a number of marsupials. The proximal colon of the lemming has been found to have a highly developed mechanism for separation of fluid and small particles, particularly microorganisms, from the larger particles of digesta (Sperber 1968; Sperber, Björnhag, and Ridderstrale 1983). The proximal colon of these animals can be divided into an ampulla coli (a short, wide segment extending from the entry of the ileum to the base of the colonic spiral), an inner spiral of about four windings around the axial structure, and an outer spiral that runs back to the base and leaves it to join the distal colon (Fig. 6.7). The mucosal surface of the proximal colon has a complicated pattern of folds, which project into the lumen (Fig. 6.8). The largest of these starts at the ampulla coli and runs as a longitudinal fold for about two windings of the inner spiral. The free edge of this fold comes close to the axial attachment of the wall, thereby dividing the lumen of the inner spiral into a narrow channel and a main channel. The narrow channel opens into the ampulla coli by a narrow slit and extends, at its distal end, as a longitudinal groove connecting to oblique furrows in the colonic wall.

The contents of the cecum, ampulla coli, and main channel of the inner spiral consist of a similar mixture of food residues and bacteria. However, the narrow channel contains dense masses of bacteria and mucus with little food residue in evidence. The main source of the mucus appears to be a thickened area of the inner spiral mucosa, distal to the narrow channel. It was concluded that digesta are mixed with mucus secreted by the middle part of the inner spiral, resulting in an aggregation of bacteria with mucus, which is transported by antiperistaltic con-

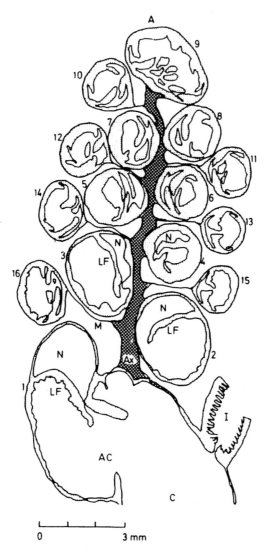

Fig. 6.8. Section through the colonic spiral. Composite from sections obtained from a 21-day-old lemming. *A*, apex; *AC*, ampulla coli; *Ax*, axial structure; *C*, cecum; *I*, ileum; *LF*, longitudinal fold; *M*, mesentery; *N*, narrow channel. Numbers *1–8* refer to sections through inner spiral. Numbers *10–16* refer to sections through outer spiral. (From Sperber, Björnhag, and Ridderstrale 1983.)

tractions into the narrow channel and back to the ampulla and cecum. This mechanism causes microorganisms to be retained in the cecum and proximal colon, and it would explain the low bacterial counts and nitrogen concentration of the feces produced by the lemming. How the bacteria aggregate with the mucus is not known.

A somewhat similar mechanism for the separation of microorganisms from fluid and particles was demonstrated in the colon of guinea pigs and chinchilla (Holtenius and Björnhag 1985). The proximal colon contains a furrow, formed by two mucosal folds about 20 cm long in the guinea pig and twice this length in the chinchilla. Contents of the furrow contained significantly higher concentrations of nitrogen and adenosine triphosphate (ATP), indicating that most of the nitrogen was present in microorganisms. Labeled bacteria, infused into the proximal colon, were transported into the cecum, and their concentration in the most proximal segment of the furrow was twice that measured in the lumen.

Sperber et al. (1983) also examined the rat for the presence of mechanisms similar to those noted in the lemming. The ampulla coli of the rat is small and difficult to distinguish. However, the main part of the proximal colon contained two rows of oblique folds, with a mucosa that was thick and composed of long, tubular glands in the luminal half of the fold. Mucus containing masses of long, fusiform bacteria were often found close to the mucosa and between the folds. Rats fed a fiber-rich diet produced two types of fecal pellets. One of these had a composition similar to that of cecal contents, whereas the other had a much lower nitrogen content. These investigators concluded that the mucus–bacteria mixture was normally returned to the cecum, and cited evidence (see Chapter 8) of abnormal accumulation of mucus in the cecum of germ-free rats. Periodic aboral transit of the bacteria–mucus mixture would account for the production of the protein-rich pellets.

In marsupials, Cork and Warner (1983) measured rate of passage in captive koalas with ^{51}Cr-EDTA and ^{103}Ru-P as fluid and particulate markers respectively. Digesta passage was measured by collection of feces in one experiment and by analysis of the contents of gut segments (compartmental analysis) in a separate experiment. The fluid marker was retained longer (MRT, 213 hours) than the particle marker (MRT, 100 hours), and both were retained longer than in any other mammal apart from the sloth, dugong, and manatee. They concluded that the principal site of digesta retention was the cecum/proximal colon, which appeared to function as a single mixing pool. Also, it appeared that the Ru-P associated mainly with small particles of digesta because of the tendency of this marker to migrate from labeled to unlabeled particles during the processes of transit and digestion (Faichney and Griffiths 1978). That is, adsorption of the Ru-P to the surface of particles is not irreversible, so the marker is in equilibrium with the available binding sites in the digesta particle pool, which contains a far greater surface area of fine than of large particles. More recently, Krockenberger (1993) measured rate of passage in free-living koalas using Co-EDTA as the fluid marker and Cr-mordanted large particles. MRTs were much shorter than those measured by Cork and Warner (1983), 99 hours and 32 hours for fluid and particles respectively.

However, selective retention of the fluid marker was still clearly evident. The shorter MRT of the particle marker is partly due to lack of migration of mordanted Cr between digesta particles, giving a truer picture of large particle passage. The shorter passage time for both markers in the wild koalas undoubtedly was due to the greater food intakes required to meet the additional energy costs of free existence, such as greater activity and thermoregulation. Nonlactating female koalas ate 35 percent more dry matter than did captive animals, and lactating koalas ate 69 percent more. The rate of passage is generally increased when food intake is increased (Warner 1981), unless there is a concomitant increase in gut volume (Karasov and Hume in press).

The same problem of particle marker migration with Ru-P just discussed was encountered by Foley and Hume (1987c) with captive greater gliders (*Petauroides volans*). With Ru-P, there was no difference in MRT between particles (46 hours) and fluid (50 hours), but large (500–1000 μm) Cr-mordanted particles were retained only half the time (23 hours). The selective retention of fluid suggested by the difference in MRTs of fluid and large Cr-mordanted particles was confirmed by analysis of digesta particle sizes along the gut; the cecum contained twice the proportion of small particles than either the stomach or distal colon. Sakaguchi and Hume (1990) have shown with ringtail possums *Pseudocheirus peregrinus*, and brushtail possums *Trichosurus vulpecula*, that small (less than 75 μm) particles labeled with Yb passed through the gut at the same rate as fluid, probably because they are swept along with the movement of fluid.

Selective retention of fluid, small food particles, and bacteria has two important nutritional consequences for small mammalian hindgut fermenters: (1) It concentrates digestive effort in the cecum by maintaining a high concentration of microbes with the potentially most digestible food particles (the small ones); and (2) it eliminates large particles from the gut relatively quickly, thereby minimizing their "gut-filling" effect and thus their inhibition of feed intake. Of the four marsupials just discussed (koala, greater glider, ringtail and brushtail possums), the first three are folivores that specialize on *Eucalyptus*, and all three species selectively retain fluid and small particles in their cecum or cecum/proximal colon and excrete larger particles relatively rapidly. This is considered to be central to their ability to specialize on such a poor quality diet as eucalypt foliage. The fourth species, the brushtail possum, has no selective retention mechanism for fluid and small particles (Wellard and Hume 1981; Foley and Hume 1987c; Sakaguchi and Hume 1990), and, probably as a consequence, is difficult to maintain in captivity on eucalypt foliage alone. In the wild, although eucalypt foliage is an important part of the brushtail possum's diet, it is invariably supplemented with foliage from other species of trees, shrubs, and grasses, even though risks of predation are high when feeding on the ground (Freeland and Winter 1975). This difference between the brushtail possum and the other three arboreal folivores is a good example of the close connection between digestive and foraging strategies of animals.

In birds without well-developed ceca, most of the total MRT is often accounted for by residence in the crop (Karasov 1990). Even in hummingbirds, the crop contributes 13 percent of total MRT (Karasov et al. 1986). Herbivorous birds usually have well-developed ceca, which increase utilization of refractory plant material by increasing total gut volume, decreasing digesta flow rate, and enhancing the extent of digestion or fermentation (Karasov 1990). Gallinaceous birds have been the most studied. Clemens, Stevens, and Southworth (1975b) and Björnhag and Sperber (1977) showed that retrograde transport of urine from the cloaca along the colon to the ceca is a normal physiological process in geese, turkeys, and guinea fowl. This carries small particles and dissolved solutes from the colonic contents into the ceca where they are fermented. Larger, potentially less digestible particles are left behind to be passed with the feces. The antiperistaltic movements of the coprodeum and colon are a continuous process except for just before and during defecation. Villi and ridges at the necklike zone of the most proximal part of the ceca also seem to help prevent large particles from entering the ceca (Fenna and Boag 1974). Consequently, in rock ptarmigan, the MRT of a fluid marker greatly exceeds that of a particle marker (Gasaway, Holleman, and White 1975). In the emu, which lacks enlarged ceca, the markers travel through the gastrointestinal tract at approximately the same rate (Herd and Dawson 1984).

The emu proves that in birds a cecum is not required for effective digestion of fiber. Because digesta retention times increase with body mass$^{0.25}$ (Karasov 1990), large birds like the emu can retain digesta long enough for effective fermentation of fiber even in a simple gut. Herd and Dawson (1984) found that the major site of microbial fermentation in the emu was the ileum.

The ostrich differs from the emu in that it has larger ceca and a very long colon, which function as effective fermentation organs (Mackie 1987), and, according to Skadhauge et al. (1984), there is no retrograde flow of urine into the ceca, or even into the colon. Instead, the urine accumulates in the coprodeum, which acts like a bladder.

The avian ceca serve other functions in addition to enhancing digestive effort; the cecal epithelium transports water and electrolytes, and this capacity increases during dehydration and/or sodium depletion (Thomas and Skadhauge 1988). Uric acid in the urine is degraded by microbial action in the ceca, and it has been shown that the cecal microbiota can utilize uric acid nitrogen for synthesis of microbial protein (Barnes and Impey 1974). Björnhag (1989) has also shown that the proportion of urine that reaches the ceca of domestic chickens increases as the nitrogen content of the food is decreased, from 6 percent on a 20 percent crude protein diet to 22 percent on an 11 percent crude protein diet. This suggests that there

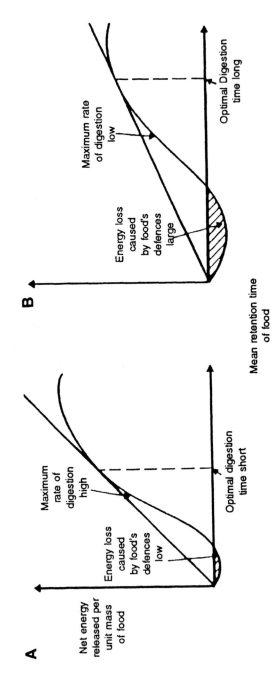

Fig. 6.9. Model of digestion in a continuous-flow system for a high-quality (**A**) and low-quality (**B**) food. (Modified from Sibly 1981 by Hume 1989.)

142

is a mechanism to compensate for a nitrogen deficiency by enhancing the recycling of uric acid nitrogen to the cecal microbes.

Digestive strategies and models of digesta flow in vertebrates

The need to match rates of digesta flow with rates of food intake and rates of nutrient absorption leads to the concept of optimal digestion time, which is modeled in Fig. 6.9. In this model, the net energy released from a food, such as plant material, is initially negative until the food's defenses, such as lignified cell walls, are overcome (for example, by chewing). This is followed by a period of rapid digestion (of cell contents), but thereafter digestion rate falls as digestion is progressively confined to cell walls. Optimal digestion time is given by the straight line from the origin tangential to the curve (Sibly 1981).

Several predictions arise from the model. First, optimal digestion time will vary among foods, being longer for poor-quality foods than for those of higher quality. Thus, animals eating poor-quality foods have larger digestive tracts, usually by being large themselves. Even within a species, the gut size of individual animals increases on poor quality diets, particularly organs of fermentation such as the cecum (Gross, Wang, and Wunder 1985; Hammond and Wunder 1991; Lee and Houston 1993).

Second, if gut capacity is limiting, the optimal strategy is to maximize digestion rate by selecting only high-quality food items. Among ruminants, those categorized by Hofmann (1973) as bulk and roughage feeders are invariably large and thus have large digestive tract capacity. Smaller antelopes have smaller gut capacities and are mixed feeders, (that is, they feed selectively on grasses until they senesce, then switch to browse species). The smallest ruminants such as the suni, dik-dik, and duiker are all concentrate selectors, taking small amounts of browse from a large number of plant species. Fermentation rates in the reticulo-rumen are highest in the smallest species (Hungate et al. 1959; Hoppe 1977).

Third, at any given level of intake, an animal should maximize retention of food to maximize the rate of obtaining energy (Sibly 1981), although clearly there must be a holding time beyond which net loss occurs (Fig. 6.9) (Penry and Jumars 1987). The negative relationships seen between level of food intake and extent of fiber digestion in ruminants (Blaxter 1962), horses (Haenlein, Holdren, and Yoon 1966), pigs (Parker and Clawson 1967), and brushtail possums (Wellard and Hume 1981) are clear demonstrations of this model prediction.

A close relationship between optimal foraging and optimal digestive strategies was recognized by Penry and Jumars (1987) in their work with invertebrate marine deposit feeders. They used principles of chemical reactor theory (Froment and Bischoff 1979) to formulate constraints that optimize digestion. From these

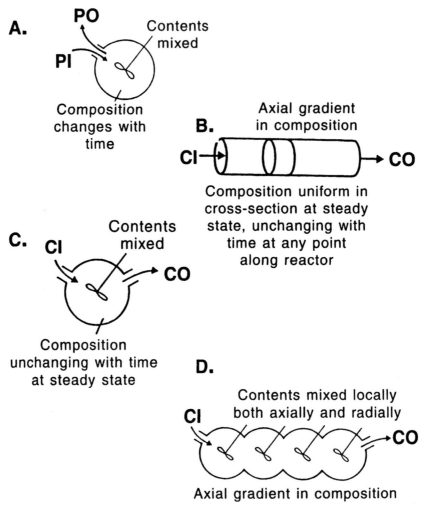

Fig. 6.10. Diagrams of four types of chemical reactors that have analogs in the digestive system, with a pulsed input (*PI*) and pulsed output (*PO*) or a continuous input (*CI*) and continuous output (*CO*). **A,** Batch reactor, which best describes digestive systems which process food in batches, such as that seen in invertebrate coelenterates. **B,** Plug-flow reactor (PFR), as characterized by the vertebrate midgut. **C,** Continuous-flow, stirred-tank reactor (CSTR) similar to that seen in gut diverticula, such as the pyloric ceca in some fish, the hindgut ceca of birds and mammals, and the cranial compartments of the ruminant and kangaroo forestomach. **D,** Modified plug-flow reactor (several stirred-tank reactors in series) similar to that seen in the haustrated sections of hindgut in many mammals and the tubiform segment of the kangaroo forestomach.

they developed models of digestion for a variety of marine deposit feeders. Because many deposit feeders digest so little of the ingested volume, the gut models developed for them were found to be readily applicable to vertebrate herbivore systems as well. Penry and Jumars' (1987) approach to modeling the digestive tract is based on the Michaelis-Menten equation to predict rates of catalytic digestion (by the animal's own enzymes) and the Monod equation to predict rates of autocatalytic digestion (by microbial fermentation). They considered three basic types of chemical reactors that could be applied to their analysis of digestion in marine deposit feeders: batch reactors; continuous-flow; stirred-tank reactors (CSTR); and plug-flow reactors (PFR). Chemical reactors are classified on the basis of (1) whether input is continuous or discontinuous, and (2) whether reactants are brought together with or without mixing (Levenspiel 1972).

Batch reactors

In a batch reactor, all reactants are added simultaneously (that is, input is discontinuous, in batches) and are well mixed (Fig. 6.10, *A*). In ideal batch reactors (that is, those that can be described accurately by simple equations), composition is uniform throughout the reactor (that is, there is perfect mixing), but composition changes with time as the reaction proceeds. The reaction is allowed to proceed for a set period, and then the reaction products and unreacted materials are all removed. The reactor may then remain empty for a period or be refilled. Extent of reaction (digestion) can be high if the time spent in the reactor is long enough, but material flow is interrupted and low overall, resulting in low production rate capabilities, unless reactor (gut) volume is very high.

In animals, batch-reactor guts may be flexible under varying food supply because they can be emptied and refilled if better quality food becomes available. Coelenterates such as *Hydra* and *Aurelia*, which have only one opening into their gastrovascular cavity, fit the description of a batch-reactor gut well. However, batch processing is not limited to digestive systems with one opening. Among vertebrates, Cochran (1987) has applied batch-reactor theory to the problem of optimal food retention times and feeding times of carnivores that partially consume individual prey. Batch reactors are not widely applicable in herbivores because of the problem of losing the microbial symbionts along with undigested food, by regurgitation from the stomach if the material is not reswallowed, for example. Nevertheless, some avian ceca may function partly as batch reactors in that they are at least partially evacuated regularly once a day (Farner 1960). Similarly, some mammalian ceca may also operate partially as batch reactors. In small hindgut fermenters that are cecotrophic (e.g., the rabbit), retrograde movement of fluid and small particles from the proximal colon into the cecum occurs throughout most of the day, but during one or two periods it ceases, and cecotrophes are formed from cecal contents (Björnhag 1987). It is during these periods of partial evacuation that the cecum operates more like a batch reactor than other types of chemical reactor.

Plug-flow reactors

In these reactors there is a continuous, orderly flow of reactants through a usually tubular vessel (Fig. 6.10, *B*). In ideal plug-flow reactors, material does not mix along the flow axis, but there is perfect radial mixing, so that, at any point along the reactor, composition is uniform in cross section and is constant with time. Reactant concentrations, and therefore reaction rates, are maintained at higher levels near the reactor entrance and decline along the length of the vessel to the outlet. Plug flow provides the greatest rate of digestive product formation in the minima of time and volume under most conditions, but the extent of digestion may be low unless reactor length (and diameter) is large.

The simple tubular guts of many marine deposit-feeding polychaetes closely meet the description of a plug-flow reactor; axial mixing of gut contents has been observed to be negligible in these species, thus fulfilling the most important assumption of the ideal plug-flow model (Penry and Jumars 1987). Among vertebrates, the small intestine best fits the plug-flow reactor model, except that some axial mixing does occur (antiperistalsis is a regular event in the avian small intestine, and perhaps in all other vertebrates as well). It is in the small intestine that catalytic digestion and absorption of the products of catalytic digestion are most efficient. There is often a direct relationship between difficulty of digestion of the usual diet of the animal and the length of its small intestine. Greater intestinal length increases retention time in the plug-flow reactor, thereby maximizing extent of digestion. Thus, in cyprinid fishes, Mohsin (1962) found that the intestinal length of carnivorous species was only 0.9 to 1.1 times the body length, but in herbivorous species this ratio was 7.2. This is one of the most consistent features of herbivorous fish (Horn 1989). In the emu, retention of digesta in an ileum that is notable for both its length and diameter allows microbial fermentation of at least the more readily digestible components of plant cell walls in a digestive tract with only a short colon and small ceca (Herd and Dawson 1984).

Continuous-flow, stirred-tank reactors

Continuous-flow, stirred-tank reactors (CSTR) are characterized by continuous material flow through a usually spherical vessel of minimal volume (Fig. 6.10, *C*). Contents are well mixed, so that in an ideal CSTR composition is uniform throughout the reactor and is unchanging with time. Reactant concentration is diluted immediately upon entry into the vessel by material recirculating in the reactor (residues from previous meals). This reduces reaction rate, but even so, extent of conversion (digestion) can be high if material (digesta) flow is low enough. Some marine deposit feeders have an expanded anterior portion of the gut in which substantial mixing occurs. This mixing chamber probably serves to overcome problems of enzyme diffusion when food is of low porosity (that is, is diluted with indigestible material), as is the case with many marine sediments (Penry and Jumars 1987). This chamber most closely resembles a CSTR. Similar mixing chambers are found in vertebrates as diverticula of the gut, such as the first and second compartments of the camelid stomach (see Fig. 4.25), the sacciform region of the macropod marsupial forestomach (see Figs. 4.10 and 4.11), and the reticulorumen of ruminants (see Figs. 4.23 and 4.29). All these examples deviate from an ideal CSTR because of incomplete mixing and differential flow of fluid and particulate digesta. Also, input is not continuous, depending on feeding patterns, but outflow is modulated so that it is more continuous than food input. The cecum also displays flow characteristics that are sometimes complicated by pulses of input and, in many cases, by selective retention mechanisms. It is probably best modeled as a CSTR, except when the organ is partially evacuated, as it is in some birds, when it displays some features of batch reactors.

Modified plug-flow reactors

Two herbivore systems at first do not appear to fit any of the three models described by Penry and Jumars (1987). These are the basically tubiform forestomach of large grazing kangaroos (see Fig. 4.10) (Langer, Dellow, and Hume 1980) and the proximal colon of large hindgut fermenters like the horse (see Fig. 4.29), which is also tubiform in gross morphology (Hume and Warner 1980). These two fermentation systems can best be modeled as a number of CSTRs in series (Fig. 6.10, *D*), as there is considerable mixing of contents both radially and axially, as a result of contractions of the gut wall, which form the haustra that are so characteristic of these two fermentation organs. As the number of CSTRs in series increases, the reactor system becomes more and more like a plug-flow reactor (Penry and Jumars 1987). Thus, for discussion and comparative purposes, tubiform fermentation chambers such as the macropod marsupial forestomach and the proximal colon of colon fermenters (for example, horses, elephants, apes) can be considered to be modified PFRs (Hume 1989; Hume and Sakaguchi 1991).

An illustration of the applicability of the modified PFR model is provided by Dellow, Nolan, and Hume (1983) who treated the forestomach of two species of wallabies as four CSTRs in series to estimate rates of fermentation and microbial protein synthesis with isotope dilution techniques. The techniques used were originally developed with ruminants, in which the infused isotopically labeled SCFA is rapidly mixed throughout the contents of the reticulorumen (Leng and Leonard 1965), where there is only one mixing pool. In the study of Dellow et al. (1983), in the wallabies, the primary mixing pool, or first CSTR, was the sacciform forestomach. The other three CSTRs were successive sections of the tubiform forestomach (Fig. 6.10, *D*). The rate of infusion of the labeled SCFA into each section of the tubiform forestomach was the rate of flow from the preceding pool, estimated with the fluid marker Cr-EDTA. Total production of SCFA was the sum of production in each of the four mixing pools, or CSTRs.

Although exemplified by the kangaroo forestomach and the equine proximal colon, any expanded haustrated tubiform section of the vertebrate gut functions as a modified plug-flow reactor (Karasov and Hume in press). This includes the tubus section of the presbytis monkey forestomach (see Fig. 4.18); the forestomach, or "anterior stomach," of the hippopotamus (see Fig. 4.24); the proximal colon of most large (>10 kg) hindgut fermenters such as wombats (see Fig. 4.10); many monkeys, apes, and humans (see Fig. 4.20); pigs (see Fig. 4.23); and rhinos and elephants (see Fig. 4.31). The proximal colon of herbivorous lizards in the families Agamidae and Iguanidae (see Fig. 3.12) also appears to function as a modified plug-flow reactor.

Applications of models

Chemical reactor theory was used by Penry and Jumars (1987) to predict the optimal configurations of stirred-tank reactor and plug-flow reactor guts in invertebrate marine deposit feeders. It was also used by Hume (1989) to predict optimal configurations of stirred-tank, plug-flow, and modified plug-flow reactor guts in mammalian herbivores. Most recently, Horn and Messer (1992) have used the approach with herbivorous fish. They described the stomach as a batch, or stirred-tank reactor, the intestine as a plug-flow reactor, the hindgut cecum as a stirred-tank reactor, and other structures where food is mechanically processed, such as gill rakers, pharyngeal mills, and muscular stomachs, as gates. They then predicted optimal gut configurations as a function of nutrient concentration and concluded that the gastrointestinal tracts of four herbivorous fish all had a plug-flow reactor component, but not all had a CSTR component, citing differences in gut length and the presence of gates as modifying factors that also had to be considered in models of the digestive tract of marine fishes.

The first vertebrate digestive system to be modeled was that of the ruminant. The approach, used by several workers (Waldo, Smith, and Cox 1972; Mertens and Ely 1979; Illius and Gordon 1991, 1992; Spalinger and Robbins 1992), was

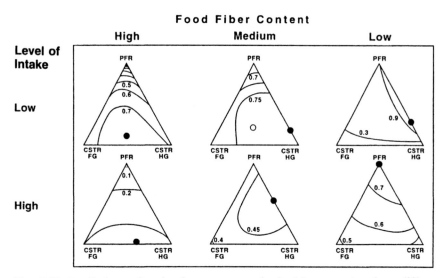

Fig. 6.11. Predictions of optimal gut structure for herbivores eating three differ-ent qualities of diet at two different levels of intake. Diets vary between high (90 percent neutral detergent fiber NDF), medium (50 percent NDF), and low (10 percent NDF) levels of fiber. Symbols represent plug-flow reactor (*PFR*); continuous-flow, stirred-tank reactor (*CSTR*); foregut (*FG*); and hindgut (*HG*). Contours represent relative rates of energy gain and are dimensionless. Solid cir-cles represent maximal rates of energy gain. The open circle represents an alter-native maximum at low rates of intake on medium-fiber diets. (Modified from Alexander 1991.)

based on first-order kinetics and the assumption that rates of fermentation are lim-ited only by the quantities of substrates present at steady state. A similar approach was taken by Justice and Smith (1992) to models of fiber digestion by small mam-malian hindgut fermenters. This confirmed that many small rodents can utilize di-ets of moderate fiber content, but there is no doubt that there is a general limit to heavy reliance on fermentative digestion of plant cell walls in small mammals (Cork and Foley 1991).

An interesting approach taken by Alexander (1991) was to treat animal diges-tive tracts as chemical reactors based only on first-order kinetics, a much simpler and limited treatment than that of Penry and Jumars (1987), but nevertheless ef-fective. The basic model was a CSTR in the foregut followed by a plug-flow re-actor (the small intestine) and a CSTR (cecum) in the hindgut. The study did not consider modified plug-flow reactor fermentation chambers in either the foregut or hindgut. Alexander's findings are summarized in Fig. 6.11. Each triangle rep-resents the set of all possible combinations of the three gut structures considered.

Solutions of the model for diets of three different fiber levels eaten at either low or high levels indicate that for high-fiber food the optimal gut design is dominated by fermentation chambers, with a foregut CSTR at least as important as a hindgut CSTR; this is the ruminant. Only a low food intake is realistic in the CSTR fermentation chamber of ruminants. Because of the flow characteristics of CSTRs, energy gains are too low at high food intakes. The evolutionary solution to the problem faced by ruminants of a limitation to flow of large particles out of the reticulorumen (Van Soest, Sniffen, and Allen 1988) is a number of CSTRs in series, as seen in the modified plug-flow reactor fermentation chamber of the kangaroo forestomach and the equine proximal colon (Hume 1989). Both have been shown to allow the flow of larger food particles and thus to maintain higher food intakes on high fiber diets than can the ruminant system (Hume and Sakaguchi 1991; Illius and Gordon 1991).

On food of medium fiber content (50 percent NDF), the optimal gut, according to Alexander (1991), is a hindgut CSTR with a short plug-flow reactor at low food intakes and a longer plug-flow reactor at higher food intakes. However, as indicated by the open symbol in Fig. 6.11, there is an alternative maximum where a foregut CSTR is at least of equal importance to the hindgut CSTR. This again represents the ruminant, and helps to emphasize the flexibility of this herbivore system at relatively low food intakes. At high food intakes the ruminant strategy is outperformed by hindgut CSTRs; these are small cecum fermenters such as lagomorphs and herbivorous rodents (Justice and Smith 1992).

On food of low fiber content, the cecum fermenter is again the optimum herbivore at low food intakes, but at high food intakes, the optimum is a simple gut (that is, a plug-flow reactor) without any fermentation chambers. These animals are omnivores rather than herbivores.

Although neither the Alexander (1991) nor the Penry and Jumars (1987) approach considers the dynamics of food particles of different sizes in the way that the model of ruminant digestion developed by Mertens and Ely (1979) does, they have the advantage in that they offer a way of enhancing future studies of correlations between digestive tract morphology and diet among species of fish, reptiles, birds, and mammals by extending the analyses beyond simple correlation.

Summary

Rate of digesta transit varies markedly among vertebrate groups, being slowest in reptiles and fastest in small birds. Rates of transit in mammals are fastest in small carnivores and slowest in large herbivores, and they are particularly slow in several arboreal folivores.

Measurement of the rate of passage of digesta is based on inert marker substances, and some of the variation in results between rate of passage studies can be attributed to the different markers used. Increased emphasis is now being placed on the use of markers that are specific to either the fluid or the particulate phase of the digesta.

Fluid digesta usually move through the gastrointestinal tract faster than particles. This is universally true for the stomach, which is the first of the two major sites of digesta retention. However, the hindgut, the other important site of digesta retention, often has mechanisms that result in greater delay in transit of fluid (and small particles, including bacteria) than of large particles. This is particularly so in the proximal colon and cecum of small hindgut fermenters (cecum fermenters). The dominant mechanism for digesta retention in the hindgut of most, if not all, vertebrates is antiperistalsis. In many terrestrial vertebrates that have a cloaca, urine is refluxed the length of the hindgut. This increases the nitrogen available for microbial growth in the ceca, thereby enhancing digestion. Digesta retention in the hindgut at some mammals is further aided by haustral contractions.

The benefits of selective retention of fluid and small particles (which tend to be washed along with the fluid) in the cecum of small hindgut fermenters are twofold. First, it eliminates large particles (the least digestible fraction) from the gut relatively quickly, thereby minimizing their "gut-filling" effect and thus their inhibition of food intake. Second, it concentrates digestive effort in the cecum by maintaining a high concentration of microorganisms together with the most digestible food particles (the small ones).

The need to match rates of digesta transit with other functions including food intake and nutrient absorption has led to the concepts of optimal digestion time and digestive strategies. Our understanding of how these are interrelated in a wide range of vertebrate systems has been aided by the modeling approach. Models of digesta transit based on chemical reactor theory have been particularly useful in the analysis of correlations between digestive tract morphology and diet among species of fish, reptiles, birds, and mammals. However, these models do not consider constraints such as detoxification of plant allelochemicals, digestion of some substrates by endogenous enzymes versus microbial fermentation, or availability of fermentation end products for absorption and metabolism. These topics are discussed in subsequent chapters.

7

Digestion of carbohydrate, lipids, and protein and the absorption of end products

The principal organic components of plants and animals are carbohydrates, lipids, proteins, and nucleic acids. Although some are ingested in a form that can be readily absorbed, most of them require hydrolysis into a limited number of simpler compounds before their absorption. This is accomplished by enzymes that are either produced by the digestive system of the host animal or by microbes indigenous to its digestive tract. This chapter concentrates on the enzymes generated by the digestive system and absorption of their end products. Vonk and Western (1984) provided a great deal of comparative information on enzymatic digestion in both vertebrates and invertebrates. A more recent review of information on variations in digestion and absorption among vertebrates is given by Karasov and Hume (in press). Information on the mechanisms of digestion and absorption has been recently reviewed by a number of authors in *Physiology of the Gastrointestinal Tract* (1994).

Table 7.1 lists the principal forms of organic compounds found in the diet of vertebrates, the major endogenous enzymes that act on them, and the absorbable end products. All of the enzymes are proteins that catalyze the hydrolysis of their substrates. This often involves the sequential breakdown of large molecules by a series of enzymes. The end products consist of a limited number of monosaccharides, fatty acids, alcohols, peptides, amino acids, and other compounds that can be absorbed. This list does not include all of the enzymes that may be present, and some of those listed are absent in many species. It also omits the vitamins, which are organic compounds that are essential in small quantities, act as coenzymes or precursors of coenzymes in the regulation of metabolism, and must be absorbed intact.

Digestion

The demonstration of endogenous digestive enzymes is complicated by the presence of enzymes ingested with the food or synthesized by indigenous microbes. Desquamation of gut epithelial cells releases additional enzymes that are involved

Table 7.1. *Principal food components and endogenous enzymes of vertebrates*

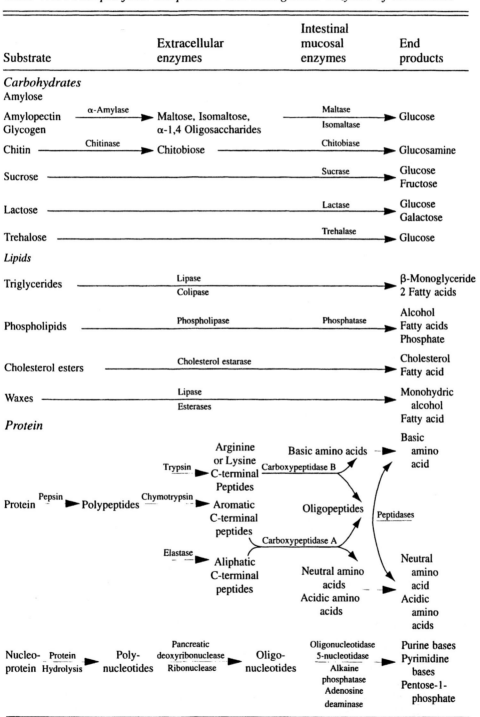

Substrate	Extracellular enzymes	Intestinal mucosal enzymes	End products
Carbohydrates			
Amylose			
Amylopectin	α-Amylase → Maltose, Isomaltose, α-1,4 Oligosaccharides	Maltase / Isomaltase	Glucose
Glycogen			
Chitin	Chitinase → Chitobiose	Chitobiase	Glucosamine
Sucrose		Sucrase	Glucose / Fructose
Lactose		Lactase	Glucose / Galactose
Trehalose		Trehalase	Glucose
Lipids			
Triglycerides	Lipase / Colipase		β-Monoglyceride / 2 Fatty acids
Phospholipids	Phospholipase	Phosphatase	Alcohol / Fatty acids / Phosphate
Cholesterol esters	Cholesterol estarase		Cholesterol / Fatty acid
Waxes	Lipase / Esterases		Monohydric alcohol / Fatty acid
Protein			Basic amino acid

Protein — Pepsin → Polypeptides
Chymotrypsin → Aromatic C-terminal peptides
Trypsin → Arginine or Lysine C-terminal Peptides — Carboxypeptidase B → Basic amino acids → Basic amino acid
Elastase → Aliphatic C-terminal peptides — Carboxypeptidase A → Neutral amino acids / Acidic amino acids
Oligopeptides → Peptidases → Neutral amino acid / Acidic amino acids

Nucleo-protein — Protein Hydrolysis → Poly-nucleotides — Pancreatic deoxyribonuclease / Ribonuclease → Oligo-nucleotides — Oligonucleotidase / 5-nucleotidase / Alkaine phosphatase / Adenosine deaminase → Purine bases / Pyrimidine bases / Pentose-1-phosphate

Source: Modified from Stevens (1977).

only with cellular metabolism. Therefore, isolation of an enzyme from homogenates of tissue from the digestive tract or its accessory glands does not prove that it is involved in normal digestive processes. The source of a digestive enzyme can be equally difficult to determine. Its presence at a given site along the tract does not necessarily indicate the site of release and even its presence in a tissue homogenate may result from adsorption to the luminal surface of cells.

Comparison of the digestive enzymes in different species is further complicated by the fact that their level of activity is modulated by changes in the diet and the time of day. Karasov and Hume (in press) concluded that pancreatic and intestinal enzyme synthesis appear to be modulated by the diet in all classes of vertebrates. Addition of carbohydrate, protein, or lipid to the diet of rats produced a 500 to 800 percent increase in the levels of pancreatic amylase, trypsinogen, chymotrypsinogen, and lipase (Brannon 1990). Addition of carbohydrate to the diet of rats and primates also increased the activity of sucrase and maltase in the brush border of small intestinal cells (Koldovský 1981). Sucrase-isomaltase and lactase activities increased along the entire length of the villus within three hours after a sucrose-containing meal, suggesting a rapid alteration in mature cells of the villus (Goda and Koldovský 1988). Disaccharidase activities in the rat intestine also show a circadian rhythm, which peaked in the dark and troughed in the light hours (Saito et al. 1975; Stevenson et al. 1975). A shift in mealtime resulted in a shift in peak activity, indicating partial dependence on food consumption as a "Zeitgeber".

An additional factor that has a major effect on the rate of digestion in ectothermic species is body temperature. As mentioned in Chapter 6, a decrease in temperature results in a reduction in rates of feeding, digesta transit, and nutrient utilization. Coulson, Coulson, and Herbert (1990) examined the effect of ambient temperatures of 31°C, 25°C, 20°C, and 10°C on protein digestion, amino acid absorption and assimilation, heart rate, and oxygen consumption in alligators. A reduction in ambient temperature was associated with a roughly proportional decrease in each of these parameters. They pointed out that rates of digestion, absorption, and assimilation must be closely integrated in animals whose mass-specific metabolic rates can differ over a range of 4,000 times between the largest crocodile and the smallest shrew. They concluded that the rate of digestion may be controlled by the rate that the villi are perfused with blood. Recall that in Chapter 1 it was suggested that the capacity of the digestive system to absorb nutrients was a possible proximate cause of "metabolic ceilings" in animals (Peterson, Nagy, and Diamond 1990).

Carbohydrates

Carbohydrates consist of individual monosaccharides or monosaccharides linked by glycosidic bonds to form chains that are two units (disaccharides), three to ten units (oligosaccharides), or more than ten units (polysaccharides) in length. The carbohydrates in plants and animals serve three main functions: structure, storage, and transport. Most of the carbohydrates in plants and animals are polysaccharides, which either provide structural support or serve for the intracellular storage of energy. The major transport carbohydrates found in plants and animals are the disaccharides sucrose, trehalose, and lactose and the monosaccharides glucose and fructose. Sucrose (glucose-fructose) is the principal transport carbohydrate in plants. Trehalose (glucose-glucose) is the blood transport carbohydrate of insects, and lactose (glucose-galactose) is the principal carbohydrate found in the milk of most mammals. Fructose is the only free monosaccharide of any quantity found in plants, and glucose is the only free monosaccharide present to any degree in animals.

Structural carbohydrates

Fig. 7.1 shows the monosaccharides and linkages for the common structural polysaccharides. The principal structural carbohydrates of terrestrial plants are cellulose, hemicelluloses, and pectins. The global terrestrial production of cellulose and hemicellulose has been estimated to be 1×10^{11} tons per year (Pel 1989). Cellulose is the major constituent of cell walls in higher plants, providing 20–40 percent of the dry matter. It consists of straight chains of glucose with a β-1,4 linkage. Hemicelluloses are a mixture of polysaccharides of varying composition. They can include xylose, galactose, arabinose, mannose, and other monosaccharides, but they often have a common core consisting of a xylose chain with a β-1,4 linkage (Kronfeld and Van Soest 1976). Pectin, the cementing substance of plant cell walls, consists primarily of an α-1,4 linked galacturonic acid chain. The cell walls of algae can contain hemicellulose and pectin along with other structural carbohydrates, but not cellulose (Vonk and Western 1984). The principal structural carbohydrate in animals is chitin, which is present in the cell walls of bacteria and fungi and in the exoskeleton of insects and many marine invertebrates (Table 7.2). Chitin constitutes 58–85 percent of the dry weight of most arthropods (Jeuniaux 1971a) and 8–60 percent of the dry weight of fungi (Berkeley 1979). Tracy (1957) estimated that marine copepods alone produced more than 10^9 tons of chitin per year. Chitin consists of an unbranched polymer of repeating units of β-1,4-linked *N*-acetyl-D-glucosamine.

Endogenous enzymes of vertebrates cannot hydrolyze the β-1,4 linkages of cellulose and hemicellulose. The α-1,4 linkage of pectin is also highly resistant to attack by α-amylase because of its axial position in the molecule (Van Soest 1994). Chitin can be hydrolyzed by chitinase to chitobiose, the dimer of β-1,4-linked *N*-acetyl-D-glucosamine, which then can be hydrolyzed to the monomer by chitobi-

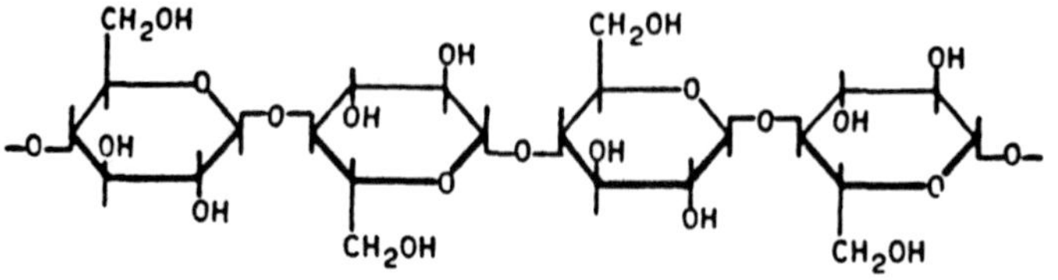

CELLULOSE

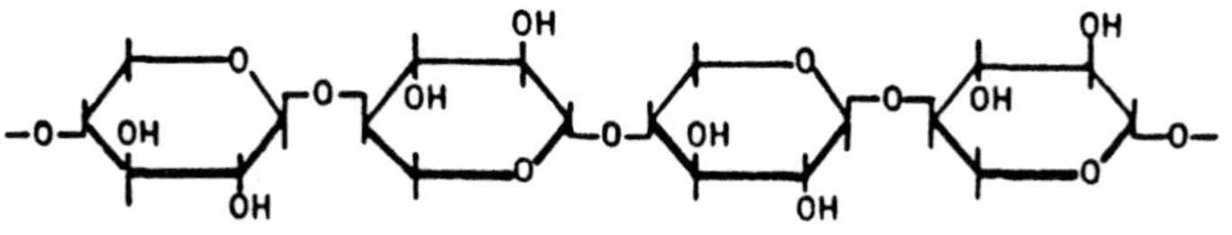

XYLAN (HEMICELLULOSE)

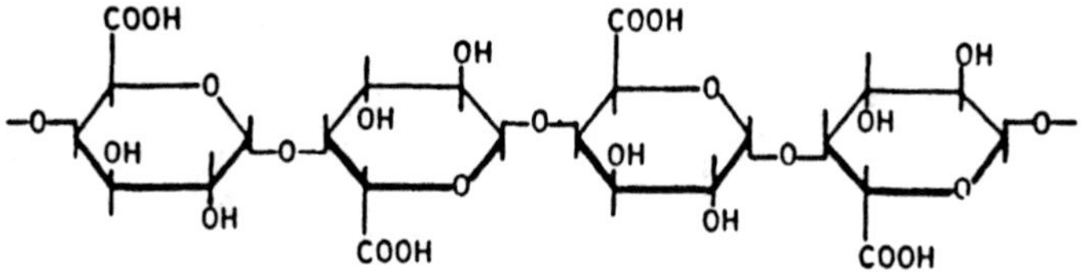

POLYGALACTURONIC ACID (PECTIN)

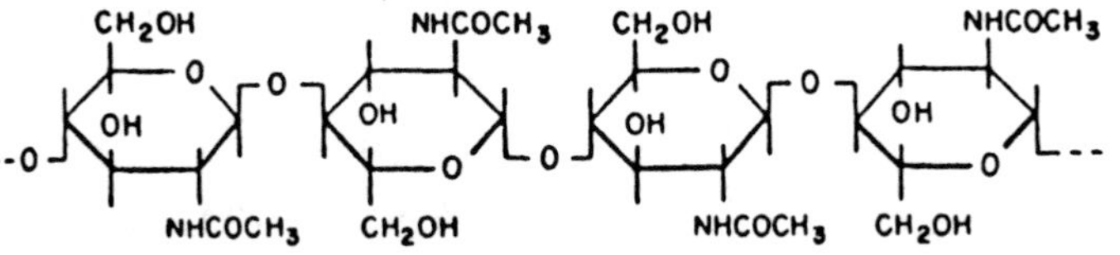

CHITIN

Fig. 7.1. Haworth projections of cellulose, xylan, pectin (Van Soest 1994), and chitin (Fruton and Simmonds 1958). The linkage in pectic acid is α-1,4. Haworth formulas are misleading in that xylan and pectin appear to have a conformation similar to cellulose, although they actually differ. Pectin, like starch, cannot exist in a linear conformation and must form kinks or coils. However, the axial position of carbon 4 in galacturonic acid results in a different configuration, as compared with starch.

Table 7.2. *The distribution of chitin in living organisms*

Phylum	Class	Structure
Fungi	Chytriodiomycetes	Cell walls
	Zygomycetes	Cell walls
	Ascomycetes	Cell walls of filamentous types
		Cell wall component of yeasts
	Basidiomycetes	Cell walls
Algae	Bacillariophyceae	Chitin fibrils of centric diatoms
Coelenterata	Hydrozoa	Perisarc of polyps, pneumatophore of some siphonophores
Aschelminthes	Nematoda	Egg shells
Phoronida		Tubes
Bryozoa		Exoskeleton, operculum
Brachiopoda		Pedicle, shell
Mollusca	Gastropoda	Radula, jaws
	Cephalopoda	Shell, dorsal shield
Annelida	Polychaeta	Jaws, chaetae, gut lining
	Oligochaeta	Chaetae
Arthropoda		Exoskeleton, lining of foregut and hindgut, apodemes, egg shells
Chaetognatha		Hooks
Pogonophora		Tubes, setae

From Vonk and Western (1984).

ase (Table 7.1). Because both enzymes are found in many indigenous gut microorganisms, their presence does not necessarily indicate endogenous enzyme production. However, Jeuniaux and coworkers found chitinase in the gastric mucosa of many mammals, birds, reptiles, and adult amphibians, as well as in the pancreas of the mole, hedgehog, pig, starling, blackbird, amphibians, and most of the reptiles examined (Table 7.3). Chicks of Leache's storm petrels, which feed on North Atlantic crustaceans, digested 35 percent of the chitin in their diet (Place 1990). Chitobiase activity also was demonstrated in gastric, pancreatic, and intestinal tissue of the insectivorous horseshoe bat (Jeuniaux 1962a).

Chitinase has been demonstrated in the digestive tract of a number of freshwater fish (Lindsey 1984). Although chitinase was absent from the digestive tract of teleosts that lacked a stomach and pyloric ceca, it has been found in the gastric mucosa of many other teleosts (Jeuniaux 1961, 1963; Dandrifosse 1963; Okutani 1966; Colin 1972; Micha, Dandrifosse, and Jeuniaux 1973b; Fänge, Lundblad, and Lind 1976; Fänge et al. 1979). It also was found in homogenates of pyloric ceca of teleosts, probably in association with disseminated pancreatic tissue. Chitinase was present in the gastric mucosa of the sharks *Squalus* and

Table 7.3. *Chitinase activity in gastric mucosa and pancreas of mammals, birds, reptiles, and amphibians*

Species	Gastric mucosa	Pancreas	Reference
Mammals			
Bat	+	0	Jeuniaux (1962a)
(*Rhinolophus ferrumeguinum*)			
Mole	+	+	Jeuniaux (1962a), Jeuniaux
(*Talpa europaea*)			and Cornelius (1978)
Hedgehog	+	+	Jeuniaux (1962a)
(*Erinaceus europaeus*)			
Dog	+	0	Cornelius, Dandrifosse,
(*Canis familiaris*)			and Jeuniaux (1975)
Fox	+	0	"
(*Vulpes vulpes*)			
Marten	0	0	"
(*Martes martes*)			
Cat	0	0	"
(*Felis silvestris*)			
Sloth	0	0	Jeuniaux (1962b)
(*Choloepus hoffmanni*)			
Mouse	+	0	Frankignoul and
(*Mus musculus*)			Jeuniaux (1965)
Rat	+	0	"
(*Rattus norvegicus*)			
Hamster	+	0	"
(*Cricetus cricetus*)			
Prosimian	+	0	Beerten-Joly, Piavaux
(*Perodicticus potto*)			and Goffart (1974)
Monkey	+	0	Jeuniaux and Cornelius (1978)
(*Cebus capucinus*)			
Human	0	0	"
(*Homo sapiens*)			
Pig	+	+	Jeuniaux (1962a)
(*Sus scrofa*)			
Sheep	0	0	"
(*Ovis aries*)			
Rabbit	0	0	"
(*Oryctolagus cuniculus*)			
Birds			
Sparrow	+	0	Jeuniaux (1962a); Jeuniaux
(*Passer domesticus*)			and Cornelius (1978)
Robin	+	0	"
(*Erithacus rebecula*)			

Table 7.3. (*cont.*)

Species	Gastric mucosa	Pancreas	Reference
Japanese nightingale (*Liothrix lutea*)	+	0	Jeuniaux and Cornelius (1978)
Starling (*Sturnus vulgaris*)	+	+	"
Blackbird (*Turdus merula*)	+	+	Jeuniaux (1962a)
Carrion crow (*Corvus corone*)	+	0	Jeuniaux and Cornelius (1978)
Chicken (*Gallus gallus*)	+	0	Jeuniaux (1962a)
Barn owl (*Tyto alba*)	+	0	Jeuniaux and Cornelius (1978)
Pigeon (*Columba palumbis*)	0	0	Jeuniaux (1962a)
Parrot (*Psittacus erithacus*)	0	0	Jeuniaux and Cornelius (1978)
Reptiles			
Terrapin (*Emys*)	+	+	Jeuniaux (1962a)
(*Clemmys*)	+	+	"
Tortoise (*Testudo hermanni*)	+	+	"
Lizard (*Lacerta*)	+	+	"
(*Uromastyx*)	+	+	"
(*Anolis*)	+	+	"
(*Chamaelo*)	+	+	"
(*Augus*)	+	+	"
Snake (*Natrix natrix*)	0	0	Micha, Dandrifosse, and Jeuniaux (1973a)
Amphibians			
Adult frog (*Rana temporaria*)	+	+	Jeuniaux (1963)
Adult toad (*Bufo marinus*)	+ +	+ +	Micha, Dandrifosse, and Jeuniaux (1973b)
Salamander (*Triturus alpestris*)	+	+	"
(*Salamandra salamandra*)	+	+	"

Scyliorhinus and rays (*Raja* spp.), but none was found in their pancreas (Micha et al. 1973b). Information on Holocephali and cyclostomes appears to be limited to the demonstration of pancreatic chitinase in chimaeras and low levels of intestinal chitinase activity in hagfish (Fänge et al. 1976, 1979). Nevertheless, the chitobiase activity of gut contents was higher than that of tissues in these species, and chitobiase was present in the gut contents of many species that lack chitinase. This, plus the fact that this chitinase does not appear to be a component of the intestinal brush border, led Vonk and Western (1984) to the conclusion that it may be of dietary or microbial origin. Despite a lack of correlation between chitinase activity and the presence of chitin in the diet of some fish (Lindsey 1984), the enzyme appears to be present in most vertebrates whose diets include substantial amounts of chitin.

Storage carbohydrates

The major storage carbohydrates of terrestrial plants are the starches amylose and amylopectin (Table 7.1). Amylose consists almost entirely of a-1,4-linked glucose units. Amylopectin consists of the same chains with branches, formed by a-1,6 linkages, usually at intervals of twenty-four glucose units. Vonk and Western (1984) listed Floridean starch, which is similar to amylopectin, and laminarin, a polymer of b-1,3-linked glucose molecules, as the principal storage carbohydrates of algae. Glycogen, the principal storage carbohydrate of animals, has a structure similar to amylopectin, except for shorter chains (ten to fourteen units) between a-1,6 branches of the glucose chain. It can comprise up to 2 percent of the fresh weight of muscle and up to 10 percent of the fresh weight of liver (Lehninger 1975).

Amylose, amylopectin, and glycogen are hydrolyzed by a-amylase to form maltose, isomaltose, maltotriose, and other a-1,4-linked and a-1,6-linked oligosaccharides (Fig. 7.2). Amylase is secreted by the pancreatic tissue of all vertebrates and the salivary glands of the echidna, grasshopper mouse *Onychomys*, bushtail possum *Trichosurus*, dog, rodents (rats, mice, guinea pig, squirrel, vole), rabbit, artiodactyls (pigs, cattle, sheep, goats, deer), and primates (Vonk and Western 1984). Salivary amylase was reported in the chicken, turkey, goose, and pigeon, but there does not appear to be conclusive evidence for this (Ziswiler and Farner 1972). Amylase activity also was demonstrated in the crop (Bolton 1962, 1965), gizzard (Hewitt and Schelkopf 1955), and bile (Farner 1943) of chickens and in the stomach of turtles (Tanaka 1942) and snakes (McGeachin and Bryan 1964). However, this may represent regurgitation of pancreatic secretions.

Pancreatic and salivary amylases are a-amylases, activated by chloride. Molecular weights and activation energies are known for some, and the amino acid

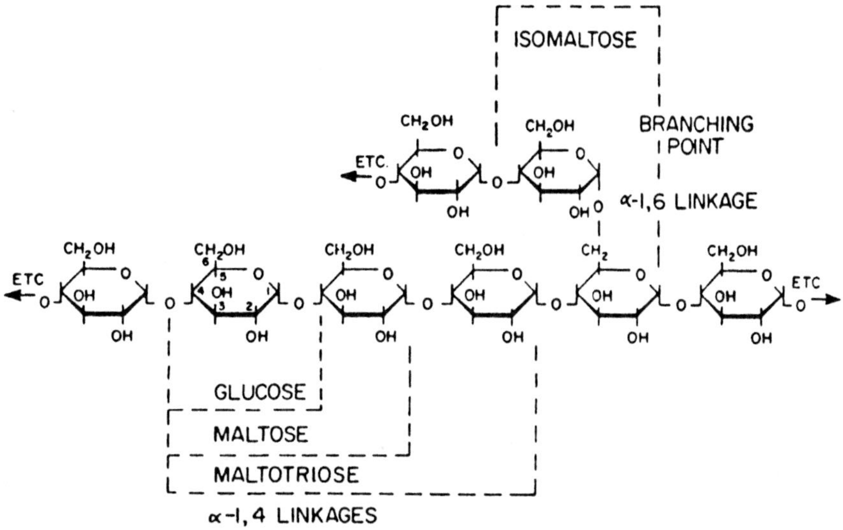

Fig. 7.2. The structure of starch. Hydrolysis catalyzed by pancreatic amylase occurs at the α-1,4-linkage, and the products of hydrolysis are straight-chain oligosaccharides. Since pancreatic amylase does not catalyze hydrolysis of the α-1,6-linkages, isomaltose is also a product of hydrolysis. Further hydrolysis is catalyzed by the maltases and isomaltase of the brush border of intestinal epithelial cells (Davenport 1982.)

composition has been determined for those of the chicken and a few mammals (Vonk and Western 1984). An intestinal γ-amylase, which did not require chloride activation and produced a stepwise release of terminal glucose from starch and oligosaccharides, also was described (Dahlqvist and Thompson 1963; Eggermont 1969; Alpers and Solin 1970). However, Vonk and Western (1984) attributed this to a lysosomal enzyme.

There is evidence for higher levels of amylase activity in herbivorous and omnivorous fish, as opposed to carnivorous species (Kapoor, Smit, and Verighina 1975), and omnivorous versus carnivorous birds (Jain 1976), but other data on birds (Bhattacharya and Ghose 1971) and reptiles (Vonk and Western 1984) conflict with this.

Oligosaccharides and disaccharides

Oligosaccharides and disaccharides are hydrolyzed to monosaccharides by enzymes located in the microvilli, or brush border, of intestinal cells (Table 7.1). Alpers (1987) listed the mammalian enzymes as isomaltase, lactase, trehalase, sucrase, glucoamylase sucrase and α-1,6-limit dextrinase. Isomaltase and the α-1,6-

Table 7.4. *Diet and intestinal disaccharidase activities of prototherian and methatherian mammals*

Species	Diet	Maltase	Isomaltase	Sucrase	Lactase	Trehalase	Cellobiase
Prototheria							
Echidna (*Tachyglossus aculeatus*)	Ants	5.5	4.4	0	0.07	2.6	0.02
Metatheria							
Brown antechinus[a] (*Antechinus stuartii*)	Insects	42.6	—	10.2	0.06	8.1	0.06
Tiger quoll (*Dasyurus maculatus*)	Carnivore	69.4	38.4	4.9	1.3	23.7	0.14
Long-nosed bandicoot (*Perameles nasuta*)	Omnivore	18.9	12.0	5.0	0.68	11.1	0.26
Short-nosed bandicoot (*Isoodon obesulus*)	Omnivore	—	—	0	—	4.0	0.22
Brushtail possum (*Trichosurus vulpecula*)	Herbivore	41.2	22.9	6.8	0.47	7.2	0.20
Ringtail possum (*Pseudocheirus peregrinus*)	Eucalyptus leaves	0.22	0.11	0.06	11.2	—	0.10
Koala (*Phascolarctos cinereus*)	Eucalyptus leaves	12.0	5.2	1.9	0.68	0.006	1.43
Eastern Grey kangaroo (*Macropus giganteus*)	Grass	0.3	0.12	0	0.05	0.12	0

Source: Modified from Kerry (1969) by Vonk and Western (1984).
All data was taken from adult specimens. Data are expressed in units (1 μmol substrate/min) per gram of mucosa (wet weight).
[a] Activities for brown autechinus expressed as units/g wet wt intestine.

limit dextrinases attack the α-1,6-linkages of amylopectin and glycogen. Because of disagreements over the number and cross-functions of these enzymes in various species, it is easier to describe their function by the specific substrate they hydrolyze.

Tables 7.4 and 7.5 list intestinal disaccharidases found in mammals. Maltase activity has been found in the intestine of most mammals. In humans, rats, and rabbits, sucrase and isomaltase are present as a single sucrase-isomaltase glycoprotein complex, but sucrase activity was absent from the intestine of the echidna, bandicoot, kangaroo, pinnipeds, cattle, sheep, and camels, and only low levels of sucrase were found in approximately 10 percent of eskimos (Kretchmer 1981). Trehalase, which would be most useful to insectivores, was found in many of the mammals examined. It was absent from the domestic cat, panther, bushtail possum, grey kangaroo, and pinnipeds, and only traces were found in the koala intestine. Cellobiase was isolated from the intestine of a number of mammals.

Activities of maltase, isomaltase, and sucrase were examined in the intestinal mucosa of the European crane and Chinese quail (Zoppi and Schmerling 1969), chicken (Brown and Moog 1967), and five species of sea birds (Kerry 1969). Maltase, isomaltase, and sucrase levels in the crane and chicken were similar to those of mammals. Quail and sea birds demonstrated lower levels of these enzymes. Only low levels of lactase were found, and trehalase activity was not clearly demonstrated in any of these birds. Lower levels of maltase activity were found in the intestine of insectivorous birds (Martinez del Rio 1990) and bats (Hernandez and Martinez del Rio 1992) than in omnivorous birds or frugivorous and insectivorous bats. Conversely, the intestines of nectarivorous species showed much higher levels of sucrase activity, as did hummingbirds, as compared with passerine species (Martinez del Rio 1990).

Maltase, isomaltase, and trehalase were demonstrated in the intestine of an herbivorous tortoise (Zoppi and Schmerling 1969) and the American alligator (Martinez del Rio and Stevens 1988). Maltase and trehalase were found in the intestinal brush border of a frog (Hourdry et al. 1979), but sucrase was absent in the adult amphibians examined (Vonk and Western 1984) and in the American alligator. Therefore, the sucrase–isomaltase complex seen in many mammals is not common to all vertebrates.

Intestinal homogenates were examined for disaccharidases in teleost fish (Kitamikado and Tachino 1960; Ushiyama et al. 1965; Kawai and Ikeda 1971; Vonk and Western 1984). Maltase activity was found in the intestine of all species, and the carp intestine showed additional sucrase and cellobiase activities, but sucrase activity was absent from the intestine of some carnivorous fish. Disaccharidase levels do not appear to have been examined in the intestine of the more primitive classes of fish.

Table 7.5. Distribution of intestinal disaccharidase activities in eutherian mammals

Order	Examples	Maltase	Isomaltase	Sucrase	Lactase	Trehalase	Cellobiase
Chiroptera	*Eidolon helvum*	+		+	–	+	+
Primates	*Tupaia, Nycticebus,*	+	+	+	+	+	+
	Perodicticus, Oedipomidas,	+	+	+	+	+	+
	Tamarinus, Actus, Saimiri,	+	+	+	+	+	+
	Cebus, Macaca, Papio, Homo	+	+	+	+	+	+
Lagomorpha	*Oryctolagus*	+	+	+	+	+	(+)
Rodentia	*Sciurus*	+		+	(+)	+	(+)
	Cricetus cricetus	+		+	(+)	+	
	Rattus norvegicus	+	+	+	(+)	+	(+)
	Cavia porcellus	+		+		+	(+)
Carnivora	*Canis familiaris*	+		+	Trace	+	+
	Ursus maritimus	+		+	+	Trace	
	Mustela	+		+		+	
	Felis silvestris	+		+		o	
	Panthera leo	+		+		o	

Pinnepedia	Zalophus californianus[a]	+		o	o	o	o
	Eumetopias jubatus[a]	+		o	o	o	o
	Arctocephalus spp.	+	+	o	Trace	o	
	Odobenus r. divergens	+	+	Trace	Trace	Trace	
	Mirounga leonina	+	+	o	Trace		
	Phoca vitulina[a]	+		o	o	o	o
Perissodactyla	Equus caballus	+	+	+	+	+	+
Proboscidea	Elephas maximus	+	+	o	+	Trace	
Artiodactyla	Tayassu tajacu	+	+	(+)	+	Trace	+
	Sus scrofa	+	+	+	+	+	+
	Bos traurus	+	(o)	+	+	+	
	Ovis aries	+		+	+	+	
	Capra hircus	+	(+)	(+)	+	+	(+)

Enzyme activity is designated + (present), *trace* or *o* (absent). Results in brackets indicate that an alternative substrate was used. These include palatinose (for isomaltose), o-nitrophenyl-β-D-glucoside (for cellobiase), the two latter substrates revealing β-galactosidase and β-glucosidase activities respectively.

[a] Suckling juveniles examined.

Source: From Vonk and Western (1984) with addition of perissodactyla data from Roberts (1975).

Lactase activity has been reported in the intestine of many mammals. It is generally found in the brush border, but in macropod marsupials milk oligosaccharides appeared to be absorbed by endocytosis and digested by an intracellular β-galactosidase (Janssens and Messer 1988; Messer, Crisp, and Czolij 1989). Unlike other carbohydrases, lactase activity is highest at birth and decreases, or may disappear, in adult animals (Koldovský 1970). Lactase is absent or present in only trace amounts in the intestine of newborn pinnipeds, which receive little lactose in their milk (Sunshine and Kretchmer 1964; Kerry and Messer 1968). Its absence from the intestine of the adult echidna, bandicoot, ringtail possum and elephant may be due to loss after weaning. Lactase appears to be absent in amphibians and reptiles, and Vonk and Western (1984) attributed the low levels found in some birds to microbial or lysosomal enzymes.

The presence and function of cellobiase is more difficult to understand. Cellobiase is a β-glucosidase that can hydrolyze the β-1,4-linkage of cellobiose, the end product of cellulose hydrolysis by C_1 and C_x cellulolytic enzymes (Vonk and Western 1984). However, the latter are not endogenous enzymes of vertebrates, and microbial fermentation of cellulose leads primarily to the release of short-chain fatty acids (SCFA), CO_2 and CH_4. Vonk and Western (1984) pointed out, however, that lactase has been shown to have both β-galactosidase and β-glucosidase activity in the pig and calf, and some of the cellobiase may be of bacterial or lysosomal origin.

Lipids

Lipids are organic compounds that are insoluble in water and soluble in organic solvents such as ether and chloroform. They serve as a major energy reserve in many plants and animals. The lipid content of plant parts varies, with seeds tending to be highest. Total lipid content also varies with species in both temperate terrestrial plants (Van Soest 1994) and marine algae (Montgomery and Gerking 1980). Those of greatest interest are the triglycerides, phospholipids, cholesterol esters, glycolipids, and waxes (Table 7.1). Each of these contains esters of alcohol and fatty acid.

Triglycerides are the major form of lipid storage in animals and the seeds of plants, and neutral esters of glycerol and three fatty acids. Phospholipids consist of a monohydric alcohol, fatty acid, phosphoric acid, and a base such as choline. The phospholipids are major components of cellular and cell organelle membranes and are, therefore, an important fraction of the lipids present in many animals. Lecithin is the principal phospholipid of animals. Cholesterol esters are the main forms in which cholesterol is found in vertebrates. Glycolipids are found mainly in photosynthetic tissue and provide much of the lipid content of pasture grasses and clover. They consist of a sugar (glucose or galactose) in a glycosidic linkage with two fatty acids and sometimes other organic acids.

Waxes are solid esters of monohydroxylic alcohols and long-chain fatty acids. They are found in the cuticle of many plants and the hives of bees, but their greatest importance comes from their presence in marine invertebrates, fish, and the spermaceti of whales (Schmidt-Nielsen 1990). Triglycerides are replaced by wax esters as the major reserve lipid of mesopelagic and bathopelagic zooplankton and fish (Lee et al. 1972). Benson, Lee, and Nevenzel (1972) found that waxes were stored or utilized by all organisms examined from depths below 1,000 m. Wax esters constituted 20 percent of the lipids found in copepods (small crustaceans) from cold, deep water (Lee et al. 1972) and Antarctic species of planktonic crustaceans at depths of 0 to 300 m (Bottino 1975). Planktonic crustaceans are an important link in the marine food chain, and Benson and Lee (1975) estimated that 50 percent of the organic material synthesized by phytoplankton is stored temporarily by marine animals as wax.

The fatty acids in lipids have a terminal carboxyl group and, because they are derived from acetate, most of them have an even number of carbon atoms. Although up to 9 percent of those formed in mammalian milk have a chain length of only four to fourteen carbons (Davenport 1982), the fatty acids in most lipids have fourteen to twenty-two carbons (Vonk and Western 1984). A large percentage of those found in animals and higher plants are C_{16} and C_{18} fatty acids. Planktonic algae contain high percentages of these, plus C_{20} and C_{22} acids.

Lipids are hydrolyzed by lipase, which can act only at a lipid–water interface, and by a variety of esterases that can only hydrolyze lipids that are in solution (Table 7.1). Place (1992b) found that oils accumulated in the cranial proventriculus of procellariiform birds (petrels, albatrosses), which regurgitated them for feeding chicks. However, lipids must be emulsified to provide the surface area required for digestion by lipase. The emulsification is accomplished principally by bile salts, phospholipids, cholesterol, and an apolipoprotein, which vary in their composition among species (Borgstrom and Patton 1991).

Information on the structure and evolution of vertebrate bile salts was reviewed by Haslewood (1978) and recently expanded to a comparison of 600 species of reptiles, birds, and mammals by Hagey (1992). All bile salts appear to be derived from cholesterol. They consist primarily of sulfated alcohols in fish, amphibians, and a few mammals and of taurine or glycine conjugates of bile acids in reptiles, birds, and most mammalian species.

Hofmann et al. (1985, 1988) described the biological and physicochemical properties of the bile acids, which allow them to function as surfactants that are immune to digestion and conserved by reabsorption from the ileum. Because of the hydrophilic and lipophilic nature of different parts of their molecules, bile salts act as detergents for the emulsification of fat. Lipase activity is inhibited by bile salts, but this is counteracted by the presence of colipase, a small protein secreted by the pancreas. Once the emulsification process has been initiated, lysolecithin, released by hydrolysis of lecithin, and the end products of triglyceride hydrolysis, also act as strong detergents. This results in the formation of micelles, with a diameter of 0.5 to 1.0 μm, which accumulate the long-chain fatty acids, monoglycerides, phospholipids, and fat-soluble vitamins released by the processes of lipid digestion.

Pancreatic lipases appear to be the most important enzymes for the digestion of fat in most vertebrates. The lipases hydrolyze triglycerides at their C_1 and C_3 ester bonds, releasing two fatty acids and a monoglyceride, although a nonspecific lipase that also attacks the C_2 position has been described in rats (Mattson and Volpenheim 1968; Patton 1975), dogs, the Northern fur seal, some teleosts (Patton, Nevenzel, and Benson 1975), and the leopard shark (Patton 1975). Lipase has been discovered in the pancreas of young chicks (Laws and Moore 1963), the western rattlesnake (Patton 1975), a frog (Scapin and Lambert-Gardini 1979), and, along with colipase, in the pyloric cecal tissue of the rainbow trout (Léger 1979). It also has been reported in the pancreas of the barndoor skate (Brockerhoff and Hoyle 1965) and the leopard shark, blue shark, dogfish, and stingray (Patton 1975). A high degree of lipase activity also was found in the intestinal epithelium of the hagfish *Myxinie glutinosa* (Adam 1963). Lipases have also been found in salivary, pharyngeal, and gastric secretions of various species. The lipase found in the gastric contents of humans has been shown to differ from that found in the pancreas.

Mammalian pancreatic enzymes include phospholipase A_2 and cholesterol esterase (Vonk and Western 1984). Phospholipase A_2, which has been demonstrated in the pancreatic juice of humans, rats, and pigs, hydrolyzes lecithin to form lysolecithin (Kidder and Manners 1978), which is then largely absorbed. This enzyme also could substitute for colipase (Bläckberg, Hernell, and Olivecrona 1981). Cholesterol esterase has been obtained from pancreatic secretions of rats and pigs. It hydrolyzes the ester to cholesterol and fatty acid and may release vitamins A, D, and E from their esters (Brockerhoff and Jensen 1974). Esterases hydrolyze monoesters more rapidly than lipase, and their specificity appears to be more dependent on the type of alcohol than on the structure of the fatty acid.

Brockerhoff and Jensen (1974) described a variety of other lipolytic esterases in the digestive tract of animals, but the specificity and origin of those found in gut contents are unclear. The fact that some of these are involved in cellular metabolism, including reformation of triglycerides, raises questions about their function as digestive enzymes in gut contents. Snake venoms contain a number of esterases, including phospholipases (Elliot 1978), but their primary function appears to be to aid in the penetration of toxins.

The importance of wax esters in the food chain of marine animals was mentioned earlier. Benson, Lee, and Nevenzel (1972) examined the digestion of wax esters and triglycerides by pancreatic lipase of the pig and by homogenates of pyloric ceca from the chum salmon and the anchovy. Porcine lipase hydrolyzed triglyceride 3.5 times more rapidly than wax esters, and the tissue homogenates from the salmon hydrolyzed triglycerides twelve times more rapidly. However, tissue homogenates from anchovies, which normally feed on copepods, hydrolyzed wax esters at twice the rate of triglycerides. A study by Patton and Benson (1975) of seven marine teleosts, showed that wax esters were hydrolyzed most rapidly by those with a stomach and pyloric ceca and that the released alcohol was converted to acid and then acyl lipids. An additional study of five marine teleosts (Patton et al. 1975) found that both triglycerides and wax esters were hydrolyzed by a nonspecific lipase that also split β-monoglycerides.

Benson and Lee (1975) concluded that all of the fish that they studied had a similar faculty to hydrolyze wax esters, oxidize the released alcohol, and incorporate the released fatty acid into acyl lipids, despite the fact that the percentage of wax ester in their normal diet varied from 70 percent in the anchovy to 5 percent in five species of tropical reef fish. However, the ability to digest wax esters appeared to correlate with digesta retention time, suggesting that their digestion by the anchovy was due to retention in its pyloric ceca.

Roby, Place, and Ricklefs (1986) examined the assimilation and deposition of wax esters by the chicks of five species of planktivorous seabirds. Up to 63 percent of the estimated digestible energy in their diet consisted of wax esters. With the use of ^{14}C-labeled acetyl palmitate, they found that the chicks efficiently hydrolyzed wax esters and assimilated the labeled fatty acids for storage, principally as triglycerides. One species, the Antarctic prion, retained much of the labeled wax ester in the stomach. In a similar study of Leach's storm petrel chicks (Place and Roby 1986), up to 38 percent of the labeled wax ester remained unhydrolyzed in the stomach after thirty-six hours. Yet the final hydrolysis of the wax ester and assimilation of the fatty acid moiety were extremely efficient, with less than 0.2 percent of the label appearing in the feces.

Place (1992b) reviewed earlier studies of wax ester digestion and assimilation by a variety of species and examined this further in six seabirds. He concluded that the more efficient digestion and assimilation of wax esters by seabirds was due to the following: (1) a higher concentration of bile salts plus triglycerides in their bile; (2) a longer retention time for lipids than aqueous compounds (123 versus 11 hours respectively in Leach's storm petrel chicks); (3) reflux and retention of intestinal contents into the gizzard; and (4) more efficient oxidation of fatty alcohol by enterocytes. Slow release from the stomach, high concentrations of bile salts, the presence of biliary triglycerides, and reflux of midgut digesta in these birds would aid in the emulsification and digestion of wax esters. Intracellular metabolism of the alcohols would aid in their absorption.

Proteins

Plant and animal proteins consist of chains of L-amino (or imino) acids linked together with peptide ($-NH \cdot CO-$) bonds. A given protein may contain up to twenty different amino acids. This, plus differences in amino acid sequence, cross-linkages, and tertiary structure, result in the wide variety of dietary proteins. The protein level in plants varies with parts (seeds, fruit, pollen, leaves, stems, roots, exudates), stage of growth, and species. The protein content of red, brown, and green marine algae varied from 3 to 13 percent of the dry matter (Montgomery and Gerking 1980). Gut contents also contain endogenous protein derived from secretions, desquamated cells and the escape of plasma protein. It has been estimated that the human digestive tract receives 10–30 g protein from secretions and 10 g from desquamated cells, plus 1.9 g of plasma albumen excreted into the tract each day (Davenport 1982).

With the exception of some newborn mammals, protein is sequentially hydrolyzed by extracellular endopeptidases, which attack peptide bonds along the protein chain, and exopeptidases, which split off terminal amino acids (Table 7.1). The action of these enzymes releases oligopeptides, dipeptides, and amino acids. Oligopeptides and dipeptides are further hydrolyzed by enzymes in the brush border or contents of intestinal epithelial cells.

Endopeptidases

The principal endopeptidases found in mammals are pepsin, chymosin (rennin), trypsin, chymotrypsin, elastase, and collagenase. Each of these enzymes is secreted as an inactive zymogen, which prevents it from attacking host tissue before its release, and is activated within the lumen of the digestive tract.

Pepsin

Secretion of pepsinogen and HCl by proper gastric glands was discussed in Chapter 2. Pepsinogen is activated by the acid media of the stomach to pepsin, which then serves as an autocatalyst for the release of additional enzyme. The enzyme is most active at a pH of 2 to 4. Pepsinogens also are secreted by tissues other than proper gastric glands. A pepsinogen originating from pyloric glands has been labeled gastricsin by some authors, and secretion of pepsinogen by the esophagus of some species of fish, adult amphibians, reptiles, and mammals was discussed in Chapter 3. Pepsin favors the hydrolysis of peptide bonds involving aromatic amino acids (Fruton and Simmonds 1958). The principal end products of pepsin digestion are large polypeptides and oligopeptides.

Pepsins of similar chemical structure and activity have been isolated from the stomachs of mammals, birds, reptiles, amphibians, and fish (Vonk and Western 1984). They appear to be present in the stomach of all mammals other than the monotremes (Griffiths 1978) and armadillos, which have an aglandular stomach. Two or three pepsinogens have been isolated from the proventriculus of chickens (Donta and Van Vunakis 1970a,b), and Herpol (1964) found that this was supplemented by pepsinogen produced by the gizzard in birds of prey (*Falco, Athene, Buteo*). Pepsin has been found in the stomach of numerous reptiles (Dandrifosse 1974) and adult amphibians, but it appears to be absent in the larvae of frogs and toads (Forte, Limlomwongse, and Forte 1969). Pepsins have been isolated from the stomach of many teleosts, including salmon, pike, perch, tuna, bluefin, albacore, cod, and hake (Vonk and Western 1984), as well as from the sleeper shark (Vonk 1927). Four pepsins have been described in the stomach of the dogfish (Bar-Eli, White, and Van Vunakis 1966; Merrett, Bar-Eli, and Van Vunakis 1969). It is absent from cyclostomes and the more advanced species of fish that lack a stomach.

Pepsin is not essential for the digestion of protein as evidenced by the absence of a stomach in some fish and larval amphibians, and the absence of pepsinogen and HCl secretion by the monotreme stomach. Humans with achlorhydria, the inability to secrete pepsinogen and HCl, can maintain their nitrogen balance (Davenport 1982). Nitrogen balance also can be maintained following surgical removal of the stomach from humans and a number of other species.

Chymosin

Chymosin, or rennin, was first described in the abomasum of young domestic cattle as the principal agent responsible for the clotting of milk. Its proteolytic activity is similar to but weaker than pepsin, and milk is clotted by pepsin as well. However, Foltmann (1981) suggested that the appearance of weakly proteolytic chymosin in the neonate, prior to peptic digestion, could serve to clot milk without damage to immunoglobulins. Chymosin is absent from the human stomach, but immunological reactions of gastric mucosal extracts from the dog, cat, rat, porcupine, kangaroo, horse, and zebra (Foltmann 1981; Foltmann et al. 1981) suggest a wide distribution among mammals.

Trypsin

Trypsinogen is secreted by the pancreas and converted to trypsin by enterokinase, an enzyme secreted by intestinal mucosal cells. Small amounts of trypsin then catalyze release of the remaining enzyme, the other pancreatic endopeptidases (chymotrypsin, elastase, and collagenase) and the exopeptidases (carboxypeptidases). Trypsin acts chiefly on peptide bonds in which the carbonyl portion is provided by basic amino acids, especially arginine. However, its function as an activator of other pancreatic proteinases makes it important to the digestion of all proteins. The distributions and contents of total proteinase, trypsin, and chymotrypsin in pancreatic tissue of various vertebrates are listed in Table 7.6. The structures of trypsinogens and trypsins have been described for a number of species (Vonk and Western 1984).

Pancreatic trypsin has been isolated from chickens (Whiteside and Prescott 1962) and turkeys (Ryan 1965). Zendzian and Barnard (1967a,b) identified trypsin in the pancreas of reptiles and amphibians. Trypsin was found in the pancreas of carp (Cohen, Gertler, and Birk 1981a,b), dogfish (Prahl and Neurath 1966), nurse shark and stingray (Zendzian and Barnard (1967a,b), and *Chimaera monstrosa*. Extracts from the pancreas of a lungfish were found to contain three trypsinogens (Reeck, Winter, and Neurath 1970). Demonstration of trypsin activity in fish that do not have a distinct pancreas has been difficult. However, trypsin was found in extracts of the hepatopancreas of goldfish, the pyloric ceca of salmon, tuna, groupers, barracuda, bass, mackerel (Vonk and Western 1984), and the intestinal mucosa of hagfish (Nilsson and Fänge 1969, 1970).

Chymotrypsin

Chymotrypsin is a pancreatic endopeptidase that preferentially attacks peptides with a C-terminal hydrophobic and, more specifically, aromatic amino acid. Because these are more common than the two basic amino acids, chymotrypsin digests a larger number of bonds than trypsin. The structure and activity of bovine chymotrypsin have received extensive study. In addition to the nonmammalian species listed in Table 7.6, chymotrypsin has been found in a lungfish, the hepatopancreas of goldfish, and the pyloric ceca of bass, mullet, mackerel, and bowfin (Vonk and Western 1984).

Elastase

Elastase is another endopeptidase found in mammalian pancreatic secretions. It attacks elastin, the fibrous protein of arteries and ligaments that is resistant to other proteinases, and is capable of hydrolyzing insulin, ribonuclease, and the proteins of some bacteria. Elastase splits bonds adjacent to uncharged, nonaromatic amino acids, with a preference for serine and alanine. Porcine elastase has been purified and crystallized, and its amino acid sequence was described by Hartley and Shotton (1971). Elastase activity has been found in the pancreas of chickens, lungfish, stingrays, and chimaeras, and the pyloric ceca of tuna, bass, and mullet (Vonk and Western 1984).

Table 7.6. *Distribution and content of proteinases in the pancreas of vertebrates*

Group	Species	Total proteinase (casein)	Trypsin (BAEE)	Chymotrypsin (BTEE)	RNase group
Group A: 20–60 mg proteinase per gram pancreatic tissue					
Turtle	*Chrysemys picta*	58	9.1	12.0	C
Turtle	*Pseudemys elegans*	53	7.6	16.0	B
Turtle	*Chelydra serpentina*	33	6.2	11.0	B
Turtle	*Chrysemys picta* (fasted at 5°C)	26	2.2	4.4	—
Frog	*Rana catesbeiana* (fed)	22	5.2	5.5	C
Group B: 10–20 mg proteinase per gram					
Turtle	*Podocnemis unifilis*	19	7.0	3.2	B
Conger eel	*Amphiuma* sp.	19	6.0	2.8	C
Horse	*Equus caballus*	14	6.0	2.8	B
Frog	*Rana pipiens*	14	1.9	1.2	C
Dog	*Canis familiaris*	12	2.2	2.0	C
Frog	*Rana catesbeiana* (fasted)	12	1.9	1.9	C
Cow	*Bas taurus*	10.3	3.2	6.4	A
Mud puppy	*Necturus maculosus*	10	2.5	2.7	C
Stingray	*Dasyatis americana*	10	1.8	1.4	C
Group C: 0–10 mg proteinase per gram					
Turkey		8.8	3.5	1.3	B
Goat	*Capra hircus*	8.7	2.9	4.1	A
Barracuda[a]	*Sphyraena barracuda*	7.3	0.5	0.4	C
Chicken	*Gallus domesticus*	6.0	0.6	1.0	B
Wallaby	*Macropus eugenii*	5.5	1.7	0.5	A
Nurse shark	*Ginglymostoma cirratum*	3.4	0.3	0.1	C
Dogfish	*Squalus suckleyi*	3.0	0.5	1.3	C
Tuna[a]	*Thunnus secundodorsalis*	1.6	0.06	0.18	C
Rabbit	*Oryctolagus cuniculus*	1.4	0.4	0.2	C

Activities of the substrates shown are expressed as the equivalent amount of bovine trypsin (BAEE and casein) or chymotrypsin (BTEE) giving the indicated activity under the same conditions.

[a] Pyloric ceca were used so the pancreatic fraction would be small and may account for their low ranking.

After Zendzian and Barnard (1967a); from Vonk and Western (1984).

Collagenase

This enzyme hydrolyzes collagen, another fibrous protein of vertebrates that is a chief constituent of the connective tissue associated with skin, tendons, and bones. Collagenase has been found in the pancreatic juice of dogs (Takahashi and Seifter 1974) and *Chimaera monstrosa* (Nilsson and Fänge 1969).

Exopeptidases

Exopeptidases remove terminal amino acids from the peptide chain. They consist of C-terminal peptidases (carboxypeptidases), effective against peptides with a free carboxyl group, and N-terminal peptidases (aminopeptidases), which attack substrates with a free amino group.

Pancreatic carboxypeptidases

Carboxypeptidases A and B are secreted by the pancreas. Carboxypeptidase A is especially active on the peptides with C-terminal aromatic or branched aliphatic residues that are end products of chymotrypsin digestion. The bovine enzyme is secreted in two forms, which are activated and then converted by trypsin to four different enzymes with essentially the same specificity. Carboxypeptidase B acts on the polypeptides with a C-terminal lysine or arginine that are end products of trypsin digestion.

Carboxypeptidase A has been demonstrated in pancreatic tissue of humans, cattle, pigs, and reptiles (turtles and terrapins), as well as the pancreas, hepatopancreas or pyloric ceca of teleosts, elasmobranchs, and cyclostomes (Vonk and Western 1984). There is also some evidence for its presence in the pancreas of chickens and adult amphibians. Carboxypeptidase B has been reported in the pancreas of humans, cattle, pigs, dogs, rats, lungfish, and dogfish and in the pyloric ceca of cod, bass, mullet, and carp (Vonk and Western 1984).

The decrease in extracellular proteolytic activity in the contents of the distal intestine of fish and other vertebrates suggested that they are reabsorbed by the posterior gut (Diamond 1977; Hofer and Schiemer 1981). However, as indicated in Chapter 8, the disappearance of extracellular endogenous enzymes could also be attributed to their digestion by indigenous microbes, with at least partial absorption of their nitrogen as ammonia and their carbon skeletons as short-chain fatty acids (SCFA).

Intestinal peptidases

The enzymes in the brush border and cytosol of the enterocytes are believed to be principally responsible for the hydrolysis of peptides rather than of intact proteins (Kim and Erickson 1985; Silk, Grimble, and Rees 1985). The enzymes involved in the final stages of peptide hydrolysis are aminopeptidases, tripeptidases, and dipeptidases. Brush border enzymes have greater activity against tetrapeptides and large oligopeptides. Much of the tripeptidase and dipeptidase activity resides in the cytosol. The concentration of these enzymes increases from duodenum to midileum and then decreases in the remainder of the small intestine of the rat (Robinson 1960), pig (Josefsson and Lindberg 1965), and sheep (Symons and Jones 1966).

Four brush border dipeptidases (Tobey et al. 1985) and seven cytosolic enzymes that hydrolyse small peptides (Rapley, Lewis, and Harris 1971) have been reported in human enterocytes. Vonk and Western (1984) listed seven exopeptidases found in the intestinal cells of mammals. Leucine aminopeptidase activity was reported in the intestinal cells of reptiles, the intestinal microvilli of teleosts, and the intestinal mucosal extracts of chimaeras and hagfish (Vonk and Western 1984). Hydrolysis of L-glycylglycine and/or other glycine dipeptides has been demonstrated with intestinal mucosal extracts from teleosts, chimaeras, and hagfish.

Nucleic acids

All cells contain ribonucleic acid (RNA) and deoxyribonucleic acid (DNA), which consist of pentose sugars (ribose or deoxyribose), phosphate, and purine and pyrimidine bases. RNA is found in the cytosol and some nuclei. DNA is found in the nucleus and mitochondria of plant and animal cells or is packaged separately in the cell contents of prokaryotes. Most of the RNA (except for bacteria and viruses) and DNA are bound to protein. Hydrolysis of nucleoproteins by gastric and pancreatic proteinases releases the nucleic acids (Table 7.1). They are then hydrolyzed by pancreatic ribonucleases and deoxyribonucleases to form polynucleotides. The polynucleotides are further hydrolyzed by intestinal nucleotidases, deaminases, and phosphorylases to form end products that are absorbed by intestinal cells (Davenport 1982).

Barnard (1969a,b) examined the distribution of pancreatic ribonucleases among vertebrates. Although ribonucleases were recorded in all classes examined, Table 7.7 shows a considerable amount of species variation in their level of activity. These levels appeared to be constant for most species and, unlike other pancreatic enzymes, independent of the diet or feeding state. The species with the highest ribonuclease levels are ruminants and macropod marsupials, which have large numbers of microbes entering the intestine from the stomach. Smith and McAllan (1971) estimated that 20 percent of the microbial nitrogen leaving the forestomach of cattle is in the form of polynucleotides, and 75–80 percent of the microbial RNA and DNA is digested in the duodenum. The polynucleotide nitrogen of bacteria may be largely converted to urea in the liver of these animals, but the

Table 7.7. *Distribution and content of RNase in the pancreas of vertebrates*

Group A: 200–1,200 µg RNase gram pancreatic tissue					
Cattle		1,200	Wallaby	*Macropus eugenii*	515
Bison	*Bison bison*	1,180	Mouse	*Mus musculus*	395
Sheep	*Ovis aries*	1,080	Lizard	*Iguana iguana*	380
Goat	*Capra hircus*	1,000	Uganda kob	*Kobus kob*	270
Kangaroo	*Macropus rufus*	600	Golden hamster	*Mesocricetus auratus*	260
Elk	*Cervus canadensis*	550	Rat	*Rattus norvegicus*	260
Kangaroo	*Macropus giganteus*	530	Guinea pig	*Cavia porcellus*	240
Group B: 20–100 µg RNase per gram					
Turtle	*Podocnemis unifillis*	90	Armadillo	*Dasypus novemcintus*	30
Pig	*Sus scrofa*	80	Horse	*Equus caballus*	25
Turtle	*Pseudemys elegans*	65	Turkey	*Meleagris* sp.	25
Hippopotamus	*H. amphibius*	62	Chicken	*Gallus domesticus*	20
Turtle	*Chelydra serpentia*	60	Opossum	*Didelphis marsupialis*	20
Caiman	*C. crocodilus*	53			
Group C: 0–20 µg RNase per gram					
Whale	*Eschrichtius robustus*	18	Monkey	*Macaca mulatta*	2
Dogfish	*Squalus suckleyi*	18	Tuna	*Thunnus secondodorsalis*	2
Frog	*Rana catesbeiana*	17	Barracuda	*Sphyraena barracuda*	2
Snake	*Bungarus fasciatus*	12	Shark	*Ginglymostoma cirratum*	2
Turtle	*Chrysemys picta*	9	Human	*Homo sapiens*	1
Mud puppy	*Necturus maculosus*	8	Elephant	*Loxodonta africana*	0.7
Grouper	*Epinephelus striatus*	8	Dog	*Canis familiaris*	0.5
Frog	*Rana pipiens*	7	Cat	*Felis silvestris*	0.5
Lungfish	*Proropterus acthiopicus*	7	Rabbit	*Oryctolagus cuniculus*	0.5
Dolphin	*Tursiops truncatus*	5	Conger eel	*Amphiuma* sp.	0.5
Stingray	*Dasyatis americana*	5	Toad	*Bufo americana*	<1
Toad	*Bufo marinus*	4	Snake	*Natrix taxi*	<1
Pigeon	*Columbia livia*	3			

Enzyme activity was determined from the rate of RNA hydrolysis at pH 7.4. Whole pyloric ceca (i.e., not pure pancreatic tissue) from tuna and barracuda were used. After Barnard (1969a,b); from Vonk and Western (1984).

phosphate released by their hydrolysis can provide an important source for both the animal and its microbes.

Absorption

Nutrients are absorbed across the intestinal epithelium by either passive diffusion or carrier-mediated transport. The rate of passive diffusion depends on the permeability to the substance, the transepithelial concentration gradient of the substance, and (for ionized substances) the transmembrane electrical gradient. Lipid-soluble substances diffuse across cell membranes at rates that also depend on their degree of lipid solubility. Water-soluble substances, such as sugars and amino acids, are absorbed to some degree by passive diffusion, and their rate of transport varies with the segment of intestine and among species (Karasov and Diamond 1985). This led to the hypothesis that the high concentrations of hexoses and amino acids that are generated in the unstirred layer adjacent to the lumen-facing (apical) membrane of intestinal cells result in their passive intracellular diffusion, along with water, by solvent drag (Pappenheimer 1990). However, most hexoses, amino acids, and water-soluble vitamins appear to be absorbed by carrier-mediated transport systems in the membranes of epithelial cells, and this is the principal means of uptake at low intraluminal concentrations.

Carrier-mediated transport across a cell membrane consists of either the "active" transfer of a substrate against its electrochemical gradient, which requires a direct or indirect investment of cellular energy, or facilitated diffusion down its electrochemical gradient. However, it is always characterized by the fact that the carrier becomes saturated at a given rate (V_{max}) of transport, regardless of the concentration gradient.

The apical membranes of absorptive cells of the midgut of all vertebrates have carriers that transfer monosaccharides, amino acids, and B-complex vitamins into the cell (Hopfer 1987). Cells lining the proximal third of the chicken ceca also absorb glucose and amino acids (Moreto and Planas 1989) and vitamin B_6 (Heard and Annison 1986) by carrier-mediated transport. Carrier-mediated uptake of glucose and amino acids has also been demonstrated in the proximal cecum of the Canada goose, sage grouse, rock dove, and red-necked phalarope (Skadhauge 1993). Cecal absorption of glucose ranged from 0.1 percent of total gut uptake in the rock dove to 49 percent in the sage grouse. Carrier-mediated transport of monosaccharides, amino acids, and water-soluble vitamins into the cell is codependent on the diffusion of Na into the cell down its concentration gradient, which is maintained by carrier-mediated secretion of Na across the basolateral membranes in exchange for K. Therefore, the energy for transport of both Na and these other nutrients is dependent on the active transport of Na across the basolateral membrane into the extracellular fluid and blood. Most of these organic nutrients are transferred across the basolateral membrane by carrier-mediated transport as well.

The mechanisms by which intestinal cells transport nutrients are difficult to study under in vivo conditions because of problems of measuring or controlling the electrical and concentration (electrochemical) gradients among lumen, cell contents, and blood. This is further complicated by the presence of an unstirred microlayer at the luminal surface of the apical membrane, which may differ from the remaining digesta in its pH and nutrient composition. The use of isolated, short-circuited epithelial tissue or vesicles prepared from the membranes of these cells to circumvent these problems is discussed in Chapter 9.

Interspecies comparisons of nutrient absorption are complicated by problems similar to those described for the comparison of enzyme activities. Absorptive capacity is dependent on the total absorptive surface of the intestinal cells. The absorption of sugars (Kinter and Wilson 1965), amino acids (King, Sepulveda, and Smith 1981), and dipeptides (Cheeseman 1986) and the reesterification of fatty acids (Shiau et al. 1980) are principally confined to the epithelial cells closer to the tip of the villus. Comparisons between mammals and lizards of equal body size showed that the mammals absorbed nutrients seven to ten times faster than the reptiles because of a 4–4.5 times greater intestinal surface area, two times greater transport activity, and a higher nocturnal body temperature. However, nutrient uptake increased with an increase in the body temperature of reptiles up to a maximum of 45°C–50°C, and, although the mammals processed food faster, with equal or higher digestive efficiency than the reptiles, there appeared to be no major differences in the mechanisms or normalized kinetics of glucose and proline absorption (Karasov and Diamond 1985; Karasov, Solberg, and Diamond 1985).

Absorptive area is the product of intestinal length and diameter (nominal surface area) times the additional contributions of the villi and microvilli. Karasov and Hume (in press) compared the nominal surface area of the small intestine of various classes of vertebrates (including the cecum of birds) as a function of body mass (Fig. 7.3). There were no significant differences among classes in the slope of this relationship, but differences in intercepts indicated that the nominal intestinal surface area of endotherms exceeded that of ectotherms, and that of the mammals exceeded that of birds. Karasov (1988) found that the multiplication factor for the villi to the nominal surface area of the small intestine ranged from 3.1 to 21.1 among sixteen mammalian species, compared with 8.1 for pigeons and 3.4 for the desert iguana. The multiplication factor for microvilli ranged from eighteen to eighty among nine species of mammals versus eighty-two in the desert iguana. Although addition of villi and microvilli to the nominal surface area increased the absorptive area of mammals by an average of 6.3 and fifty-three times,

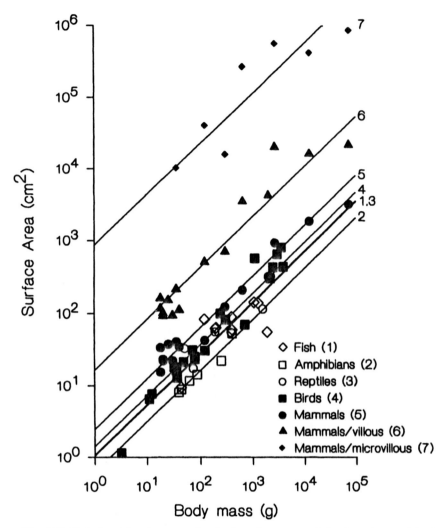

Fig. 7.3. Relationships between luminal surface area of the intestine and body mass for each class of vertebrate. Surface areas are nominal (length × diameter) unless otherwise indicated, and include ceca when present in fish and birds. There are no significant differences in slope (0.71). The calculated proportionality coefficients (intercept at zero) are: amphibians, 0.63; fish 1.06; reptiles 1.08; birds 1.43; mammals 2.47. The intercept for mammalian villus area is 16.44 and the intercept for mammalian microvillus area is 867. (With permission of W. H. Karasov, to be published in Karasov and Hume in press.)

respectively, there was no significant change in the slope of the relationship between surface area and body mass.

The absorption of individual substrates also is affected by their levels in the diet. The relative rate of sugar versus amino acid absorption appears to be high-

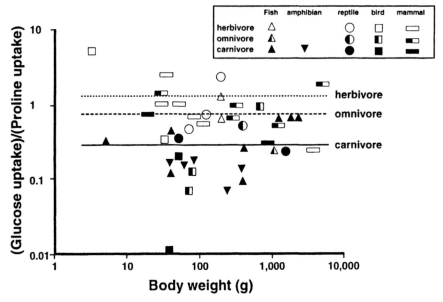

Fig. 7.4. Relative rates of sugar and amino acid transport in herbivorous, omnivorous, and carnivorous species of vertebrates. Each species was eating either its natural diet or one of similar nutrient composition. The ordinate is the ratio of the uptake capacity for D-glucose/L-proline for the midgut or total intestine. These ratios appeared to be independent of body mass. Therefore horizontal lines depict average values. Note that this ratio is highest for herbivores and lowest for carnivores in all classes. Glucose uptake showed a greater variation (herbivores > omnivores > carnivores) than proline. (Karasov and Diamond 1988).

est in herbivores and lowest in carnivores in all classes of vertebrates (Fig. 7.4). This relationship between diet and the rates of absorption of various nutrients also applies to individual animals. Ferraris and Diamond (1989) examined the effect of diet on the intestinal uptake of sugars, dipeptides, amino acids, vitamins, and minerals in laboratory rodents. The uptake of sugars, amino acids, and dipeptides increased with an increase in the percentage of each substrate in the diet. However, the uptake of iron, calcium, copper, and phosphate was reduced with an increase in their dietary levels, as they predicted for substances that are toxic in high concentrations. The transport of vitamins also tended to be reduced with an increase in substrate concentration, and the uptake of choline and ascorbic acid were not even increased in animals deficient in these vitamins (Stein and Diamond 1989; Karasov 1992b). These alterations in nutrient uptake appear to be due to changes in the number of carriers in the apical and basolateral membranes (Ferraris and Diamond 1989; Cheeseman and Harley 1991; Said, Mock, and Collins 1989; Shirazi-Beechey et al. 1991; Titus, Karasov, and Ahearn 1991; Said and Mohammadkhani 1993).

Carbohydrates

D-glucose and D-galactose are transported into the absorptive cells of the midgut of all vertebrates (Storelli, Villela and Cassano 1986; Obst and Diamond 1989; Ahearn et al. 1992) and the proximal ceca of birds (Planas et al. 1986; Planas, Ferrer, and Moreto 1987; Vinardell and Lopera 1987; Moreto and Planas 1989) against their concentration gradients by a carrier-mediated mechanism that is dependent on the passive diffusion of Na into the cell (see Fig. 9.8). The characteristics of this carrier appear to be similar in all classes of vertebrates, except for differences in its relative affinity for these substrates and evidence for multiple glucose carriers in some mammals (Karasov 1988; Pajor, Hirayama and Wright 1992). A separate mechanism for the Na-independent facilitated diffusion of fructose has been demonstrated in the apical membrane of many vertebrates. Glucose was transported across the basolateral membrane of the small intestinal absorptive cells of mammals (Stevens, Kaunitz, and Wright 1984), chickens (Kimmich 1981), and fish (Reshkin and Ahearn 1987; Reshkin et al. 1988) by carrier-mediated facilitated diffusion. The carrier in the basolateral membrane of the chicken enterocyte and in the rat jejunum (Cheeseman 1993) transferred both glucose and fructose.

Alliot (1967) reported that the intestine of the cat shark absorbed N-acetyl-D-glucosamine more rapidly than glucose. However, Place (1990) found that N-acetyl-D-glucosamine and glucosamine end products of chitin digestion were absorbed from the intestine of Leach's storm petrel at approximately 25 percent and 50 percent of the efficiency of glucose absorption. Karasov (1988) also found that the carrier-mediated mechanism for N-acetyl-D-glucosamine across the apical membrane of the mouse small intestine had an affinity much lower than that of the D-glucose carrier.

Lipids

Most end products of lipid digestion are incorporated into micelles before their release for passive diffusion when they come into contact with the apical membrane of intestinal cells. Their rate of diffusion into the enterocyte is dependent on their degree of lipid solubility. For example, the rate of fatty acid diffusion increases with chain length (Westergaard 1987; Meddings 1989), and the protonated (undissociated) form of the acids permeates faster than the anionic (dissociated) form. There also are species differences in the relative permeability to fatty acids (Thomson, Hotke, and O'Brien 1983). Following their uptake by the cell, these end products are resynthesized into triglycerides and phospholipids and assembled into chylomicrons, which are small spheres 0.1–3.5 μm in diameter and coated with a mixture of protein, cholesterol, triglyceride, and phospholipid. The chylomicrons pass through the cell and are released into intercellular spaces for removal by the lymphatic system.

Free short-chain organic acids and bases, which are both water- and lipid-soluble, need not be emulsified before their passive diffusion across epithelial membranes and can pass across the enterocytes unchanged. This includes the short-chain fatty acids (SCFA) and ammonia that are end products of microbial fermentation (see Chapter 8). It also includes numerous drugs and plant alkaloids such as barbituric acid, caffeine, strychnine, and morphine, which explains their early discovery and use as therapeutic agents. Conjugated and ionized bile salts are absorbed by carrier-mediated transport from the terminal ileum for recirculation and secretion in the bile.

Peptides and amino acids

Tetrapeptides are shortened by brush border enzymes before entering the cell, but tripeptides and dipeptides can be absorbed from the lumen faster than their constituent amino acids and independent of Na transport. Carrier-mediated, Na-independent, facilitated transport of dipeptides has been demonstrated in the intestinal brush border membrane of the rat (Ganapathy and Leibach 1982, 1985) and a number of other mammals, as well as in the chicken (Calonge, Ilundain, and Bolufer 1990) and the tilapia fish (Reshkin and Ahearn 1991). There is evidence for multiple peptide transport systems, including one that cotransports peptides with two protons (Webb, Matthews, and DiRienzo 1992). Peptides can compete with one another for absorption, but they do not compete with free amino acids. A carrier-mediated pathway for dipeptide transport also was reported in the basolateral membrane of the rabbit (Dyer et al. 1990) and tilapia (Reshkin and Ahearn 1991). However, with the exception of small amounts of peptides such as glycylglycine, only amino acids are transported into the blood.

Amino acids are transported by apical membranes into intestinal cells of the midgut of all vertebrates and the proximal cecum of the chicken by a number of Na-dependent, carrier-mediated systems that transport different groups of amino acids (Stevens, Kaunitz, and Wright 1984; Christensen 1985; Storelli, Vilella, and Cassano 1986; Storelli et al. 1989; Moreto and Planas 1989). Some carriers are dependent on Cl as well (Munck and Munck 1990). Mammals, birds (Lerner 1984), reptiles (Moin-Un-Nisa, Zain, and Zain-Ul-Abedin 1970), and fish (King et al. 1986; Storelli et al. 1989) appear to have separate carriers for the neutral, basic, acidic, and imino acids. However, these carriers show some over-lapping specificities, and some species appear to have additional mechanisms for the transport of specific amino acids (Munck 1983; Stevens, Kaunitz, and Wright 1984;

Karasov, Solberg, and Diamond 1985; Storelli et al. 1989). Carrier-mediated transport of acidic and neutral amino acids has been demonstrated in the basolateral membrane of the rat (Mircheff, van Os, and Wright 1980) and eel (Reshkin et al. 1988).

Vitamins

Fat-soluble vitamins (A, D, E, and K) can be absorbed by passive diffusion in the same fashion as other lipid-soluble substances. Water-soluble vitamins (biotin, choline, nicotinic acid, riboflavin, and pantothenic acid) are absorbed by mammals against their concentration gradient by carriers that are generally Na-dependent (Rose 1987). Where examined, this appears to be true for nonmammalian vertebrates as well (Karasov and Hume in press). Ascorbic acid (vitamin C) is required by fish, some passerine birds (Chaudhuri and Chatterjee 1969), bats (Birney, Jenness, and Ayaz 1976), guinea pigs, and some primates, which lack the ability to synthesize it. Na-dependent, carrier-mediated ascorbic acid transport has been demonstrated in primates and guinea pigs (Rose 1987). A similar mechanism that involved cotransport of 2 Na:1 L-ascorbate was described in intestinal brush border vesicles of the European eel (Maffia et al. 1993). Vitamin B_6 was found to be absorbed by the small intestine and, to a lesser extent, by the crop and ceca of chickens (Heard and Annison 1986). However, this appeared to be dependent on passive diffusion. Facilitated diffusion of inositol was reported in basolateral membrane vesicles from intestinal cells of the herbivorous tilapia *Oreochromis mossambicus*, and the carnivorous eel *Anguilla anguilla* (Reshkin et al. 1989).

Ontogeny

The diet and digestive system can undergo major changes during the early development of fish (Tanaka, Kawai, and Yamamoto 1972; Kawai and Ikeda 1973a,b; Dabrowski 1979; Lauff and Hofer 1984; Buddington 1992; Buddington and Doroshov 1986) and the metamorphosis of amphibians (Andrew 1959; Reeder 1964). However, the most extensive studies of neonatal development have been conducted on mammals. As mentioned earlier, mammals differ from other vertebrates in that they suckle their young, which are born at an early stage of development. The composition of milk can vary a great deal among different species. Oftedal (1984) listed the milk composition of fifty-seven species belonging to nine mammalian orders. On a dry matter basis, the fat content ranged from 2 percent in the black rhinoceros to 82 percent in the harp seal. Protein content ranged from 7 percent in humans to 40 percent in the dog and raccoon. Sugar content ranged from 5 to 8 percent in lagomorphs to 66 percent in the horse. In most species, the major carbohydrate is lactose, and the diet following weaning is lower in fat and higher in carbohydrates, with little or no lactose. These changes in the diet are preceded by the eruption of teeth and major changes in digestive enzymes and absorption of nutrients.

Although there is considerable species variation in the prenatal and neonatal development of the digestive system, much of our information comes from the extensive studies conducted on the rat, which have been reviewed by Henning (1985, 1987) and Henning, Rubin, and Shulman (1994). The gastrointestinal tract of the rat is functionally immature at birth and during the first two postnatal weeks. Lipids are digested by a lingual and possibly a gastric lipase, but the stomach does not secrete HCl or pepsinogen, and there is very little secretion of bile or pancreatic enzymes. Digestion by enzymes in the brush border of intestinal cells is largely restricted to the hydrolysis of lactose. Most macromolecules, including immunoglobulins and nutrient proteins in the milk, are absorbed by pinocytosis. Immunoglobulins are specifically absorbed by jejunal epithelium and transported intact into the blood. Other proteins are absorbed by nonspecific mechanisms in the apical membrane of ileal epithelium and are digested by lysosomal cathepsins and peptidases in the cytosol of these cells before the transport of amino acids and peptides into the circulatory system.

During the third and fourth week of postnatal development, there is a dramatic increase in the secretion of salivary amylase, gastric pepsinogen, HCl, bile, pancreatic amylase, lipase, trypsin, and chymotrypsin. Microvillus lactase activity decreases, with a reciprocal increase in maltase and sucrase activity, and pinocytosis and cathepsin activities are reduced to adult levels. These developments are relatively unaffected by a delay in weaning. They are preprogramed before the change in diet, and the major stimulus is the rise in the corticosterone and thyroxin levels. The weaning of suckling rats begins at seventeen days and is complete by day twenty-six, at which time the adult mode of digestion is fully operational.

Although this general pattern of gastrointestinal development seems to apply to most mammals (Koldovský 1981; Janssens and Messer 1988), there are some major differences among species. A lipase secreted by the lingual salivary glands also has been demonstrated in newborn humans and cattle (Hamosh 1979). However, the human digestive system is much more mature at the time of birth (Henning 1987). Salivary amylase secretion and intestinal sucrase and maltase activity are at adult levels. Although pancreatic secretions of amylase and lipase are low or absent, pepsinogen, HCl, and pancreatic proteases are secreted at 50 percent of the adult level. Pancreatic amylase also was present in prenatal horses, cattle, and sheep, and it increased most rapidly after weaning (Koldovský 1970).

The mechanisms for carrier-mediated absorption of glucose and amino acids appear to be present at the time of birth in all mammals (Buddington and Diamond 1989, 1990; Buddington 1992). The rates of glucose and amino acid uptake per mg of rabbit intestine were maximal at birth and declined to a much lower level in the adult, but the rate of fructose uptake showed a marked increase at the time of weaning. In lambs, however, glucose transport increased to a maximum two weeks after birth and then declined to negligible amounts at weaning (Shirazi-Beechey et al. 1991a,b). The fact that a four-day infusion of glucose into the proximal intestine of adult sheep resulted in a 40–80 percent increase in glucose absorption indicated that the normal decline was due to the low amounts of glucose presented to the intestine following the development of microbial fermentation in the forestomach.

The transfer of immunity from dam to progeny is limited by their ability to survive digestion as well as the intestine's ability to absorb these proteins. For example, gastric secretion of HCl is well developed in the guinea pig before birth and, although gastric contents of humans are usually neutral at birth, the pH decreases rapidly the first day. Trypsin activity increases in the first week after birth in human infants and calves and over the first five weeks after birth in pigs, although a trypsin inhibitor is present in human pancreatic juice and in the colostrum of calves. Intact protein can be absorbed from the intestine into the lymphatic system during the first twenty-four to thirty-six hours following the birth of cats, pigs, goats, sheep, cattle, and horses (Koldovský 1970). It may be absorbed over a longer period by dogs, for up to three weeks by mice and rats, and for as long as six weeks following the birth of hedgehogs. Brambell (1970) noted an inverse relationship between the ability of a species to transfer immunoglobulins via the placenta versus intestinal absorption (Table 7.8). Sanderson and Walker (1993) reviewed information on the mechanisms for uptake and transport of macromolecules.

The survival of orphan neonatal mammals is often dependent on the use of substitute milk of composition similar to that produced by the species. For example, the feeding of lactose to neonate or adult animals that have little or no lactase in their small intestine can result in diarrhea caused by microbial fermentation of lactose in their hindgut. Cow's milk is often used with dilution or supplementation. However, the neonates of some species such as rabbits can be easily killed by overfeeding, and the milk of many mammals that nurse their young for long periods shows major changes in its composition (Oftedal 1984).

Buddington (1992) compared information on the ontogeny of intestinal nutrient transport in eight species of mammals with that of a bird, amphibian, and two species of fish. Age-related changes in the uptake of nutrients corresponded to shifts in the composition of the diet and the need to absorb greater quantities of

Table 7.8. *Transmission of passive immunity*[a]

Species	Prenatal	Postnatal
Horse	0	+++ (24 h)
Pig	0	+++ (24–36 h)
Ox, goat, sheep	0	+++ (24 h)
Wallaby (*Setonix*)	0	+++ (180 d)
Dog, cat	+	++ (1–2 d)
Fowl	++	++ (<5 d)
Hedgehog	+	++ (40 d)
Mouse	+	++ (16 d)
Rat	+	++ (20 d)
Guinea pig	+++	0
Rabbit	+++	0
Human, monkey	+++	0

[a] 0, No absorption or transfer; + to +++, degrees of absorption or transfer. (From Brambell 1970.)

food. Although the time and types of transport for nutrients varied with species, the transport of some nutrients such as sugars and amino acids appeared before the onset of their appearance in the diet of all species. The mechanisms of response included changes in the density, distribution, and types of transport carriers for different nutrients, and changes in the physicochemical characteristics of the brush border membrane. These changes appeared to be signaled by both internal preprogramed and external sources.

Summary

Most of the endogenous, extracellular digestive enzymes appear to be present in all vertebrates. Some, such as chitinase and pepsin, are present in each class of vertebrate but not in all species. Others, such as phospholipase A_2 and cholesterol esterase, elastase, and carboxypeptidase B, may be absent in some vertebrate classes but not others. Digestive enzymes present in the brush border or cytosol of intestinal cells have been examined in only a few mammals and in even fewer other vertebrates. Maltase appears to be common to all vertebrates. Sucrase and trehalase are absent from the intestine of some mammals and many lower vertebrates. Lactase is absent from the intestinal mucosa of some neonate mammals, many adult mammals, and most lower vertebrates.

The levels of enzyme activity can vary with species, the stage of an animal's neonatal development or its diet. Carnivores tend to have higher levels of proteinases and lower carbohydrase levels than omnivores or herbivores. Animals that have large numbers of indigenous microbes in their foregut tend to have higher levels of pancreatic nuclease and lower levels of carbohydrases. The complement of digestive enzymes can undergo marked changes during the early development of fish, amphibians, and mammals, and much of this appears to be genetically programed before hatching or birth. Enzyme levels also can be modulated by dietary changes following maturation of individual animals.

Most of the mechanisms for transport of nutrients across intestinal epithelium appear to be common to all vertebrates. Phagocytosis or pinocytosis seems to be confined mostly to newborn mammals. Lipid-soluble substances can be absorbed by passive diffusion across intestinal cell membranes, but absorption of water-soluble organic compounds is largely limited to carrier-mediated transport across intestinal cell membranes and passive diffusion between the cells. Carrier-mediated transport involves facilitated diffusion of fructose and Na-dependent, concentrative transport of glucose, galactose, peptides, amino acids, and water-soluble vitamins.

The mechanisms for carrier-mediated transport of most nutrients appear to be present at birth but can increase in the number of carriers and their affinity for the substrate with development of neonates and changes in the diet of adults. The rate of passive diffusion increases with an increase in the intraluminal concentration of the nutrient, but its relative contribution to the uptake of water-soluble nutrients remains uncertain.

The presence of most of these digestive enzymes and absorptive mechanisms in all classes of vertebrates indicates their early appearance and conservation. The appearance of other enzymes and changes in the chemistry of bile salts and acids provide clues to the evolution of these animals. Although the assimilation of nutrients by vertebrates is generally dependent on the hydrolysis of food by endogenous enzymes, indigenous microbes can play an equally important role, and their contribution is discussed in Chapter 8.

8

Microbial fermentation and synthesis of nutrients and the absorption of end products

One of the most interesting and least understood characteristics of the digestive system is the symbiotic relationship between animals and the microorganisms that normally inhabit their digestive tract. Shortly after birth or hatching, the lumen and epithelial surfaces of the gastrointestinal tract become colonized with bacteria. The digestive tract provides the nutrients and environment required by these microorganisms, and they provide protection and nutrients in return. The type and number of microorganisms can be affected by major changes in the diet, such as those associated with weaning in mammals, but they tend to consist of populations that are stable and characteristic of the species (Dubos 1966). The lowest numbers of bacteria are generally found in the stomach because of its low pH and proteolytic activity. Rapid transit of digesta limits the number in the midgut, as well. The largest populations are found in the hindgut of terrestrial vertebrates and the foregut of some herbivores, where they are joined by colonies of indigenous protozoa and fungi in some species.

These indigenous microorganisms can be critical to the growth and even the survival of some species. Studies of gnotobiotic (germ-free) animals show that indigenous bacteria help protect the gut from disease by stimulation of immune mechanisms and competition with pathogenic organisms. Their ability to convert otherwise undigestible compounds into absorbable nutrients is especially important to herbivores. Much of the information on gut microbiology was derived from studies of the ruminant forestomach, which have been extensively reviewed by Phillipson (1977), Church (1988), and Van Soest (1994). Subsequent studies have shown that microbial fermentation provides an equally important contribution to the conservation of nitrogen, Na, and water by the hindgut of most vertebrates.

The ruminant forestomach

Microbes

During the first weeks after birth, the ruminant forestomach becomes colonized with *E. coli aerogenes* and streptococci, which are joined by lactobacilli and streptococci in the suckling animal (Eadie and Mann 1970). Weaning is followed by development of the extremely complex group of flora and fauna that are characteristic of adult animals (Hungate 1966, 1968; Bryant 1977; Wolin 1979; Allison 1984). Culture counts give estimates of 10^{10} to 10^{11} of predominantly anaerobic bacteria per gram of fluid in rumen contents (Table 8.1). Microscopic counts, which include organisms that are dead or require specific culture media, give higher numbers. Table 8.2 lists the principal species found in the rumen of sheep and cattle, along with their fermentative properties. They ferment carbohydrate into absorbable nutrients. They degrade protein and other nitrogenous compounds and use the end products for synthesis of microbial protein. They also hydrolyze lipids and hydrogenate fatty acids (Tamminga and Doreau 1991), which accounts for the high levels of saturated fatty acids found in the meat of these animals, and synthesize B-complex vitamins.

Although the metabolic characteristics of individual species of bacteria vary, the end results of their combined fermentation activities are shown in Fig. 8.1. They produce extracellular enzymes that digest polysaccharides into monosaccharides and convert protein into peptides, which are absorbed by the bacterial cells. Monosaccharides are converted by intracellular enzymes to pyruvate, as they are in the cells of vertebrates. However, rather than entering the Krebs' Cycle for aerobic metabolism to CO_2 and H_2O, pyruvate is converted by anaerobic metabolism to short-chain fatty acids (SCFA), mainly acetate, propionate, and butyrate, plus CO_2, H_2, CH_4 and H_2O. The SCFA were called *volatile fatty acids* (*VFA*) in earlier literature because of their ready separation from other components of digesta by steam distillation, but they are now generally referred to as *short-chain fatty acids* (*SCFA*).

The rate of total SCFA production is dependent on the substrate–soluble carbohydrate (starches and sugars) > pectin > cellulose (Fig. 8.2, *A*). The acetate/propionate ratio is reduced by either an increase in the concentration of soluble carbohydrates or a reduction in rumen pH, and both of these SCFA are replaced with lactate at a pH lower than 5.0 (Fig. 8.2, *B*). CH_4 production is directly proportional to acetate production and inversely proportional to the production of butyrate, but it depends on other factors that affect the growth and replication of methanogenic organisms as well.

Rumen bacteria can either utilize peptides for synthesis of bacterial protein, or deaminate amino acids to form carbon skeletons plus ammonia. The carbon skele-

Table 8.1. *Short-chain fatty acid concentrations and microbial counts in the digestive tract of vertebrates*

Gut segment and species	Percent body weight	SCFA mmole/l	Bacteria g digesta	Protozoa' g digesta	References
Forestomach					
MAMMALS					
Cattle and sheep	10–25	60–120	10^{10}–10^{11}	10^4–10^5	Hungate (1966); Phillipson (1977); Allison (1984); Höller et al. (1989)
Camel	10–17	80–180	10^6–10^{11}	10^3–10^6	Hungate et al. (1959); Williams (1963)
Llama	—	94–186	—	—	Vallenas and Stevens (1971b); Engelhardt, Ali, and Wipper (1979)
Hippopotamus	—	60–180	—	+	Thurston, Noirot-Timothee, and Arman (1968)
Collared peccary	—	89	—	10^6	Carl and Brown (1983); Lochmiller et al. (1989)
Eastern grey kangaroo	10–15	101–102	—	10^4	Dellow and Hume (1982); Dellow et al. (1988)
Red-necked wallaby	9–15	103–129	—	—	Hume (1977); Dellow et al. (1988)
Tammer wallaby	—	119	10^{11}	10^4	Dellow et al. (1988)
Swamp wallaby	12–17	56–62	10^{11}	10^4	Dellow et al. (1988)
Red-necked pademelon	8–11	91–93	—	—	Hume (1977); Dellow et al. (1988)
Quokka	—	82–137	10^{10}	10^6	Moir (1965); Moir (1968)
Three-toed sloth	20–30	37–95	—	0	Britton (1941); Denis et al. (1967)
Colobus monkey	17	181–219	10^6–10^8	0	Kuhn (1964); Ohwaki et al. (1974)
Langur monkey	—	103–165	10^{10}–10^{11}	0	Bauchop and Matucci (1968)
Gray whale	—	26–558	10^7–10^{8d}	—	Herwig et al. (1984)
Bowhead whale	—	8–1060	10^8–10^{10}	—	Herwig et al. (1984)
Minke whale	6.7	—	10^{10}	—	Mathiesen, Aagees and Sørmo (1990); Olsen et al. (1994)

Hyrax	—	50–140	—	—	Clemens, (1977); Leon (1980); Rübsamen, Hume, and Engelhardt (1982)
Hamster	—	81	—	—	Sakaguchi and Matsumoto (1985)
Dugong	5	18	—	—	Murray et al. (1977)
BIRDS					
Hoatzin	12	115–143	10^9	10^4	Grajal et al. (1989); Dominguez–Bello (1993)
Distal midgut					
MAMMALS					
Raccoon	—	10	—	—	Clemens and Stevens (1979)
Dog	—	18	—	—	Banta et al. (1979)
Bush baby	—	30	—	—	Clemens (1980)
Domestic pig	—	30	—	—	Clemens, Stevens, and Southworth (1975a)
Vervet monkey	—	35	—	—	Clemens (1980)
Horse					
Pelleted diet	—	5	—	—	Argenzio, Southworth, and Stevens (1974)
Grass diet	—	45	10^8	—	Mackie and Wilkins (1988)
BIRDS					
Emu	—	92	—	—	Herd and Dawson (1984)
REPTILES					
Florida red-bellied turtle	—	60–120	—	—	Björndal and Bolten (1990)
FISH					
Rainbow trout	—	15	10^7	—	Trust et al. (1979); Paris, Murat, and Castilla (1977)
Common carp	—	19	—	—	Trust et al. (1979); Paris, Murat, and Castilla (1977)
Grass carp	—	7	10^7–10^9	10^3–10^6	
Sea carp	—	27	—	—	Clements, Gleeson, and Slayton (1994)
Herring cale	—	35	—	—	Clements, Gleeson, and Slayton (1994)
Butterfish	—	37	—	—	Clements, Gleeson, and Slayton (1994)
Tilapia	—	18	—	—	Titus and Ahearn (1988)

Table 8.1. (*Cont.*)

Hindgut

Gut segment and species	Percent body weight	SCFA mmole/l	Bacteria g digesta	Protozoa g digesta	References
MAMMALS					
Raccoon	—	15–70	—	—	Clemens and Stevens (1979)
Dog	—	90–190	—	—	Banta et al. (1979)
Domestic pig	—	180–200	—	—	Clemens, Stevens, and Southworth (1975a)
Pony	—	70–90	—	—	Argenzio, Southworth, and Stevens (1974)
Elephant	—	70–106	—	+	Clemens and Maloiy (1982); Van Hoven, Prins, and Lankhorst (1981)
Rhinocerous	—	—	—	+	Clemens and Maloiy (1982)
Dugong	—	183–236	—	—	Murray et al. (1977)
Bush baby	—	130–160	—	—	Clemens (1980)
Vervet monkey	—	110–229	10^{10}	—	Clemens (1980); Bruoton, Davis, and Perrin (1991)
Sykes monkey	—	140–180	—	—	Clemens and Phillips (1980)
Baboon	—	100–170	—	—	Clemens and Phillips (1980)
Samango monkey	—	174–199	10^7–10^{10}	—	Bruoton, Davis, and Perrin (1991)
Human	—	99–135	10^{11}–10^{12}	—	Simon and Gorbach (1987); Savage (1986); Cummings et al. (1987)
Rabbit	—	89–90	—	—	Hoover and Heitmann (1972)
Norwegian rat	6 (Cecum)	100	—	—	Sakata (1987)
Naked mole-rat	7–8 (Cecum)	65–115	10^9	10^8	Buffenstein and Yahav (1991)
Guinea pig	—	46–130	—	—	Hagen and Robinson (1953); Rechkemmer and Engelhardt (1993)
Hamster	—	95	—	—	Sakaguchi and Matsumoto (1985)

				References
Porcupine	6 (Cecum)	71	—	—	Johnson and McBee (1967); Vispo and Hume (1995)
Beaver	4 (Cecum)	—	—	—	Hoover and Clarke (1972)
Capybara	7 (Cecum)	26–44	10^6–10^8	10^6	Baldizan, Dixon, and Parra (1983); Borges, Dominquez–Bello, and Herreia (1995)
Rock hyrax	8–12 (Cecum + appendages)	55–105	—	—	Clemens (1977); Rübsamen, Hume, and Engelhardt (1982); Eloff and Van Hoven (1985)
Greater glider	—	30–40	—	—	Foley, Hume, and Cork (1989)
Koala	10	18–36	—	—	Cork and Hume (1983)
Brushtail possum	—	60–75	—	—	Foley, Hume, and Cork (1989)
Wombat	16	33–126	—	—	Barboza and Hume (1992b)
BIRDS					
Chicken	—	—	10^{10}	—	Salanitro et al. (1978)
Goose	—	10–70	—	—	Clemens, Stevens, and Southworth (1975b)
Ptarmigan	5	100	—	—	Gasaway (1976a,b); McBee and West (1969)
REPTILES					
Caiman	—	15	—	—	Guard (1980)
Geochelone carbonaria	—	24	—	—	Guard (1980)
Chrysemys picta	—	23–63	—	—	Guard (1980)
Gopher tortoise	—	100	—	—	Björndal (1987)
Green sea turtle	—	156–206	10^{10}	—	Björndal (1979, 1991)
Green iguana	—	16–51	10^{10}	+	Guard (1980); Troyer (1984a,b); Garza and Hernandez (1986)
AMPHIBIANS					
Leopard frog	—	—	10^9	—	Gossling, Loeche, and Mace (1982)
FISH					
Kyphosus cornelii	—	16	—	+	Rimmer and Wiebe (1987)
Kyphosus sydneanus	—	18	—	+	Rimmer and Wiebe (1987)

Values are the range of levels found in the gut segment. Dashes indicate an absence of information. The presence of protozoa is indicated by a +, and their absence is indicated by a 0 where counts were made.

Table 8.2. *Fermentative properties of ruminal bacteria*

Species	Function[a]	Products[b]
Bacteroides succinogenes	C,A	F,A,S
Ruminococcus albus	C,X	F,A,E,H,C
Ruminococcus flavefaciens	C,X	F,A,S,H
Butyrivibrio fibrisolvens	C,X,PR	F,A,L,B,E,H,C
Clostridium lochheadii	C,PR	F,A,B,E,H,C
Streptococcus bovis	A,SS,PR	L,A,F
Bacteroides amylophilus	A,P,PR	F,A,S
Bacteroides ruminicola	A,X,P,PR	F,A,P,S
Succinimonas amylolytica	A,D	A,S
Selenomonas ruminantium	A,SS,GU,LU,PR	A,L,P,H,C
Lachnospira multiparus	P,PR,A	F,A,E,L,H,S
Succinivibrio dextrinosolvens	P,D	F,A,L,S
Methanobrevibacter ruminantium	M,HU	M
Methanosarcina barkeri	M,HU	M,C
Spirochete sp.	P,SS	F,A,L,S,E
Megasphaera elsdenii	SS,LU	A,P,B,V,CP,H,C
Lactobacillus sp.	SS	L
Anaerovibrio lipolytica	L,GU	A,P,S
Eubacterium ruminantium	SS	F,A,B,C

[a]C = cellulolytic, X = xylanolytic, A = amylolytic, D = dextrinolytic, P = pectinolytic, PR = proteolytic, L = lipolytic, M = methanogenic, GU = glycerol-utilizing, LU = lactate-utilizing, SS = major soluble sugar-fermenting, HU = hydrogen-utilizing.
[b]F = formate, A = acetate, E = ethanol, P = propionate, L = lactate, B = butyrate, S = succinate, V = valerate, CP = caproate, H = hydrogen, C = carbon dioxide, M = methane.
Source: Modified by Allison (1984) from Hespell (1981.)

tons can be metabolized to acetate, propionate, or butyrate, or, in the case of the branched-chain amino acids valine, leucine, and isoleucine, they can be converted to isobutyrate, isovalerate, and 2-methylbutyrate, respectively. Ammonia nitrogen also can be used for protein synthesis, and some of this ammonia is produced by the breakdown of dietary and endogenous nonprotein nitrogenous compounds. The major endogenous compound is urea, which is produced in the liver and enters the rumen via saliva and diffusion through the rumen wall. Ammonia in excess to that used for microbial protein production is reabsorbed and converted to urea by the liver, or nonessential amino acids. Therefore, there is a continuous circulation of urea into and ammonia out of the rumen, with the net flow of nitrogen dependent on its utilization for microbial production of protein.

Carbohydrate and protein metabolism are interdependent. Therefore, limitations in carbohydrate availability limit microbial protein synthesis and replication, with

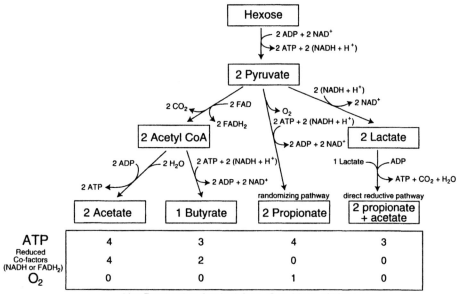

Fig. 8.1. Pathways of SCFA production by the rumen or colonic bacterial biomass. The production of methane is necessary for the production of oxidized cofactors in the pathways leading to acetate and butyrate production but not in the pathways leading to propionate production. (Modified from Herdt 1992.)

a greater shunt of deaminated peptides into SCFA production. Insufficient nitrogen in the diet limits microbial protein synthesis and increases the utilization of urea nitrogen for this purpose, with an overall reduction in SCFA production.

Although protozoa are much less numerous than bacteria, they can occupy an almost equal volume of rumen contents. The predominant species are anaerobic ciliates that belong to the families Isotrichidae and Ophryoscolecidae (Hungate 1966; Ogimoto and Imai 1981). Rumen protozoa are capable of fermenting carbohydrate, storing starch, digesting protein, hydrogenating fatty acids, and regulating the numbers of bacteria (Clarke 1977; Coleman 1980; Hobson and Wallace 1982; Prins 1991). They contribute relatively little to carbohydrate fermentation, but they store starch and produce protein for digestion and the subsequent absorption of amino acids during passage through the abomasum and midgut. In ad-

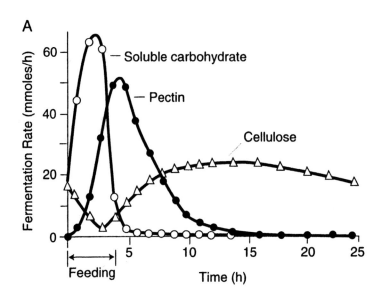

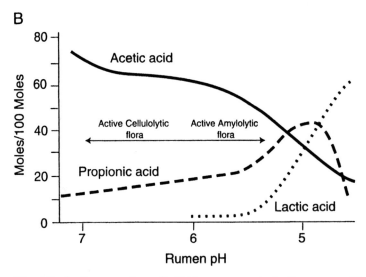

Fig. 8.2. **A,** Fermentation of alfalfa components in the rumen. (From Baldwin, Koong, and Ulyatt, 1977.) **B,** Relationship of ruminal pH to proportions of acetic, propionic, and lactic acid production. (From Kaufmann, Hagmeister, and Dirksen, 1980.)

dition to the bacteria and protozoa, anaerobic fungi can be found at concentrations of 10^3 to 10^5 organisms per gram in animals on high-fiber diets (Fonty 1991). Although present in small numbers, they may play an important role in the digestion of poor quality roughage.

Fermentation of carbohydrates

Rumen microorganisms ferment sugars, starches, cellulose, hemicelluloses, and pectins to short-chain organic acids, CO_2, CH_4, and H_2. The principal organic acids, regardless of carbohydrate substrate, are the SCFA acetic, propionic, and butyric acids. The total concentration of SCFA in the forestomach of sheep and cattle varies between 60 and 120 mM, depending on diet and time after feeding (Table 8.1). When animals are fed hay or other roughage, they consist of 60 to 70 percent acetate, 15 to 20 percent propionate, and 10 to 15 percent butyrate. Addition of grain or other readily fermentable starches or sugars to the diet increases their rate of production and reduces the acetate/propionate ratio. Although fermentation of starch to SCFA produces less energy than its conversion to glucose by endogenous enzymes, microbial fermentation of structural carbohydrates is a tremendous advantage to animals on a high-fiber diet.

Most of the SCFA produced in the forestomach are absorbed directly (Barcroft, McAnally, and Phillipson 1944; Engelhardt and Hauffe 1975). Substantial portions of the three main fatty acids are metabolized by ruminal epithelium (Pennington 1952; Pennington and Sutherland 1956; Stevens and Stettler 1966a). Their rate of metabolism increases with chain length (acetate < propionate << butyrate), and part of the acetate and most of the butyrate are metabolized to ketone bodies. The remainder is transported into the blood and provides a substantial source of the maintenance energy (Table 8.3). The forestomach of grazers, such as sheep, cattle, and wildebeest, contributes 10 to 15 percent of the animal's body weight and provides more energy than the smaller forestomach of browsers, such as the black-tailed deer, moose, and a gray duiker (Table 8.3).

Because their pK is approximately 4.8, the production of SCFA tends to decrease the pH of rumen digesta. The pH is normally maintained within the range of 5.5 to 7.0 by HCO_3 and PO_4 secreted by the salivary glands, HCO_3 secreted by the rumen, and by absorption of SCFA, which increases with a reduction in digesta pH. However, engorgement on grain, fruit, or other foods containing high levels of rapidly fermentable carbohydrate can precipitate a series of events referred to as *carbohydrate engorgement or overload* (Dirksen 1970). Rapid production of SCFA depresses the pH of rumen contents to levels that result in the replacement of some of the normal bacteria with higher populations of lactobacilli and production of lactic acid (Fig. 8.2, *B*). High levels of SCFA and lactic acid result in hypertonic rumen digesta, which draws water into the rumen and produces systemic dehydration. The rapid absorption of these organic acids also can produce ulceration of the forestomach epithelium and systemic acidosis. Various degrees of this condition are seen in cattle that are converted rapidly to a high-concentrate diet for the production of meat or milk (Dougherty 1977).

Table 8.3. Contribution of SCFA produced in different segments of the digestive tract to maintenance energy requirement

Species	Segment	Percent of maintenance energy requirement[a]	References
Foregut fermenters			
Cattle	Reticulorumen	63 A	Siciliano-Jones,
Bos taurus	Total hindgut	9 B	and Murphy (1989)
Sheep	Reticulorumen	53–79 A	Leng and Leonard 1965;
Ovis aries	Reticulorumen	29–42	Leng, Corbett and Brett
	Cecum	5–7	(1968); Faichney (1969);
			Hume (1977)
Wildebeest	Reticulorumen	67	Van Hoven and
Connochaetes gnou	Cecum	6	Boomker (1981)
Moose	Reticulorumen	40	Gasaway and
Alces alces			Coady (1974)
Black-tailed deer	Reticulorumen	23	Allo et al. (1973)
Odocolleus columbianus	Total hindgut	1	
Gray duiker	Reticulorumen	18–38	Boomker (1983)
Sylvicapra grimmia	Total hindgut	10–15	
Red-necked wallaby	Forestomach	42	Hume (1977)
Macropus rufogriseus	Total hindgut	1	
Red-necked pademelon	Forestomach	21	Hume (1977)
Thylogale thetis	Total hindgut	2	
Hindgut fermenters			
MAMMALS			
Rock hyrax	Forestomach + cecum	44	Rübsamen, Hume,
Procavia habessinica	+ colonic appendages		and Engelhardt (1982)
Cape hyrax	Cecum +	31–36	Eloff and van
Procavia capensis	Colonic appendages		Hoven (1985)

198

Species	Region	Value	Reference
Rabbit *Oryctolagus cuniculus*	Cecum	6 30 A 38 A	Hoover and Heitman (1972); Parker (1976); Marty and Vernay; (1984)
Hairy-nosed wombat *Lasiorhinus latifrons*	Total colon	33	Barboza and Hume (1992b)
Common wombat *Vombatus ursinus*	Total colon	30	Barboza and Hume (1992b)
Howler monkey *Alouatta palliata*	Cecum	31	Milton and McBee (1983)
Pony *Equus caballus*	Cecum	30	Glinsky et al. (1976)
Naked mole-rat *Heterocephalus glaber*	Cecum	10–22	Buffenstein and Yahav (1991)
Pig *Sus scrofa*	Total hindgut	10–12 9–23 A 31	Imoto and Namioka (1978); Kennelly et al. (1981); Rose, Hume, and Farrell (1987)
Guinea pig *Cavia porcellus*	Cecum + upper proximal colon	17	Sakaguchi et al. (1985)
Brushtail possum *Trichosurus vulpecula*	Total hindgut	16	Foley, Hume, and Cork (1989)
Beaver *Castor canadensis*	Cecum + proximal colon	10	Hoover and Clarke (1972)
Koala *Phascolarctos cinereus*	Cecum + proximal colon	9	Cork and Hume (1983)
Greater glider *Petauroides volans*	Cecum	8	Foley, Hume, and Cork (1989)
Porcupine *Erethizon dorsatum*	Cecum	8	Johnson and McBee (1967)
Human *Homo sapiens*	Total hindgut	6–9 B	McNeil (1984)
Laboratory rat *Rattus noryvegicus*	Cecum	5	Yang, Manoharan, and Michelsen (1970)

Table 8.3. (Cont.)

Species	Segment	Percent of maintenance energy requirement[a]	References
Birds			
Emu	Ileum + colon	11	Herd and Dawson (1984)
Dromaius novaehollandiae			
Rock ptarmigan	Ceca	9	Gasaway (1976)
Lagopus mutus			
Sharp-tailed grouse	Ceca	7	"
Typmpanuchus phasianellus			
Willow ptarmigan	Ceca	4–9	"
Lagopus lagopus			
Reptiles			
Florida red-bellied turtle	Small intestine + hindgut	100	Björndal and Bolten (1990)
Pseudemys nelsoni			
Lizard	Cecum + proximal colon	47	Foley et al. (1992)
Uromastyx aegyptius			
Green iguana	Cecum	30, 38	McBee and McBee (1982)
Iguana iguana			
Green turtle	Cecum	15	Björndal (1979)
Chelonia mydas			

[a]Maintenance energy is calculated as twice the BMR or is assumed to be equivalent to ad libitum digestible energy intakes of captive nonreproducing adult animals. Rates of SCFA production were measured in vitro by incubation or in vivo either by isotope dilution (A) or from measurement of digesta flow (B).

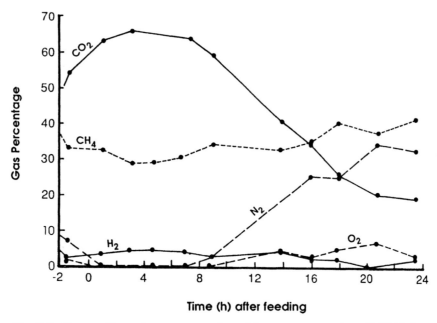

Fig. 8.3. Composition of rumen gases in a dairy cow on a ration of alfalfa hay and grain. (From Washburn and Brody 1937.)

Rumen gases vary in both their rate of production and their composition with time after feeding (Fig. 8.3). Carbon dioxide is derived from fermentation of carbohydrate and the neutralization of SCFA with HCO_3. Methane appears to be almost totally derived from the reduction of CO_2 by formate, succinate, and H_2. The latter reaction accounts for the low concentrations of H_2 in the rumen, except for the first few days of a fasting period. Nitrogen and O_2 are added from swallowed air, and N_2 can diffuse into the rumen from the blood as well. Oxygen is rapidly reduced by rumen microorganisms, and some of the CO_2 is directly absorbed into the bloodstream, but much of the CO_2 and most of the CH_4 produced in the rumen are removed by eructation, as described in Chapter 5. Kleiber (1961) found that an adult cow on a diet of 5 kg of hay lost 191 liters of CH_4, which was equivalent to 10 percent of the daily digestible food energy.

The low pH and high proteolytic activity of abomasal contents inhibit further production of SCFA by rumen microbes, and this, plus their rapid absorption (Engelhardt and Hauffe 1975) and dilution with abomasal secretions, results in relatively low levels of these fatty acids in the abomasum. Svendsen (1969) found that the abomasum of cattle fed a hay diet contained 6 mM SCFA and released gas into the rumen (via the omasal canal) at a rate of 0.8 liters per hour. The gases released from abomasal contents were composed of 17 percent CO_2, 65 percent CH_4, and 17 percent N_2. The N_2 was assumed to arise from the swallowing of air and diffusion from blood, and the CO_2 could result from the acidification of the HCO_3 arriving in the forestomach digesta. Although the CH_4 may have derived from reduction of CO_2, this would require relatively strong H donors plus enzymes. Therefore, it appeared that some of the CH_4 was produced by microbial fermentation in the abomasum itself. The subsequent feeding of a high-grain diet resulted in a doubling of the SCFA concentration and gas production and a reduction in abomasal motility. This led Svendsen to conclude that a combination of decreased motility and increased gas production may be responsible for the abomasal distension and displacement seen in cattle on high-concentrate diets. An increase in the rate of SCFA absorption also may account for the abomasal ulcers found in calves and adult cattle fed this type of diet.

Nitrogen recycling and synthesis of microbial protein

Rumen bacteria can convert both dietary protein and other dietary and endogenous nitrogenous compounds into peptides, amino acids, and ammonia. As mentioned above, one of the major sources of nonprotein nitrogen is urea, the waste product of protein metabolism in the liver. Urea enters the forestomach via salivary secretions and diffusion across forestomach epithelium and is rapidly converted to ammonia by bacteria attached to its epithelial surface (Houpt and Houpt 1968; Egan, Boda, and Varady 1986). Some of the ammonia is used for microbial synthesis of essential and nonessential amino acids, which are incorporated into microbial protein. These amino acids are absorbed from the small intestine, following the digestion of microbes by endogenous enzymes. Most of the remaining ammonia is absorbed from the ruminant forestomach and converted to nonessential amino acids and urea in the liver. The rumen of sheep on a chopped orchard grass diet secreted 2.1 g of urea and absorbed 3.8 g of ammonia N per day (Remond et al. 1992). Therefore, the rumen bacteria can increase both the quality and quantity of protein available to animals on a diet that contains low levels of protein.

The utilization of nitrogen and the fermentation of carbohydrate by rumen bacteria are interdependent. A shortage of dietary nitrogen can hamper the digestion of poor quality roughage such as straw (Balch and Campling 1965). This may be due to the marked effect of nitrogen deficiency on bacterial numbers (Teather et al. 1980). The rate that urea is recycled by rumen microorganisms is similarly dependent on an adequate supply of carbohydrates (Engelhardt, Hinderer, and Wipper 1978; Obara, Dellow, and Nolan 1991).

The recycling of urea nitrogen also aids in the conservation of nitrogen and water. It increased the efficiency of nitrogen utilization by sheep (Cocimano and Leng 1967), camels (Schmidt-Nielsen et al. 1957), and llamas (Hinderer 1978).

Restriction of water increased the rate at which urea was recycled into the digestive tract of temperate, tropical, and desert ruminants (Phillips 1961a,b; Livingston, Payne and Friend 1962; Utley, Bradley, and Boling 1970; Mousa, Ali, and Hume 1983; Freudenberger and Hume 1993). Therefore, it reduced the amount of water required for the renal excretion of urea.

Synthesis of vitamins

Rumen microorganisms can synthesize all of the B-complex vitamins required by cattle and sheep, and a deficiency of B_{12} can be produced only on diets deficient in cobalt, which is required for its microbial synthesis (Phillipson 1970). Much of the thiamin, pantothenic acid, pyridoxine, and biotin are present in solution, and some of the B vitamins appear to be directly absorbed from the rumen (Dziuk 1984).

Digestive tracts of other species

Various degrees of microbial digestion similar to that seen in the ruminant forestomach have been demonstrated along the gastrointestinal tract of all classes of vertebrates. Because of the low levels of SCFA in food and their rapid absorption from the gastrointestinal tract, their presence in digesta is a sign of microbial fermentation. Table 8.1 lists the concentrations of SCFA found along the gastrointestinal tract of a number of mammals, birds, reptiles, and fish, and the microbial counts, where measured. It shows similar SCFA concentrations in the gut of carnivores, omnivores, and herbivores. However, the quantity produced and absorbed varies with gut capacity and retention time. The contribution of SCFA to the maintenance energy requirement of some of these species is listed in Table 8.3.

Foregut

Bacterial counts similar to those of the ruminant forestomach have been demonstrated in the forestomach of the camelids, hippopotamus, sloth, macropod marsupials, peccaries, colobus and langur monkeys, and whales and in the crop and distal esophagus of the hoatzin, a folivorous South American bird (Table 8.1). These were accompanied by substantial numbers of protozoa in the forestomach of the camel, hippopotamus, hoatzin, and some macropods, but protozoa appeared to be absent from the forestomach of the sloth and the colobus and langur monkeys. Dellow et al. (1988) found additional colonies of fungal sporangia, like those of the rumen, in the forestomach of four species of kangaroos and wallabies.

The levels of SCFA also were similar to those found in the forestomach of ruminants. The extremely high levels found in some gray and bowhead whales were probably from samples collected long after death. Where measured, the proportions of acetate, propionate, and butyrate were similar to those in the ruminant forestomach as well. The principal gasses in the forestomach of the langur monkey were CO_2 and CH_4 (Bauchop and Martucci 1968). Although Kuhn (1964) reported similar findings for the colobus monkey, Ohwaki et al. (1974) found no CH_4 in the stomach of these animals. Carbon dioxide and CH_4 also were present in the forestomachs of numerous species of macropod marsupials (Kempton, Murray, and Leng 1976; Engelhardt et al. 1978; Dellow et al. 1988). The forestomach of the quokka and tammar wallaby contained higher levels of H_2 and lower levels of CH_4 than those of ruminants, and only negligible amounts of CH_4 were present in the forestomach of the grey kangaroo. Hume (1982) suggested that the more rapid transit of digesta may inhibit the establishment of the slow-growing methanogenic bacteria. The forestomach of macropod marsupials also demonstrated a similar process of nitrogen recycling (Dellow and Hume 1982; Dellow, Nolan, and Hume 1983). Urea increased the efficiency of nitrogen utilization by the euro (Brown 1969) and tammar wallaby (Lintern-Moore 1973; Kinnear and Main 1975; Kennedy and Hume 1978) fed low-protein diets.

Forestomach contents constituted from 10 to 30 percent of the body weight of the camel, macropods, colobus monkey, and sloth (Table 8.1) and 75 percent of the total gut content of the hippopotamus. The crop and distal esophagus of the hoatzin contained a quantity of digesta equivalent to 12 percent of its body weight. Therefore, the SCFA generated in the foregut of these animals should provide an important source of their total energy requirements. The contribution to the maintenance energy requirement of two macropod marsupials is shown in Table 8.3.

The contribution of SCFA to the energy economy of whales (Table 8.1) is unknown. However, Ellis (1981) estimated that the body weight of bowhead whales is three tons per meter of body length. Therefore, the 53 liters of digesta found in the forestomach of a bowhead whale 9 meters long (Tarply et al. 1987) would contribute little to the nutritional requirements of the animal. SCFA also have been found in the forestomach of small, toothed whales (Morii and Kanazu 1972; Morii 1979). The major substrates available for SCFA production in the baleen whales appear to be chitin and small amounts of kelp. The forestomach of minke whales, which feed largely on krill, contained 10^{10} bacterias per gram of fluid digesta, including chitinase-producing organisms (Mathiesen, Aagees and Sørmo 1990; Olsen et al. 1994), and Martensson, Nordøy, and Blix (1994) showed that the ability of these whales to digest krill was superior to that of simple-stomached crabeater seals. The low levels of SCFA and relatively small volume of the dugong stomach (Figs. 4.31, 4.32) indicate that it plays only a minor role in the fermentation of carbohydrate by these marine herbivores.

High concentrations of SCFA were found in the proximal compartment of the stomach of the hyrax and hamster (Table 8.1). However, Rübsamen, Hume, and Engelhardt (1982) found that acetate was the predominant SCFA (87–98%) in the hyrax stomach, and this was accompanied by relatively high levels of lactic acid. Higher levels of starch in the stomach should reduce the acetate/propionate ratio, as they do in the rumen and the hindgut of pigs, but some types of bacteria can synthesize acetate from CO_2 and a wide range of other substrates (Daniel and Drake 1993). Therefore, a predominance of acetate could be due to differences in the bacteria that populate these segments of the gut. The presence of lactic acid indicates colonization by lactobacilli, which Savage (1977) found attached to the stratified squamous epithelium lining the forestomach of many rodents. Rübsamen, Hume, and Engelhardt (1982) concluded that the SCFA produced in the stomach of the hyrax provided only a small fraction of total-tract production.

Low levels of SCFA were found in the simple stomachs of the raccoon, pig, bush baby, vervet monkey, and pony (Fig. 8.4), as well as the baboon and Sykes monkey (Clemens and Phillips 1980), the koala (Cork and Hume 1983) and the elephant (Clemens and Maloiy 1982). As in the hyrax, this was accompanied by substantial concentrations of lactic acid in the stomachs of the dog, pig, and pony (see Figs. 9.5–9.7). Although lactic acid may be released from gastric mucosa, the high concentrations seen after feeding suggested that it was produced by indigenous microbes. The quantity of SCFA produced in the stomach of these animals would be of little nutritional significance, but it shows that bacterial fermentation occurs in the stomach of these animals.

Midgut

The numbers of bacteria in the mammalian midgut are generally much lower than those in the rumen. The human small intestine contains 10^4 to 10^6 viable, predominantly anaerobic organisms per gram of digesta (Savage 1986). However, Mackie and Wilkins (1988) found that the counts of anaerobic bacteria in grass-fed horses ranged from 10^6 per gram in the duodenum to 10^8 per gram in the ileum (Table 8.1). The presence of a high percentage of proteolytic organisms, especially in the duodenum, raised the question of whether they digest or compete with the endogenous enzymes. There appears to be little information available on bacterial counts in the midgut of birds, reptiles, or amphibians. However, interest in the nutrition of herbivorous fish has stimulated studies of their indigenous bacteria. Because the intestinal tract of most fish is composed principally of midgut, most of the counts are confined to this segment.

Aerobic and facultative anaerobic bacteria, at 10^3 to 10^8/g of digesta, were found in the midgut of a variety of teleosts (Trust and Sparrow 1974; Hamid, Sakata, and Kakimoto 1976; Horsley 1977; Fishelson, Montgomery, and Myrberg 1985;

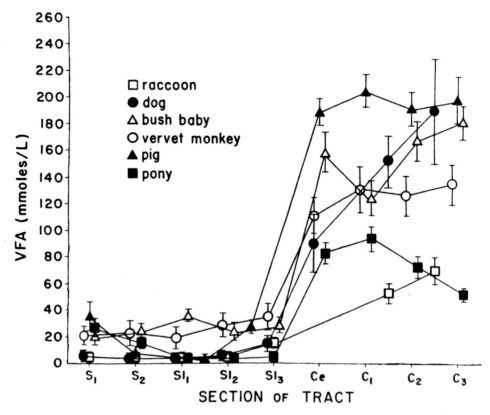

Fig. 8.4. Mean (± SE) values for concentrations of VFA (SCFA) along the gas-trointestinal tract of raccoon (*Procyon lotor*), dog (*Canis familiaris*), pig (*Sus scrofa*), bush baby (*Galago crassicaudatus*), vervet monkey (*Cercopithecus pygerythrus*), and pony (*Equus caballus*). All animals were fed at twelve-hour intervals during a three- to four-week period before the study. Raccoons, pigs, and ponies were given the same pelleted diet. Each value represents the mean from twelve animals, killed in groups of three at two, four, eight, and twelve hours after feeding. Sections of tract were oral (S_1) and aboral (S_2) halves of the stomach, two or three equal segments of small intestine (SI_1, SI_2, SI_3), cecum (Ce), and two or three segments of colon (C_1, C_2, C_3). (Modified from Argen-zio, Southworth, and Stevens 1974; Clemens, Stevens, and Southworth 1975a; Banta et al. 1979; Clements and Stevens 1979; Clemens 1980.)

Lesel, Fromageot, and Lesel 1986). Obligate anaerobes, which predominate in the forestomach and intestine of mammals, have received less study. However, Trust et al. (1979) measured the number of strictly obligate anaerobes cultured from homogenates of the gastrointestinal tract of the grass carp *Ctenopharyngodon idella*; goldfish *Carassius auratus*; and rainbow trout *Salmo gairdneri*. Although the number of anaerobes in the intestinal tract of the rainbow trout was insignificant, they found 10^7 to 10^9 obligate anaerobes/g of homogenate in the intestine of the grass carp and 10^8 obligate anaerobes/g in the goldfish intestine. In a review of bacterial flora of fishes, Cahill (1990) concluded that large populations of facultative and obligate anaerobes were present in the gut of a number of species.

Chitinolytic bacteria, in counts of 10^7 to 10^9 per gram, were found in the intestine of a number of marine fish (Seki and Taga 1963; Goodrich and Morita 1977; Danulat 1986), but there seems to be no information on microbial digestion of chitin in fish. A group of extremely large indigenous bacteria, 600×80 μm in size, was found in the intestinal contents of herbivorous surgeonfish in the Red Sea (Fishelson, Montgomery, and Myrberg 1985) and on the Great Barrier Reef (Clements, Sutton, and Choat 1989) and was subsequently identified as gram-positive bacteria related to the clostridia (Angert, Clements, and Pace 1993). Substantial numbers of these microorganisms were consistently present in the intestine of these fish. A study of the brown surgeonfish *Acanthurus nigrofuscus* indicated that they reduced the amylase, protease, and lipase activity of host enzymes but may contribute to the digestion of lipid (Pollak and Montgomery 1994). Protozoa, from 10^3 to 10^6 per gram, also were reported in the midgut of the herbivorous herring cale *Odax cyanomelas* (Clements 1991).

SCFA are found along the midgut of many vertebrates (Table 8.1 and Fig. 8.4), with the highest concentrations generally in the distal segments. Low levels were found in the small intestine of dogs, raccoons, and ponies fed commercial, pelleted diets, but mean concentrations of 30 to 45 mM were present in the terminal midgut of the pig, bush baby, vervet monkey, and grass-fed horses. The concentrations found in the terminal midgut of the emu *Dromius novaehollandiae*, and Florida red-bellied turtle *Pseudemys nelsoni*, were equivalent to those of the rumen. Herd and Dawson (1984) found that the ileum of the emu contained 37 percent of the total gut contents, as compared with only 5 percent in the hindgut. They concluded that production of SCFA (principally acetate) in this segment accounted for most of the energy derived from microbial fermentation in this species (Table 8.3). Björndal and Bolten (1990) found that the SCFA in the midgut of the Florida red-bellied turtle, which contained 68 percent of the total gut contents (versus 17 percent in their hindgut), were similar to those of the rumen in both composition and concentration. From the rate of SCFA production, they estimated that the absorption of SCFA from the midgut could account for the mean daily energy requirement of these animals (Table 8.3). Therefore, it appears that substantial amounts of carbohydrate are subjected to microbial fermentation in the midgut of some species, despite the potential problems of microbial competition with and digestion by endogenous enzymes.

Microbial fermentation has been examined in a number of fish. Cellulase activity has been reported in the gastrointestinal tract of a number of freshwater fish

(Prejs and Plaszesyk 1977). Davies (1985) concluded that the inclusion of plant fiber, cellulose, or chitin in the diet provided no nutritional advantage to the rainbow trout. Cellulose digestion has been reported in the gut of the carp *Cyprinus carpio* (Shcherbina, Mochul'skaya, and Erman 1970; Shcherbina and Kazlauskene 1971), the freshwater catfish *Ictaluris punctatus*; and sixteen species of marine detritus feeders (Stickney and Shumway 1974). No evidence of cellulolytic activity was found in the intestine of the algae-feeding *Tilapia esqulenta* (Fish 1951). However, Clements, Gleeson, and Slaytor (1994) pointed out that algae contain little cellulose, so cellulolytic activity may not be a good measure of microbial fermentation in fish.

SCFA in concentration of 7 to 37 mM have been reported in the intestine of the carnivorous rainbow trout, omnivorous common carp, and herbivorous grass carp; tilapia *Oreochromis mossambicus*; butterfish *O. pullus*; sea carp *Crynodus lophodon*; and herring cale (Table 8.1). Highest concentrations were usually reported in the terminal midgut. Acetate was the predominant fatty acid in all of these species. Although these SCFA levels are low, Clements, Gleeson, and Slaytor (1994) suggested that they may provide a significant source of the energy required to fuel the much lower rates of metabolism of these ectotherms (see Chapter 1).

Hindgut

Microbes

The bacteria inhabiting the large intestine have been described for a number of mammals (Savage 1977, 1986; Hoogkamp-Korstanje et al. 1979; Robinson, Allison, and Bucklin 1981; Finegold, Sutter, and Mathisen 1983; Allison 1984; Mikel'Saar et al. 1984). They are generally similar to those found in the rumen in both total numbers and species (Wolin 1981; Allison 1984). At least 400 species, representing 40 genera of bacteria, were isolated from human feces (Savage 1986). The populations associated with epithelial tissue differ from those found in the lumen, and fecal populations do not necessarily represent the luminal contents of the hindgut. Bacterial populations can also vary between segments. For example, the epithelial surface of the cecum and proximal colon of the koala contained much greater numbers of bacteria capable of degrading tannin–protein complexes in their normal diet of *Eucalyptus* foliage (Osawa et al. 1993). Methanogenic bacteria appear to preferentially colonize the distal colon of humans that are CH_4 excretors (Pochart et al. 1993). However, total bacterial counts of 10^8 to 10^{12}/g have been measured in the hindgut of mammals, birds, reptiles, and amphibians, and protozoa have been reported in the hindgut of a number of mammals, the green iguana and two species of marine herbivorous fish (Table 8.1).

Fermentation of carbohydrates

Studies conducted in the 1940s demonstrated the presence of SCFA in the large intestine of sheep, cattle, deer, pigs, rats, rabbits, and dogs (Barcroft, McAnally, and Phillipson 1944; Elsden, Hitchcock, Marshall, and Phillipson 1946; Phillipson 1947). A series of studies conducted in the 1970s showed that SCFA concentrations in the hindgut of mammalian carnivores, omnivores, and herbivores were similar to those seen in the rumen (Fig. 8.4) and that they were absorbed at similar rates per unit surface area (Table 8.4). The total concentrations of SCFA in the hindgut of these animals were relatively unaffected by time after feeding, by changes in the diet, or a 48-hour period of starvation in the dog. The acetate/propionate/butyrate ratio was similar to that of the ruminant forestomach, and the addition of fiber to the diet had a similar effect on the acetate/propionate ratio in the pig (Argenzio and Southworth 1974; Imoto and Namioka 1978). However, the addition of fiber to the diet increased the total volume (and SCFA content) of hindgut contents. Furthermore, compartmental analysis of the pony hindgut showed marked cyclic changes in the rate of SCFA production with time after feeding (see Fig. 9.16). Therefore, the relatively constant levels of SCFA in hindgut contents were maintained by their absorption and the secretion and absorption of water.

Calloway (1968) reviewed information on the composition of gases in the large intestine of dogs, rats, pigs, cattle, horses, and humans. The gasses were the same as those found in the rumen (CO_2, CH_4, H_2, N_2, and O_2), but there was considerable variation among species and with changes in diet. It was estimated that the large intestine accounts for approximately 13 percent of the CH_4 produced in the gastrointestinal tract of sheep (Murray, Bryant, and Leng 1976). However, the human large intestine contains higher percentages of H_2 and N_2 and lower percentages of CO_2 and CH_4 than the rumen, and CH_4 was absent in about two thirds of the human population (Levitt and Bond 1970).

Subsequent studies have shown that this process of microbial production of SCFA is common to the hindgut of all vertebrates that have been examined (Table 8.1). Relatively high concentrations have been found in the hindgut of mammals. Concentrations vary among segments of hindgut and species. The lower levels of SCFA found in the hindgut of some mammals such as the raccoon can be attributed to its small capacity and short digesta retention time (see Figs. 4.8 and 6.1). The lower levels in marsupial folivores have been attributed to the tannins in their diet of *Eucalyptus* forage. Where examined, the ratio of the three fatty acids are similar to those found in the rumen.

Substantial concentrations of SCFA have been reported in the hindgut of birds, reptiles, and fish (Table 8.1). Substantial amounts of SCFA were demonstrated in the ceca of omnivorous and herbivorous birds. Annison, Hill, and Kenworthy

Table 8.4. *Transport of SCFA across isolated, short-circuited gastrointestinal epithelium*

Tissue	Short-chain volatile fatty acids mμmol/cm² × h							
	Ox		Pig		Pony		Dog	
	Loss lumen side	Gain blood side	Loss lumen side	Gain blood side	Loss lumen side	Gain blood side	Loss lumen side	Gain blood side
Stratified squamous epithelium	10.5 ± 1.4	2.4 ± 0.4	6.1 ± 2.4	0.4 ± 0.1	1.4 ± 1.4	0.02 ± 0.02	—	—
Cardiac mucosa	—	—	8.3 ± 1.8	1.1 ± 0.3	—	—	—	—
Proper gastric mucosa	—	—	5.0 ± 1.0	0.3 ± 0.04	9.4 ± 3.8	0.2 ± 0.1	—	—
Pyloric mucosa	—	—	5.8 ± 1.1	0.5 ± 0.02	5.5 ± 1.4	0.4 ± 0.1	—	—
Cecal mucosa	—	—	10.3 ± 2.9	4.3 ± 0.6	8.2 ± 0.8	1.6 ± 0.2	—	—
Proximal colon mucosa	—	—	8.0 ± 0.8	3.7 ± 0.6	9.8 ± 1.4	1.8 ± 0.2	8.8 ± 0.3	4.0 ± 0.2
Distal colon mucosa	—	—	9.8 ± 1.0	3.1 ± 0.2	6.5 ± 1.5	1.5 ± 0.2	8.8 ± 0.3	4.0 ± 0.2

Values are means ± SE for the rate of SCFA transport (lumen to blood) obtained during 2.5-hour experimental periods. Ringer solution, containing a 90-mM equimolar mixture of acetate, propionate, and butyrate, was used to bathe lumen surface of tissue, and normal Ringer was used to bathe blood surface; both solutions were buffered at pH 7.4 with bicarbonate. Transmucosal electrical PD was clamped at zero. (Modified from Stevens and Stettler 1966b; Argenzio, Southworth, and Stevens 1974; Argenzio and Southworth 1974; and Herschel et al. 1981.)

(1968) estimated that cecal production of acetate provided 11 percent of the energy required by chickens. Guard (1980) found substantial concentrations of SCFA in the hindgut of the carnivorous caiman (*Caiman crocodillus*) and an omnivorous terrapin (*Chrysemys picta*) and high levels in the cecum of the herbivorous green iguana *Iguana iguana* and tortoise *Geochelone carbonaria*. Björndal (1979, 1991) found that the green turtle *Chelonia mydas*, which feeds on algae and sea grasses, digested about 83 percent of the cellulose in its diet. Their hindgut contained 55 percent of the digesta in the gastrointestinal tract, with high levels of SCFA and H_2 and low levels of CH_4. She estimated that 15 percent of the daily energy balance was provided by absorption of the SCFA produced in the cecum alone. Substantial degrees of cellulose digestion were demonstrated in five species of herbivorous tortoises (Hamilton and Coe 1980: Nagy and Medica 1986; Bjorndal 1987, 1989). The contribution of these fatty acids to the total energy requirements of herbivorous reptiles and birds are listed in Table 8.3.

Although there appears to be no information on microbial fermentation in the hindgut of adult amphibians, the large populations of bacteria in the hindgut of the leopard frog (Table 8.1) included both acetogenic and butyricogenic organisms (Gossling, Loesche, and Nace 1982). Because of the short and indistinct hindgut of fish, most of the data on intestinal SCFA levels in them refer to either the entire intestinal tract or its proximal and distal segments. However, the hindgut of the Western Australian kyphosids *Kyphosus cornelii* and *K. sydneyanus*, which is separated from the midgut by a sphincter, contained acetate at levels of 16 and 38 mM respectively (Rimmer and Wiebe 1987).

The substrates for SCFA production in the hindgut are shown in Fig. 8.5, *A*. Dietary starches and endogenous carbohydrates serve as the principal substrates in carnivores and omnivores and as a substantial source in herbivores. Significant quantities of dietary starch were shown to reach the hindgut of humans, rats, mice, hamsters, guinea pigs, and rabbits (Baker et al. 1950); pigs (Keys and DeBarthe 1974); cattle (Karr, Little, and Mitchell 1966); sheep (Orskov, Fraser, and McDonald 1971); and ponies (Hintz et al. 1971). The amount of starch that escaped digestion in the small intestine depended on the dietary source, and it was reduced by either boiling or grinding. Endogenous carbohydrates appeared to be the principal substrates for SCFA production in the human hindgut (Ehle, Robertson, and Van Soest 1982).

Mucus, which is 80 percent polysaccharide, may be a major substrate (Vercellotti, Salyers, and Wilkins 1978). This could account for the large quantities that accumulate in the cecum of germ-free rats and guinea pigs (Gordon and Bruckner 1984). Carbohydrates and amino acids in epithelial cells that are sloughed into the intestine can serve as additional substrates for SCFA production (el-Shazly 1952a,b).

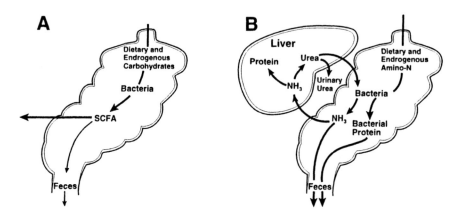

Fig. 8.5. Bacterial fermentation of carbohydrates and metabolism of nitrogen in the hindgut of vertebrates. **A,** Bacterial fermentation of dietary and endogenous carbohydrates to SCFA, which are largely absorbed prior to defecation. The dietary carbohydrates consist principally of starch that has escaped digestion in the upper gastrointestinal tract and structural carbohydrates, which cannot be digested by endogenous enzymes. The endogenous carbohydrates include those found in mucus and sloughed epithelial cells. **B,** Resident bacteria also convert protein and other nitrogen-containing compounds into ammonia and bacterial protein. The principal endogenous substrates are the enzymes secreted into the upper digestive tract and urea, which passively diffuses into the intestine. Some of the ammonia that results from metabolism of endogenous urea is incorporated into microbial protein or excreted in the feces, but substantial amounts are absorbed and returned to the liver for either the production of more urea or the synthesis of protein. Bacteria in the hindgut of birds and reptiles appear to serve similar functions, except that the urea is replaced by uric acid excreted into the cloaca by the urinary system. Microbial protein is lost in the feces of species that do not practice coprophagy. (Modified from Wrong and Vince 1984.)

Cellulose, hemicellulose, and pectin are major substrates in the hindgut of herbivores. Microbial fermentation in the hindgut accounted for 63–73 percent of the neutral detergent fiber digested by ponies (Hintz et al. 1971) and 12 percent of the total cellulose digested by sheep (Goodall and Kay 1965).

SCFA stimulate the growth of enterocytes (Sakata and Yajima 1984) and provide energy for other body functions. Henning and Hird (1972b) found that butyrate was metabolized by ileal, cecal, and colonic mucosa of the rabbit. All of these tissues metabolized butyrate to CO_2. Metabolism to ketone bodies was low in the ileum and distal colon but high in the cecum and proximal colon. Ketogenesis provides an additional source of energy to epithelial cells. Colonic epithelial cells metabolize lesser quantities of propionate and acetate, as well, but butyrate appears to provide their major source of energy (Clausen and Mortensen 1994). Absorption of butyrate from the hindgut also stimulates the growth of jejunal mucosal cells, apparently via the autonomic nervous system (Frankel et al. 1994). Absorption of SCFA from the hindgut provides an additional source of energy for other body functions and a substantial part of the maintenance energy of hindgut-fermenting herbivores (Table 8.3). The absorption of SCFA from the hindgut also provides a substantial contribution to the absorption of water (see Chapter 9).

Nitrogen recycling and synthesis of microbial protein

Bacteria in the hindgut recycle nitrogen in a manner similar to that of the ruminant forestomach (Wrong 1988). Fig. 8.5, *B* shows the major substrates for microbial digestion and synthesis of nitrogenous compounds. Wrong and Vince (1984) concluded that most of the nitrogen that enters the human large intestine from the ileum consists of urea and creatinine at levels equivalent to that of blood, and protein derived from intestinal secretions, sloughed epithelial cells, and dietary residues. Significant amounts of amino acids entered the large intestine from the ileum of pigs (Holmes et al. 1974). They appeared to be derived predominantly from mucin and the amino acids that are least efficiently absorbed by the small intestine (Taverner, Hume, and Farrell 1981 a,b). The quantity of endogenous amino-acid nitrogen in the ileum of pigs was markedly increased by an increase in dietary fiber (Sauer, Stothers, and Parker 1977). This may be due to the more rapid rate of digesta transit through the small intestine of animals on high-fiber diets.

High concentrations of urea were found in the gut of rats following the feeding of antibiotics (Chao and Tarver 1953) or the establishment of germ-free conditions (Levenson et al. 1959). The levels found in the experimentally cleansed large intestine of humans (Wrong, Edmonds, and Chadwick 1981), and the perfused colon of sheep and goats (Engelhardt et al. 1984) indicated that most of the urea is hydrolyzed to ammonia by bacteria attached to the epithelial surface. Much of the ammonia is either incorporated into microbial protein or absorbed and recycled for the synthesis of nonessential amino acids or urea in the liver (Fig. 8.5, *B*).

Urea was extensively recycled by the large intestine of the rabbit (Regoeczi et al. 1965), pony (Prior et al. 1974), greater glider and brushtail possum (Foley and Hume 1987b), rock hyrax (Hume, Rübsamen, and Engelhardt 1980), and donkey (Izraely et al. 1989). Nitrogen balance was maintained in rabbits on an 8 percent crude protein diet by infusion of urea into the cecum (Salse, Crampes, and Raynaud 1977), and utilization of urea by the hyrax large intestine increased with reduction of dietary protein or restriction of water (Hume, Rübsamen, and Engelhardt 1980), in a manner similar to that of the ruminant forestomach. Urea recycling is also important in animals that hibernate for long periods. The black bear *Ursus americanus*, which hibernates five months without eating, drinking, or defecating, can recycle 20 percent of its endogenous urea through the digestive tract (Guppy 1986). Uric acid, the principal waste product of protein metabolism in reptiles and birds, also was recycled into ammonia and microbial protein following its excretion into the cloaca of chickens (Bell and Bird 1966; Mead 1974; Mead and Adams 1975) and willow ptarmigan (Mortensen and Tindall 1981). Björnhag (1989) showed that the amount of urine refluxed into the avian ceca increased with a reduction in dietary protein. Karasawa and Maeda (1994) found that the nitrogen of uric acid and of urea added to the feed was utilized by cecal bacteria of the chicken and, following absorption of ammonia, was used for synthesis of nonessential amino acids.

Carnivores tend to produce larger quantities of urea. Oxidation of amino acids can account for 41 to 85 percent of the total energy production of fish (Mommsen and Walsh 1992). Marine elasmobranchs and a few other species maintain high plasma urea levels, but freshwater elasmobranchs and all teleosts maintain levels similar to those of mammals (see Chapter 9). Recycling of urea has been demonstrated in the gut of sharks (Knight, Grimes, and Colwell 1988) and the gulf toadfish *Opsanus beta* (Walsh, Dannulat, and Mommsen 1990).

The concentration of nitrogen decreased as digesta progressed through the large intestine of the sheep (Hogan and Phillipson 1960; Goodall and Kay 1965), pony (Hintz et al. 1971; Wootton and Argenzio 1975), rabbit (Rérat 1978), and wombat (Barboza and Hume 1992a). Most of the nitrogen in the digesta and feces is contained in bacterial protein (Wootton and Argenzio 1975; Wrong and Vince 1984; Wrong, Vince, and Waterlow 1985). Substantial amounts of microbial protein were digested in the colon of the rabbit (Bonnafous and Raynaud 1968), pony (Wootton and Argenzio 1975), and human (Wrong 1988). Although this may be partially due to pancreatic trypsin activity, most of it appears to result from bacterial activity (Wrong 1988).

Active absorption of amino acids by the hindgut has been demonstrated in the newborn rat (Batt and Schachter 1969) and pig (James and Smith 1976; Jarvis et al. 1977), but only passive absorption was shown in the adult rat (Binder 1970), prairie vole (Hume, Harasov, and Darken 1993), and tortoise (Baillien and Schoffeniels 1961). Although introduction of isotopically labeled microbial cells into the horse cecum resulted in the appearance of isotopically labeled amino acids in the blood (Slade et al. 1971), this was probably due to absorption of the isotope in other forms. For example, the infusion of isotopically labeled alanine into the cecum of rabbits resulted in considerable breakdown and little absorption (Hoover and Heitmann 1975). Infusion of isotopically labeled leucine and isoleucine into the cecum of pigs also failed to demonstrate absorption of either amino acid (Krawielitzki et al. 1983, 1984). Thus, there appears to be no substantial evidence for active absorption of amino acids by the hindgut of adult mammals. However, as stated in Chapter 7, substantial amounts of amino acids are absorbed by the cecum of birds and by the proximal colon of chickens on a high-sodium diet. The possible contribution of amino acids synthesized by the hindgut bacteria to the nutrition of other nonmammalian species needs to be explored.

Synthesis of vitamins

The presence of B vitamins in rat feces was recognized early (Steenbock, Seel, and Nelson 1923), and subsequent studies showed that rats fed a diet deficient in biotin, riboflavin, pyridoxine, pantothenic acid, B_{12}, folic acid, or vitamin K showed no vitamin deficiencies if they were allowed to ingest their feces. However, germ-free rats were deficient in thiamin (Wostmann, Knight, and Reyniers 1958), pantothenic acid (Daft et al. 1963), folic acid and biotin (Luckey et al. 1955), and vitamin K (Gustafsson 1948), and they were susceptible to riboflavin deficiency (Luckey et al. 1955) when placed on diets deficient in these vitamins. Most of the thiamin, riboflavin, and niacin in the rat cecum were present in bacteria (Mitchell and Isbel 1942), and B_{12} was taken up by a variety of bacteria (Giannella, Broitman, and Zamcheck 1971). Thiamin was absorbed from the rat cecum (Kasper 1962), and B_{12}, pyridoxine, and pantothenic acid were absorbed equally well following their oral or large intestinal administration to human patients (Sorrell et al. 1971). Wrong, Edmonds, and Chadwick (1981) concluded that there was good evidence that thiamin, nicotinic acid, riboflavin, pantothenic acid, biotin, pyridoxine, folic acid, B_{12}, and vitamin K are synthesized by microbes in the human large intestine and that all of these, except nicotinic acid, riboflavin, and pantothenic acid, are subject to some degree of absorption.

Coprophagy

Coprophagy, the ingestion of feces, provides another avenue by which some species recover nutrients produced in the hindgut. Some animals are only coprophagic at certain stages in their development or during nutritional deficiency. Ingestion of maternal feces was observed in young golden hamsters (Dieterlein 1959), rats (Galef 1979), thoroughbred foals (Francis-Smith and Wood-Gush 1977), and koalas (Minchin 1937). When they first emerge from the pouch, juvenile koalas consume a special maternal feces called "pap." The pap has a higher percentage of water, higher pH, and much higher counts of tannin-protein-complex degrading bacteria, which suggests that it is of cecal origin (Osawa, Blanchard, and O'Callaghan 1993). Hatchling green iguanas regularly consumed the feces of adult animals, which helps microbial colonization of their gut and may provide additional nutrients (Troyer 1984 a,b). The nursing females of many mammalian species lick the anus of their young to elicit reflex defecation and then ingest the feces. Some marsupial females follow a similar process by periodically cleaning their pouch. Ingestion of urine and feces from the offspring has been shown to conserve 50–80 percent of the water that otherwise would be lost by lactating rodents and dingos (Baverstock, Watts, and Spencer 1979).

Working sled dogs (Kronfeld 1973) and domestic cats (McCuistion 1966) on high-carbohydrate diets became hypoglycemic and coprophagic after an overnight fast. Horses became coprophagic within three months after they were placed on a 6.2 percent crude protein diet, and they continued to ingest their feces until a week after the protein level was raised to 10 percent (Schurg et al. 1977). Coprophagy also has been noted in shrews (Crowcroft 1952; Baxter and Meester 1982).

In contrast, coprophagy is practiced on a regular basis by the adults of many species of small herbivorous mammals, which include the rabbits (Olsen and Madsen 1943; Hamilton 1955; Geis 1957), hares (Watson and Taylor 1955; Bookhout 1959; Pehrson 1983), and pika (Haga 1960). It has been reported in numerous rodents, including the naked mole rat (Hill et al. 1957), beaver (Richard 1959), mountain beaver (Ingles 1961), ground squirrel (Turcek 1963), rat (Leon 1974; Galef 1979), Scandinavian lemming (Sperber, Björnhag, and Ridderstrale 1983), guinea pig and chinchilla (Björnhag and Sjöblom 1977), capybara (Herrara 1985), and nutria (Snipes et al. 1988). It appeared to be absent in the North American porcupine (McBee 1977). Coprophagy also was reported in the folivorous lemur *Lepilemur mustelinus leucopus* (Hladik et al. 1971), and the ringtail possum (Chilcott and Hume 1985).

The nutritional significance of coprophagy was reviewed by Hörnicke and Björnhag (1980). Osborne and Mendel (1911) made early mention of the fact that prevention of coprophagy depresses the growth rate of rats. It resulted in a 15 to 25 percent depression in growth rate of rats on rations adequate in dietary nutrients, and they maintained their growth rate only if allowed to ingest feces directly from the anus (Barnes et al. 1957; Barnes and Fiala 1958; Barnes, Fiala, and Kwong 1963). It also produced changes in fecal microbes (Barnes 1962; Fitzgerald, Gustafsson, and McDaniel 1964). Coprophagy appeared capable of providing rats with all of their requirements for vitamin K and at least part of their requirements for the B-complex vitamins.

Björnhag (1994) pointed out that a 25-gram vole needs about twelve times more energy per kg of body mass than a 650-kg horse. Therefore, the vole would need to eat at least twelve times as much food to extract an equivalent amount of energy per unit body weight. Because ingested feces displace food that would normally be taken into the stomach, the advantages of coprophagy depend on the relative nutrient content of the feces and diet. As mentioned in Chapter 6, some animals selectively ingest special, high-nutrient feces that are derived from cecal contents, which Harder (1950) called *cecotrophy*. Rabbits produce two distinct types of feces; a dry, hard fecal pellet, which is discarded, and a soft fecal pellet, which is contained within a strong mucous envelope (Eden 1940; Heisinger 1965; Henning and Hird 1972a; Jilge 1979). The soft feces are produced during one or two periods each day. They attach to hairs around the anus and are ingested directly and swallowed without mastication. They remain within the membrane for up to six hours after ingestion, where their contents are maintained at a relatively neutral pH because of high levels of PO_4 buffers (Griffiths and Davies 1963). The stimulus for ingestion of cecotrophs may be SCFA due to their strong odor and taste and high concentration in soft feces (Henning and Hird 1972a). Germ-free rabbits consume neither their hard nor soft fecal pellets (Yoshida et al. 1968).

Cecotrophy can be extremely important to the nutrition of rabbits. The soft feces contain high levels of SCFA, microbial protein, B vitamins, Na, K, and water (Table 8.5). Cecotrophs are a rich source of nitrogen, providing up to 30 percent of the total nitrogen intake of rabbits, and the microbial protein showed a high content of essential amino acids (Hörnicke 1981). Therefore, the practice of cecotrophy can greatly improve both the quantity and biological value of protein in the diet. The SCFA in these fecal pellets also provides an additional source of energy and the B vitamins provided can be in excess of the animal's needs. For example, B_{12} was synthesized at 100 times the daily requirement (Simnet and Spray 1961). Chilcott and Hume (1985) estimated that cecotrophy recycled energy equivalent to 58 percent of the ringtail possum's intake of digestible energy. It also recycled nitrogen equivalent to 81 percent of their daily intake, which could account for the low maintenance nitrogen requirement of these marsupial folivores.

Evacuation of nitrogen-rich cecotrophs also has been demonstrated in the mountain hare and a number of rodents, as well as the ringtail possum (Table 8.6). Cecotrophs are produced by chinchillas during the day time and by guinea pigs in short periods dispersed throughout the day and night (Holtenius and Björnhag 1985). Some animals that practice cecotrophy show differences in the chemical composition of the two types of feces but less difference in their size, shape, or consistency. Cecotrophs are well formed in the naked mole rat, mountain beaver, ground squirrel, and ringtail possum but are pulpy in the beaver and lepilemur. Pika produce soft feces at random during the day or night (Haga 1960). They consist of an amorphous paste with no mucous envelope, and they are only partly ingested. The remainder are pasted to stones or other surfaces for later ingestion.

Robertson (1982) described widespread coprophagy among herbivorous and detritivorous fish on a Pacific coral reef. However, autocoprophagy and intraspecies coprophagy were rare, and most of the interactions consisted of herbivores eating

Table 8.5. Some components of rabbit feces and cecotrophs[a]

Component	Concentration	Food	Feces Mean	SD	Cecotrophs Mean	SD	Ratio Cecotrophs to feces	Reference
H_2O	%	9.73	51.6	± 3.4	70.1	± 1.8	1.36	Kandatsu, Yoshihara, and Yoshida (1959)
Crude protein	%	14.90	18.15	± 0.57	28.15	± 2.84	1.52	
True protein	%	10.94	6.97	± 0.52	18.55	± 1.83	2.66	
Non-protein nitrogen	%	0.63	0.19	± 0.04	1.60	± 0.20	8.42	
Crude fiber	%	19.7	29.6	± 1.6	17.8	± 2.4	0.60	
SiO_2	%	2.92	4.99	± 0.79	3.75	± 0.65	0.75	
SO_3	1%	0.68	0.50	± 0.07	1.28	± 0.21	2.56	
MgO	%	0.72	0.87	± 0.06	1.28	± 0.12	1.47	
CaO	%	1.19	1.80	± 0.37	1.35	± 0.22	0.75	
Fe_2O_3	%	0.01	0.25	± 0.06	0.26	± 0.06	1.04	
Total P	%	0.83	0.95	± 0.17	1.54	± 0.11	1.62	
Inorganic P	%	0.16	0.60	± 0.19	1.04	± 0.26	1.73	
Organic P	%	0.65	0.35	± 0.06	0.50	± 0.15	1.43	
Cl	mmole/kg DM		33	± 4.5	55	± 4	1.67	Bonnafous (1973)
Na	mmole/kg DM		37.6	± 8.2	104.7	± 21.2	2.78	
K	mmole/kg DM		84	± 11.6	259.6	± 29.5	3.09	
Total SCFA	mmole/kg DM		45		180		4.0	
Acetate	mmole/kg DM		39.7	± 8.8	123	± 3	3.10	
Propionate	mmole/kg DM		2.8	± 1.7	10.7	± 3	3.82	
Butyrate	mmole/kg DM		2.7	± 2.0	44.1	± 5.0	16.33	
pH			7.91	± 0.16	6.38	± 0.10	0.811	
Cholesterol	%		0.71	± 0.13	0.40	± 0.04	0.56	
Bacteria	10^{10} g DM		31	± 1.5	142	± 12	4.58	

[a]Values are given on dry matter (DM) basis; means with standard deviation (SD).
From Hörnicke and Björnhag (1980).

Table 8.6. *The concentrations of nitrogen in cecal contents, cecotrophs, and feces in comparison to the food in twelve animal species with colonic separation mechanisms*

	Nitrogen content (mg g^{-1} DM)				
Species	Food	Cecal contents	Cecotrophs	Feces	References
Rabbit *Oryctolagus cuniculus*	24[a]	42	46	15	1,2
Mountain hare *Lepus timidus*	10[b]	nm	45	11	3
Rat *Rattus norvegicus*	21[a]	33	35	20	2,4
Scandinavian lemming *Lemmus lemmus*	18[c]	52	nm	17	4
Water vole *Arvicola terrestris*	nm[d]	60	63	19	5
Kangaroo rat *Dipodomys microps*	nm[e]	nm	40	27	6
Guinea pig *Cavia porcellus*	19[a]	31	47	17	7
Chinchilla *Chinchilla laniger*	20[a]	27	34	18	7
Nutria *Myocaster coypus*	27[f]	33	34	22	8
Ringtail possum *Pseudocheirus peregrinus*	13[g]	39	46	12	9
Donkey, horse *Equus asinus, E. caballus*	11[h]	12*	—	8	2,10
Turkey *Meleagris gallopavo*	21[i]	55	—	16	11

nm, not measured; DM, dry matter.

Food: *a*, commercial rabbit food; *b*, willow twigs; *c*, mosses and oats; *d*, fresh herbs; *e*, saltbush leaves (*Atriplex confertifolia*); *f*, commercial nutria pellets, oats, beet pulp, hay; *g*, *Eucalyptus* foliae; *h*, straw and oats; *i*, whole oats.

*Nitrogen concentration in the dorsal colon 16 mg g^{-1} DM.

References: 1, Björnhag (1972); 2, G. Björnhag unpublished results; 3, Pehrson (1983a); 4, Sperber et al. (1983); 5, I. Sperber and Y. Ridderstråle personal communication; 6, Kenagy and Hoyt (1980); 7, Holtenius and Björnhag (1985); 8, Snipes et al. (1988); 9, Chilcott and Hume (1985); 10, Sperber et al. (1992); 11, Björnhag and Sperber (1977). (From Björnhag 1994.)

the feces of zooplanktivores and other carnivorous fish. In some species, this occurred to such a degree that fecal material constituted an important component of their diet.

Herbivores

The ability to utilize the leaves, petioles, and stems of plants as a significant source of nutrients expanded the diversity and distribution of vertebrates. Herbivory requires the reduction of large quantities of plant material into particles that are either rapidly passed through the gut for the extraction of nutrients from the cell contents or that are retained for the more lengthy process of microbial fermentation. Although the first strategy appears to have been adopted by some fish and a few mammals, such as the giant panda, microbial fermentation was adopted by most terrestrial herbivores. Microbial fermentation seems to be common to the gut of all vertebrates. However, the quantity of nutrients produced in the digestive tract is dependent on its capacity for the retention of substrates and microorganisms. Assimilation of these nutrients is determined principally by their retention time and the mucosal surface area. Therefore, the difference between most herbivores and other species is a greater capacity for and retention of digesta.

The rarity of herbivorous fish has been attributed to the fact that they are ectotherms with a relatively small gut capacity. A reduction in water temperature decreases the rates of digesta transit and food utilization at all stages of fish growth (Kaushik 1986) and reduces the rates of bacterial metabolism and generation. As noted in Chapter 6, the information available on digesta transit in fish is sparse and confined largely to the time required for the first appearance of food. It shows that the rate of passage is greatly affected by body temperature and can vary considerably among species. Although the herbivores tended to have a longer digestive tract than omnivores and carnivores, the freshwater herbivores, which have been predominantly studied, tended to show a more rapid transit of food (two to ten hours).

Among the herbivorous species, the freshwater grass carp appears to have received the most study. The adult fish can consume its weight in vegetation each day and reduce it to particles 3 mm in size with its pharyngeal teeth (Hickling 1966; Sutton 1977). Yet they have neither a stomach nor pyloric ceca, and their intestinal tract is only two to three times their body length. Although SCFA were found in their intestinal tract (Table 8.1), there is disagreement over their ability to digest cellulose (Cross 1969; Law and Jamalullail 1981; Van Dyke and Sutton 1977), and an increase in dietary fiber content reduced their digestion of dry matter, protein, and fat (Mgbenka and Lovell 1986; Hajra et al. 1987). Straganov (1963) found that the feeding activity of grass carp was intermittent and very selective at water temperatures below 16°C but intensive and inclusive of a broad range of plants at higher temperatures. Transit of food through the digestive tract required eighteen hours at 10°C (Barrington 1957), but this was reduced to seven to eight hours at 27°C (Hickling 1966). Trust et al. (1979) pointed out that although a higher body temperature would increase the rate of bacterial regeneration and metabolism, it also increases the rate of passage through the lumen. Therefore, it would appear that at higher body temperatures and rates of fiber intake, the grass carp obtains most of its nutrients by the rapid passage of large quantities of plant material through the gut rather than from microbial fermentation.

Microbial fermentation may play a more important role in the digestion of plants by marine herbivores that feed on planktonic algae, algal turfs, and detrital drift associated with coral reefs (Choat 1991). Lobel (1981) found that marine herbivores that did not triturate plant material with pharyngeal teeth or a gizzardlike stomach demonstrated a low gastric acidity, which could release the cell contents from the plant material. Horn (1992) concluded that herbivorous fish demonstrated the following four different types of plant digestion: (1) gastric acid digestion in the stomach, (2) mechanical reduction with a gizzardlike stomach, (3) mechanical digestion with a pharyngeal mill, or (4) hindgut fermentation. The first group, which are mainly browsers with a long intestine, includes a few cyclids and several marine species. The second group, with a gizzardlike stomach and long intestine, are mainly grazers, including the muglids and some acanthurians. The third group, those with a pharyngeal mill, no stomach, and a moderately long intestine, are grazers, or browsers and include the acaids, odacids, and hemiramphids. The fourth group was represented by the kyphosids (sea chubs), which are browsers with a hindgut that is separated from the midgut by a valve and includes a pair of small ceca (see Fig. 3.10).

Despite evidence suggesting that most herbivorous fish do not depend on gut microbes for digestion of plant material, the presence of SCFA (Table 8.1) indicates microbial fermentation. Although the concentrations and quantities in the gut of these fish are low, SCFA may provide a significant amount of energy. Much attention has been given to microbial fermentation of cellulose. As pointed out by Clements, Gleeson, and Slaytor (1994), many of the herbivores feed on algae, which contain little cellulose. The gut microbes may be fermenting other carbohydrates such as laminarin. The concentration on carbohydrate fermentation also fails to consider other contributions that these bacteria may be making to the conservation of nitrogen or water or to the digestion of wax esters.

The larval forms of many amphibians are herbivores with complex mouthparts for the trituration or filtering of food. However, little is known about microbial digestion at this stage of development, and all adult amphibians are carnivores, with weak dentition and a limited gut capacity.

Among the over 6,000 species of extant reptiles, only about 50 species of lizards, a few tortoises, and the marine green turtle are herbivores. The paucity of reptilian herbivores has been attributed to the lack of an efficient masticatory apparatus, the inability to maintain a high body temperature, and the lack of sufficient gut capacity in most species. The upper and lower jaws of reptiles are the same width, and their articulation allows only a vertical, scissorlike movement, unsuitable for the efficient mastication of food (Ostrom 1963). A drop in the ambient temperature increased digesta transit time and reduced the rate of cellulose and hemicellulose digestion by the green turtle (Bjorndal 1979; 1987). Herbivorous hatchlings of the green iguana (Troyer 1984a) and freshwater turtle *Pseudemys nelsoni* (Bjorndal and Bolten 1992) digested plant material as well or better than the adults. This was attributed to selective feeding, smaller bite size, and a more rapid rate of digesta transit. However, the high rate of food intake, digesta transit, and digestion seen in hatchlings of the algae-eating marine iguana *Amblyrynchus cristatus* was attributed to their ability to maintain a higher body temperature than the adults (Wikelski, Gall, and Trillmich 1993).

The small body size of most reptiles restricts their relative gut capacity, as it does in small mammals. Pough (1973) found that whereas adult herbivorous lizards weighed more than 300 grams, their young and those species that weigh less than 50–100 grams tend to be carnivores or omnivores. The hindgut of herbivorous reptiles tends to be longer and more voluminous than that of carnivores or omnivores (Skoczylas 1978), and the relationship between hindgut capacity and body weight in the adult green iguana is similar to that of mammalian herbivores (Troyer 1984b). An increase in body size increased the efficiency of fiber digestion by the green turtle (Bjorndal 1979). The fact that body size had no effect on digestive efficiency when food intake was held constant in studies of the gopher tortoise *Gopherus polyphemus* (Bjorndal 1987), indicated that digestive efficiency was dependent on gut capacity rather than on other factors associated with the growth of these animals.

Although birds are endotherms, and the gizzard can provide an efficient masticatory apparatus, there are relatively few species that can subsist largely on the fibrous portion of plants. The restrictions for flight, which would account for the relocation of their masticatory apparatus, may have restricted the distribution and weight of gut contents as well. Most of the herbivorous species have either a limited ability or no ability to fly. The ceca are the principal site of microbial fermentation in most birds. Use of the colon by the ostrich, the small intestine by the emu, and the crop by the hoatzin for this purpose appear to be rare exceptions.

Despite evidence that the earliest mammals were carnivores, herbivorous species are now found in eleven of the twenty mammalian orders, and five of these orders are composed only of herbivores (see Table 4.1). The herbivores are the most abundant in number and species, show the widest ecological distribution, and include the largest species of terrestrial animals. It has been estimated that they constitute over 90 percent of the mammalian biomass (Björnhag 1987). The success of mammalian herbivores is partly due to the ability to maintain a high body temperature (Crompton, Taylor, and Jagger 1978) and, thus, high rates of microbial activity. However, it also is due to an extremely efficient masticatory apparatus and some major adaptations of the gastrointestinal tract.

As mentioned earlier, most mammals have a very efficient apparatus for the mastication of food (Davis 1961; Crompton and Parker 1978). The articulation and muscles of the mammalian jaw provide for either transverse movement of the mandible, as seen in ungulates and most other species (see Fig. 11.4, *A*), or the fore-and-aft movement characteristic of rodents and elephants. Mastication is further aided by large premolars and molars with uneven occluding surfaces, which are especially well developed in herbivores (Janis and Fortelius 1988), and by the use of muscles in their cheeks and mobile tongue to guide food between these grinding surfaces.

The gastrointestinal tract of mammalian herbivores demonstrates a wide range of adaptations for microbial digestion. The giant panda appears to obtain its nutrients by selective feeding and rapid transit of plant material through a relatively simple and short gastrointestinal tract (Dierenfeld et al. 1982), with little help from gut microbes. However, the digestive tract of most mammalian herbivores includes a voluminous cecum, colon, or forestomach that retain this material for prolonged periods.

Parra (1978) showed that the relative volume of gut contents in mammalian herbivores decreased with a decrease in the body size of the species. This, coupled with the fact that the rate of metabolism per unit of body mass increases with a decrease in species size (Chapter 1), places limitations on the extent to which small herbivores can delay digesta transit. Small herbivores appear to have handled this restriction by rapid passage of larger food particles through their gut and the selective retention of small digesta particles, fluid, and bacteria in a voluminous cecum (Chapter 6). The efficiency of this system is enhanced in many of these species by coprophagy and, especially, cecotrophy.

Among the larger mammalian herbivores, adaptations of the digestive tract took two other major directions. In some, such as the equids, tapirs, rhinoceros, elephants, wombats, and large apes, an expanded proximal colon provides the major site for digesta retention and microbial digestion. Digesta are retained with the aid of haustra, plus permanent compartmentalization in perissodactyls and elephants. However, an expanded foregut provides a minor, supplementary site in the hyrax, sirenians, and some rodents, and the principal site for microbial digestion in most of the artiodactyls, the macropod marsupials, some edentates and primates, and the hoatzin. Use of the hindgut as the major site allows for the more efficient extraction of nutrients by endogenous enzymes and absorptive mechanisms of the midgut. However, use of the foregut for this purpose allows microbial degradation of plant toxins and recovery of the amino acids synthesized by these bacteria.

Dietary diseases associated with microbial fermentation

As mentioned earlier, rapid fermentation of starches or sugars can lower the pH of rumen contents and stimulate growth of lactobacilli, production of lactic acid, and a marked reduction in pH. This can result in damage to the forestomach epithelium and systemic dehydration and acidosis. The relatively high levels of lactic acid in the forestomach of the hyrax and some rodents, and the stomach of dogs, pigs, and ponies also could be due to dietary starch and a low pH. Infusion of acetic acid inhibited HCl secretion and increased the permeability of gastric mucosa in the dog (Babkin, Hebb, and Kreuger 1941) and cat (Flemström and Frenning 1968).

Ulceration of the stratified squamous epithelium of the stomach is a serious problem in pigs subjected to intensive husbandry practices. In a postmortem survey of 48,000 pigs from twenty countries, O'Brian (1986) found a 13 percent incidence of ulcers in this region of the stomach, which can result in anemia, anorexia, and weight loss and an estimated two percent rate of mortality. Argenzio and Southworth (1975) recorded acetate plus lactate concentrations of 60 mmoles/l in the proximal (stratified squamous epithelial and cardiac mucosal) regions of the stomach of pigs four hours after a meal containing high levels of ground corn. In vitro studies by Argenzio and Eisemann (in press) showed that a lumen bath containing a 60 mM acetate solution at a pH of 2.5 or less resulted in both structural and functional damage to stratified squamous, cardiac, and proper gastric epithelium. They concluded that ulceration was confined to stratified squamous epithelium under in vivo conditions because of the lack of HCO_3 secretion into a mucous barrier and slower restitution of this tissue following injury.

Although the upper digestive tract tends to dampen the effects of a change in diet, the hindgut also can be affected. High levels of dietary starch were associated with high concentrations of SCFA and lactic acid and atony in the cecum of cattle (Svendsen 1969). The effects of diet on the hindgut of horses were reviewed by Clarke, Argenzio, and Roberts (1990). High-performance horses are often fed a high-concentrate meal once or twice a day for additional energy and convenience in feeding. The equine stomach is relatively small and shows little receptive relaxation for the storage of food (Phaneuf and Ruckebusch 1983), and digesta are passed rapidly through the small intestine. Abrupt change from an all-hay to an all-concentrate diet resulted in a marked increase in the total number of cecal bacteria and the number of lactic acid-producing microorganisms and a marked reduction in pH. Cecal contents of animals fed a high-concentrate diet at 12-hour intervals showed a twenty-fivefold increase in lactate concentration and a one-unit reduction in pH. The pH remained low for five to six hours after feeding, even after prolonged adaptation of animals to the diet. Hindgut digesta are normally maintained at a relatively neutral pH (see Chapter 9). However, the epithelium can be severely damaged or destroyed by SCFA or lactic acid at a pH of 5.5 or less (Argenzio and Meuton 1991). Rapid production of SCFA and gas could account for the high incidence of large intestinal torsion, impaction, and colic in horses fed high-concentrate diets (White 1990).

The early studies of Burkitt (1971) showed a relationship between the low-fiber, high-protein, high-fat diets in affluent Western civilizations and the higher incidence of colorectal cancer and other diseases of the human large intestine. The protective effect of cereal fiber against colorectal cancer has been confirmed by numerous epidemiological and case-control studies. However, the reasons for this beneficial effect are unknown. Insoluble fiber (cellulose, lignin, and some hemicelluloses) can reduce digesta transit time (see Chapter 6) and increase the volume of digesta in the large intestine. The more soluble fiber, which includes pectins, gums, and other hemicelluloses (Van Soest 1994), has a greater viscosity and forms gels that can sequester substances that are potentially carcinogenic. All but lignin can be fermented to SCFA.

Jacobs (1988) concluded that wheat bran and cellulose had a protective effect against colonic cancer in humans, but the more rapidly fermentable carbohydrates of fiber (corn bran, pectin, carrageenan, agar, and metamucil) enhanced the development of chemically induced tumors in laboratory animals. A lack of enhancement in germ-free animals suggested that the end products of fermentation served as stimulatory agents. There also appears to be a relationship between colonic lumen pH, cell proliferation, and carcinogenesis (Jacobs and Lupton 1986). However, other observations from carcinogen studies suggested that the more fermentable carbohydrates are the most protective (Van Soest 1994), and it is becoming increasingly evident that SCFA, particularly butyrate, play important roles in the maintenance and repair of hindgut epithelium.

The low-fiber diets of affluent societies also tend to be high in protein and fat. A series of studies by Visek and coworkers suggests that excessive microbial production of ammonia caused by these diets promotes the development of colonic cancer. Administration of ammonium acetate increased the incidence of N-methyl-N-nitro-N-nitrosoguanidine–induced colonic cancer in rats (Clinton et al. 1988).

Perfusion of the colon with ammonium acetate or ammonium chloride resulted in histological damage and a decreased life span of colonic epithelial cells (Hsi-Chiang Lin and Visek 1991b). An increase in dietary protein increased ammonia concentrations and an increase in dietary fat raised the pH of colonic digesta (Hsi-Chiang Lin and Visek 1991a), suggesting that the damage was due to absorption of the more lipid-soluble NH_3.

Ammonia absorption also has been implicated as a cause of gastric ulcers associated with chronic infections of *Helicobacter pylori* (initially classified as *Campylobacter pyloridis*) in humans. Patients with gastric and duodenal ulcers showed a high incidence of infection with these bacteria (Marshall et al. 1985; Graham 1989). Inoculation with these bacteria resulted in a rapid onset of gastritis (Morris and Nicholson 1987) and hypochlorhydria (Marshall 1987), and their eradication with antibiotics induced the healing of ulcers (Graham 1989; Rauws, Langenberg and Houthoff 1988). Hazell et al. (1987) suggested that this may be due to the toxic effects of ammonia produced by the urease activity of these organisms. This was supported by the finding that prolonged feeding of ammonium acetate to rats resulted in mild gastritis and an increase in both mucosal thickness and antral gastrin levels (Lichtenberger et al. 1995).

There also is evidence that dietary fat may promote colonic carcinogenesis by influencing the amount and type of microbial metabolism of bile acids and cholesterol (Reddy 1983; Hill 1983). Therefore, the advantages of a high-fiber, low-protein, low-fat diet may be due to the effects of fiber on digesta transit, volume, and microbial populations, or to limitations in the substrates available for microbial production of SCFA, ammonia, or other agents toxic to the epithelial cells.

Summary

Bacteria colonize the gastrointestinal tract of all vertebrates. Their numbers and species appear to be determined principally by digesta retention time and pH. The shorter retention time of the stomach and midgut of most species results in lower numbers of bacteria, and the low pH of gastric contents encourages the growth of lactobacilli. The longer retention time and favorable pH of digesta in the hindgut of terrestrial vertebrates, the foregut of a number of herbivores, and the midgut of a few species results in greater numbers of bacteria. These bacteria are accompanied by colonies of protozoa in the foregut and hindgut of some herbivores, and by colonies of fungi in the foregut of a few species.

Microbes indigenous to the gastrointestinal tract provide nutrients required by the animal and protection from invasion by pathogenic organisms. Microbial fermentation of carbohydrates produces substantial levels of SCFA in the hindgut of terrestrial vertebrates, the midgut of the emu and Florida red-bellied turtle, the crop of the hoatzin, and the forestomach of a number of mammalian herbivores. The SCFA in the hindgut and forestomach of most mammals have similar acetate/propionate/butyrate ratios. However, acetate proportions are particularly high in the forestomach of the hyrax and some rodents and in the midgut of most species examined.

Absorption of SCFA provides energy for the gut epithelium and other tissues of the body. Gut microbes convert nitrogenous compounds into ammonia and microbial protein and synthesize B-complex vitamins. The absorption of ammonia conserves the nitrogen secreted into the digestive tract as urea and endogenous enzymes, and absorption of the vitamins can satisfy some of the requirements for these nutrients. The greater capacity and digesta retention time in the hindgut, midgut, or foregut of herbivores allows the additional fermentation of structural carbohydrates and provides SCFA in quantities that can satisfy a significant fraction of their energy requirements. The contribution of gut microbes to the nutrition of herbivorous fish is uncertain. This would be limited by their gut capacity, the inhibition of digesta transit and microbial metabolism at low ambient temperatures, and by rapid digesta transit at high ambient temperatures. However, because of their lower rate of metabolism, the low levels of SCFA found in the intestine of fish may provide a significant source of energy, and the contribution of gut microbes to nitrogen conservation and other functions in fish needs to be investigated.

The small number of herbivorous reptiles can be attributed to ectothermy, a relatively inefficient masticatory apparatus and limited gut capacity of present-day species. Explanations given for the high percentage of herbivorous dinosaurs are discussed in Chapter 11. The relatively low number of herbivorous birds can be attributed to the restrictions of flight on the capacity and distribution of gut contents.

The success of mammalian herbivores can be attributed to endothermy, an efficient masticatory apparatus, and an expansion in gut capacity. The cecum is the principal site of microbial fermentation in most small herbivores. This allows recovery of nutrients from food digested by endogenous enzymes prior to its fermentation by bacteria. Selective retention of bacteria and small particles of plant material in the cecum and rapid passage of larger particles through the colon provides an efficient system for animals with a relatively low gut capacity and high rate of metabolism. Coprophagy, particularly cecotrophy, provides many small herbivorous mammals with the additional advantage of cycling microbial protein and B-vitamins through the upper digestive tract.

A larger body size and lower metabolic rate allows for a greater gut capacity and retention time in larger animals. The principal sites of microbial fermentation in large mammalian herbivores are either the proximal colon or foregut. Microbial fermentation in the proximal colon requires retention of water and eliminates cecotrophy. However, colon-fermenting herbivores can increase the quantity of nutrients extracted from high-fiber, low-protein diets by an increase in food intake and rate of digesta passage. Microbial fermentation in the foregut tends to limit the intake and transit of food, but it allows the recovery of microbially synthesized protein and vitamins and the microbial degradation of plant toxins. For reasons given in Chapter 9, it can also reduce the quantities of Na and water that need to be absorbed by the distal colon.

The advantages of microbial fermentation in the forestomach of cetaceans are difficult to understand. It may aid in the digestion of chitin or wax esters by some baleen whales, but it appears to provide little direct nutritional advantage to either these animals or other species that feed on fish or marine mammals. This may be an arrangement that was conserved from their terrestrial ancestors, which are believed to have been closely related to the ungulates (Romer 1966). The advantages of microbial digestion in the crop and distal esophagus of the hoatzin or the midgut of the emu and Florida red-bellied turtle are equally difficult to understand. Use of the crop and distal esophagus places the fermentation chamber ahead of the masticatory apparatus. Use of the midgut would seem to result in bacterial competition with and digestion of endogenous enzymes.

Many of the gastrointestinal diseases of domesticated, cultured, or captive animals can be attributed to feeding them the wrong diet. The digestive tract of most herbivores is constructed for almost continuous feeding on a diet high in plant fiber and relatively low in protein, starches, and sugars. Therefore, major problems can come from the intermittent feeding of a diet that contains low levels of fiber and high levels of rapidly fermentable starch or sugar. The hindgut of many omnivores also appears to require a minimal amount of plant fiber for normal function, as evidenced by the higher incidence of cancer and other diseases in the colon of humans on low-fiber diets. Problems that can arise from the relationships between microbial fermentation and the absorption of electrolytes and water are discussed in Chapter 9.

9

Secretion and absorption of electrolytes and water

Electrolytes play a major role in the absorption, distribution, secretion, and excretion of water, and their distribution determines the ionic composition and pH of the body fluid compartments. Large quantities of electrolytes and water are secreted into the digestive tract to aid in the digestion and transit of food. Most of this must be reabsorbed for maintenance of the electrolyte–water and acid–base balance of body fluids. Therefore, secretion and absorption of electrolytes is one of the most important functions of the digestive system, and it consumes a considerable amount of the energy required by animals.

The earliest vertebrates are believed to have been freshwater fish, whose major problem is the elimination of excess water. Freshwater fish do not drink water (Smith, Farinacci, and Breitweiser 1930). Excess water is excreted by the kidneys following glomerular filtration of plasma and resorption of ions and other required substances from the renal tubules. Marine fish have the opposite problem. The osmolality and NaCl content of sea water is three to four times that found in the body fluids of most vertebrates (Fig. 9.1). This is resolved in the hagfish, sharks, rays, coelacanth, and marine frog *Rana cancrivora* by body fluids that are either isotonic or hypertonic to sea water (Bentley 1982). This is principally due to the retention of urea (Fig. 9.2). Marine elasmobranchs have serum urea levels that range from 209 to 453 mM, compared with 81 to 180 mM for euryhaline species and 1 to 2 mM for freshwater river rays (Griffiths et al. 1973). However, the body fluids of marine teleosts have an osmolality only slightly higher than that of freshwater species.

Although marine teleosts drink sea water, the osmolality and NaCl levels of their gastric contents are only slightly above that of their plasma (Smith et al. 1930). Studies of euryhaline fish such as eels and flounder, which can adapt to a wide range of salinities, showed that the movement from freshwater to sea water was accompanied by replacement of the stratified epithelium in their esophagus

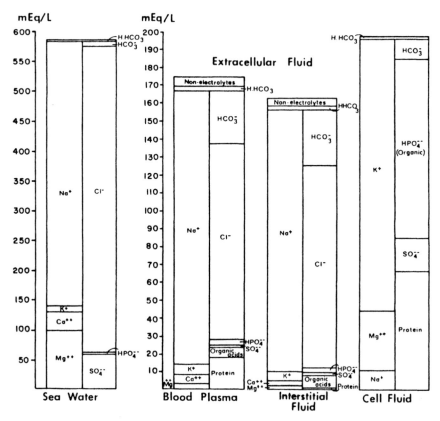

Fig. 9.1. Electrolyte composition of sea water and human body fluid compartments. (From Gamble 1954.)

with a mitochondria-rich columnar epithelium (Yamamoto and Hirano 1978; Humbert, Kirsch, and Meister 1984). The columnar epithelium absorbed Na and Cl, reducing NaCl levels and osmotic activity of gastric contents (Hirano and Mayer-Gostan 1976; Kirsch 1978; Parmelee and Renfro 1983). The glomerulus is absent from the kidney of many species and functions intermittently in others, which reduces the volume of water excreted in the urine. Excess Na and Cl are actively secreted by the gills and salt glands in the rectum of some cartilaginous fish and marine catfish (van Lennep and Lanzing 1967). Salt glands are found in many species of marine reptiles and birds as well (van Lennep and Young 1979).

Terrestrial animals need to conserve both electrolytes and water. Adult amphibians do not drink water, but they absorb Na, Cl, and water via their skin, kidney, urinary bladder, and hindgut. The kidneys of amphibians, reptiles, and most birds are limited in their ability to concentrate urine, as compared with most ter-

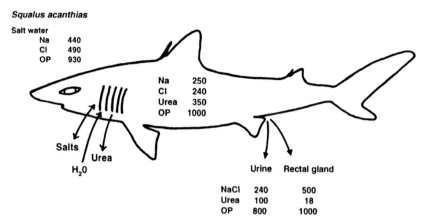

Fig. 9.2. Summary of osmotic regulation in a typical elasmobranch. Values for NaCl, Na, Cl, and urea are given in mM or mEq/L. Osmotic pressures (*OP*) of sea water, body fluids, and urine are given in mOsm/L. (Modified from Kormanik 1992.)

restrial mammals. However, as mentioned in Chapter 5, reflux of urine from the cloaca of reptiles and birds allows hindgut resorption of electrolytes and water from both their urinary excretions and the secretions of their digestive systems (Fig. 9.3). Minnich (1970) concluded that much of the urinary sodium and water excreted by the kidneys of the desert iguana were reabsorbed by the cloaca. Junqueira, Malnic, and Monge (1966) concluded that the colon also was responsible for absorption of urinary ions and water in snakes (*Xenodon, Philodryas,* and *Crotalus* sp.) The urinary and digestive systems exit separately from the body of most mammals. However, the loop of Henle in the mammalian nephron provides for the more efficient concentration of urine, and an increase in the relative length of the colon in most mammals facilitates the resorption of electrolytes and water secreted into the digestive tract.

Electrolyte composition of body fluid compartments

The electrolyte composition and osmolality of the cellular and extracellular (interstitial and blood plasma) compartments in mammals are illustrated in Fig. 9.1. The major cation in cells is K, with much lower concentrations of Mg and Na. The principal anions are HPO_4, SO_4, HCO_3 and proteins. Extracellular fluid has a much different ionic composition, with Na as the principal cation, and Cl and HCO_3 the major anions. Despite these differences, each compartment maintains an equal concentration of cations and anions, and its osmolality is approximately equivalent to the sum of the osmotic activity of these electrolytes. The digestive tract provides an additional body compartment that can be very large in herbi-

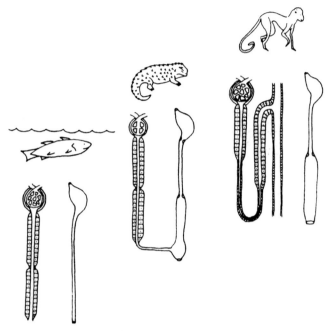

Fig. 9.3. Development of the nephron and hindgut in relation to habitat. The nephrons of fish, amphibian, reptile, and bird kidneys are limited in their ability to concentrate urine. However, urine is excreted into the cloaca of adult amphibians, reptiles, and birds, and recovery of urinary electrolytes and water is aided by the cloaca and hindgut of reptiles and birds. Microbial digestion of uric acid also aids in the conservation of nitrogen. A few mammals retain the cloaca, but the majority excrete their digesta and urine separately. Recovery of urinary electrolytes and water is aided by the kidney's loop of Henle, and nitrogen conservation is aided by microbial production of ammonia. (Modified from Smith 1943 by Stevens 1977.)

vores. Table 9.1 gives the distribution of body fluids among the intracellular, extracellular, and transcellular compartments of humans and sheep. Although the relative distribution of fluids between intracellular and extracellular space is similar, the gut content of sheep is almost eight times that of humans per 100 kg body weight. The principal electrolytes found in the gastrointestinal tract of most vertebrates are Na, K, H, NH_4, Cl, HCO_3 and the anions of SCFA, but the composition and osmotic activity varies with diet and the segment of tract.

Fig. 9.4 shows the osmolality and major electrolytes in the gastrointestinal contents of a pony. Gastric contents had a higher osmotic activity than body fluids because of a release of electrolytes by dissolution and digestion of food. The principal cation was Na, derived from the diet and salivary secretions. The K in gas-

Table 9.1. *Body fluid compartments*

	Volume	
	L/100 kg body weight	
Compartment	Human	Sheep
Intracellular	50	31.1
Extracellular	20	15.5
Plasma	(5)	(4.9)
Interstitial	(15)	(10.7)
Transcellular	2.8	22.2
Alimentary tract	(1.4)	(20)
Other[a]	(1.4)	(2.2)
TOTAL	72.8	68.8

[a]Cerebrospinal fluid, synovial fluid, aqueous humor, and urine.

tric contents was derived largely from the diet, and ammonia was probably derived principally from microbial hydrolysis of endogenous urea. The concentrtion of H (not shown) secreted by oxyntic cells of the stomach is generally 1 mEq/L or less. The osmotic activity of digesta in the small intestine was reduced to levels equivalent to those of plasma, because of the absorption of nutrients and the secretion of water. Concentrations of Na remained high throughout most of the intestinal tract because of its addition in the pancreatic and biliary secretions and the absorption of Na and water at a relatively constant ratio. The more distal segments of the colon showed a reduction in the level of Na and a reciprocal increase in K levels because of more efficient absorption of Na and both the release and secretion of K. Ammonia levels decreased along the small intestine and then increased in the proximal colon.

The principal anion in the stomach was Cl, derived from food and the secretion of HCl. The concentration of Cl was reduced as digesta passed through the small intestine, with a reciprocal increase in HCO_3, which is the major anion in pancreatic, biliary, and ileal secretions. Although HCO_3 is the predominant anion found in the ileum of the horse, pig, and human (Alexander 1962; Fordtran 1973), PO_4 was the principal anion in the ileum of the dog, cat, rabbit, and guinea pig (Alexander 1965). The marked reduction in HCO_3 levels in the pony cecum and colon was associated with a reciprocal increase in SCFA produced and in PO_4 released by microbial digestion.

Many functions of the digestive system are dependent on the pH of gastrointestinal contents. Figs. 9.5–9.7 show pH and organic acid levels measured along

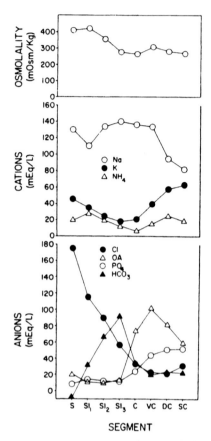

Fig. 9.4. Digesta osmolality and ion concentrations along the gastrointestinal tract of the pony. Data represent mean values obtained from four measurements taken over a twelve-hour period between meals in ponies fed a conventional, pelleted diet. Segments represent the stomach (*S*), three equal segments of the small intestine (*SI*), the cecum (*C*), and the ventral colon (*VC*), dorsal colon (*DC*), and small colon (*SC*). Concentrations of PO_4 were calculated on the basis of a pKa of 6.8 for NaH_2PO_4 and the mean pH of digesta in each segment. At the pH of large intestinal contents, ammonia and the organic acids (*OA*), which consist of SCFA and lactic acid, exist principally in their ionized form. Concentrations of HCO_3 were calculated as the difference in concentration of measured cations and anions. (Modified from Argenzio 1975.)

the gastrointestinal tracts of the dog, pig, and pony two, four, eight and twelve hours after a meal. The pH in the proximal and distal halves of the dog and pony stomachs and the proximal half of the pig stomach underwent marked cyclic changes with time after a meal. These were accompanied by cyclic changes in the concentrations of SCFA and lactic acid, but the highest concentrations of these organic acids tended to correlate with the periods of highest pH. Alexander and

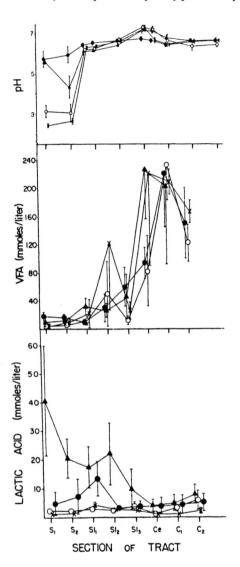

Fig. 9.5. Mean (±SE) values for pH, VFA (SCFA), and lactic acid concentrations in segments of the gastrointestinal tract of dogs fed meat at 12-hour intervals. Symbols correspond to the following four time periods after feeding: 0 or twelve hours (*closed circle*); two hours (*closed triangle*), four hours (*open circle*), eight hours (*x*). Segments of tract are cranial (S_1) and caudal (S_2) halves of the stomach; proximal (SI_1), middle (SI_2), and distal (SI_3) thirds of the small intestine; cecum (*Ce*); and proximal (C_1) and distal (C_2) colon. (From Banta et al. 1979.)

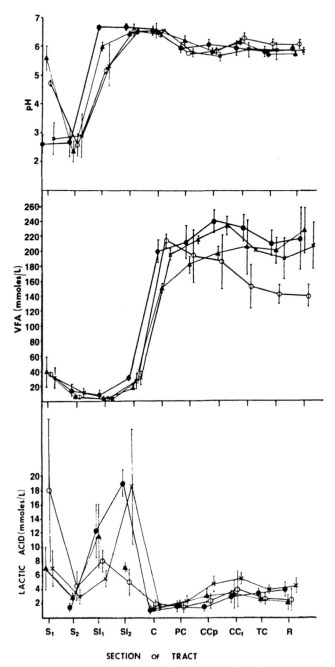

Fig. 9.6. Mean (±SE) values for pH, VFA (SCFA) concentration and lactic acid concentrations in segments of the gastrointestinal tract of pigs fed a conventional, high concentrate diet at 12-hour intervals. Symbols correspond to those in Fig. 9.5. Segments of tract are cranial (S_1) and caudal (S_2) halves of the stomach; proximal (SI_1) and distal (SI_2) small intestine; cecum (C); proximal colon (PC), centripetal colon (CCp), centrifugal colon (CCf), and terminal colon (TC); and rectum (R). (From Argenzio and Southworth 1974.)

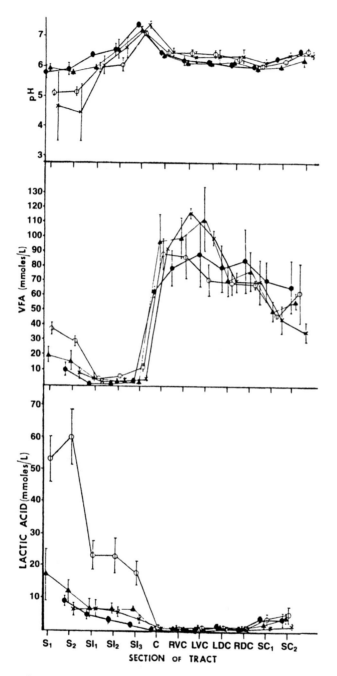

Fig. 9.7. Mean (±SE) values for pH, VFA (SCFA) concentration, and lactic acid concentration in various segments of the gastrointestinal tract of ponies fed a conventional, pelleted horse diet at 12-hour intervals. Symbols correspond to those in the previous two figures. Segments of tract are cranial (S_1) and caudal (S_2) halves of the stomach; proximal (SI_1), middle (SI_2), and distal (SI_3) thirds of the small intestine; cecum (C); right ventral colon (RVC) and left ventral colon (LVC); left dorsal colon (LDC) and right dorsal colon (RDC); and proximal (SC_1) and distal (SC_2) small colon. (From Argenzio et al. 1974.)

Davies (1963) noted a similar inverse relationship between the concentrations of lactic acid and HCl in the stomach contents of pigs. Therefore, the cyclic changes in the pH of gastric contents in the dog and pony can be attributed principally to an increase in the rate of HCl secretion with feeding. The pH in the cranial segment of the pig stomach was lowest prior to the meal, suggesting a more continuous secretion of HCl. It then increased following feeding because of the secretion of HCO_3 by cardiac mucosa. Surgical removal of the proper gastric glandular mucosa from the stomach of pigs resulted in significantly higher concentrations of SCFA and lactic acid (Borch-Madsen 1946).

The pH of intestinal contents also varied with time after feeding. The pH in the small intestine varied from 6.0 to 7.4 in the dog and pony, and from 5.3 to 6.8 in the pig. It was highest just prior to feeding in the proximal segment of the small intestine and decreased two hours after the meal because of the entrance of gastric chyme. The ileal contents of the dog and pony demonstrated a pH significantly higher than that of any other segment of the gastrointestinal tract because of the secretion of HCO_3. The pH of large intestinal contents varied from 6.0 to 7.3 in the dog, 5.5 to 6.5 in the pig, and 6.0 to 6.5 in the pony. This could be due to changes in the rate of HCO_3 appearance in the ileal effluent, HCO_3 secretion by large intestinal mucosa, and the microbial production of SCFA.

The pH profile of digesta along the gastrointestinal tract of the nurse shark *Ginglymostoma cirratum*, was similar to that of the dog and pony (Caira 1989). The proximal portion of the stomach and the intestinal tract of sharks with an empty stomach had an average pH between 6.8 and 7.2. The mean pH of the distal portion of the stomach was higher than this, approximately 7.8. Sharks with food in their stomachs showed a much lower pH (2.0 to 3.4) in both stomach segments and a slightly lower pH in the cranial segments of the intestine.

Electrolyte transport mechanisms

Mechanisms of transport

Electrolytes can be transported across gut epithelium by passive diffusion through paracellular channels or by carrier-mediated transport across the membranes of epithelial cells. The earliest model of ion transport across epithelial cells was proposed by Koefed-Johnson and Ussing (1958) to explain the absorption of Na by frog skin. Frog skin was placed between two chambers containing identical solutions of frog Ringers. The transepithelial electrical potential of approximately 50 mV (blood side positive), which is normally present, was short-circuited by the application of an external current. An isotope of Na was then added to one or the

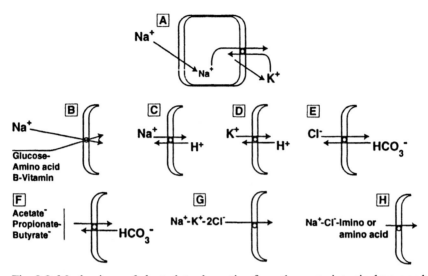

Fig. 9.8. Mechanisms of electrolyte absorption from the gastrointestinal tract and the salivary, pancreatic, and hepatic ducts. Model A was originally proposed for electrogenic transport of Na across the frog skin. Except for the number of Na and K ions exchanged by the basolateral membrane, it remains the accepted model for electrogenic Na transport by many epithelial cells. The remainder of the figure shows other mechanisms proposed for electrolyte transport across the lumen-facing or apical membrane of epithelial cells. Cotransport of Na with other nutrients (**B**) has been described for the midgut of all vertebrates, the proximal cecum of birds, and the colon of chickens on high-sodium diets. Na–H (**C**) and Cl–HCO$_3$ (**E**) exchange have been described in a wide range of gastrointestinal and glandular duct epithelia. Exchange of K for H (**D**) has been demonstrated in gastric oxyntic cells and distal colonic epithelium, and SCFA anion–HCO$_3$ exchange (**F**) has been proposed for epithelium of the midgut of fish and mammalian hindgut. Cotransport of Na-K-2 Cl (**G**) has been described in the fish intestine and in the gallbladder epithelium of some mammals. Cotransport of Na, Cl, and imino or amino acids (**H**) has been described in the intestine of fish and ileum of mammals. With the exceptions of electrogenic Na transport alone (**A**) or in cotransport with other organic nutrients (**B**), each of these mechanisms has been described in the basolateral membrane of some epithelial cells, as well.

other bathing solution for the measurement of Na flux in each direction. Results showed a net transport of Na from the outside to the inside of the skin in the absence of an electrochemical gradient. Furthermore, the electrical current generated by Na transport accounted for the electrical potential in the nonshorted tissue. They proposed that Na was absorbed into the epithelial cell by passive diffusion down an electrochemical gradient, which was maintained by its active transport across the basolateral membranes of the cell in exchange for K (Fig. 9.8,

A). Permeability of the basolateral membrane to K allowed its rapid back-diffusion out of the cell, resulting in the net transport of Na and its positive charge. The transepithelial electrical potential generated by Na transport favored the passive diffusion of Cl, and the osmotic gradient produced by the absorption of these ions resulted in the passive diffusion of water.

The frog skin model has been expanded, but it remains as the basic model for one type of Na transport, and the Ussing chamber technique has provided much of our understanding of ion transport across epithelial cells. Techniques more recently developed for the preparation of vesicles formed from the disrupted lumen-facing (apical) and serosal-facing (basolateral) membranes of cells, have provided another approach to these studies. Carrier-mediated exchange of Na for K is seen in the basolateral membranes of all epithelial cells. This ion pump consists of membrane Na/K ATPase, which may be fueled by lipids, glucose, glutamine, ketone bodies, or SCFA. The RQ values for rabbit and dog intestine suggested that ketone bodies were a major fuel during interdigestive periods (Frizzell, Markscheid-Kaspi, and Schultz 1974; Lester and Grim 1975) and that SCFA provided a major energy source in the colon of rabbits and humans.

Transport of electrolytes across the epithelium of the digestive system proved to be much more complicated, involving more electrolytes that can be transported in opposite directions by different tissues. Although electrolytes can diffuse by paracellular passage through some types of epithelial tissue, they are transported mainly across the membranes of epithelial cells. Sodium, K, and Cl can passively diffuse through ion-selective channels in the membranes of some cells. This is one of the means by which Na is transported electrogenically (i.e. with a net movement of charge) across some epithelial cells. However, most of the transport of electrolytes is mediated by membrane carriers. The major mechanisms for carrier-mediated transport across the lumen-facing membrane are shown in Fig. 9.8. Cotransport of Na with other nutrients, such as glucose or amino acids, across the apical membrane results in its electrogenic transport across midgut epithelial cells. The molecular physiology of sodium-glucose cotransport was reviewed by Hedigar and Rhoads (1994). Electrolytes also are transported by nonelectrogenic mechanisms. Mechanisms for nonelectrogenic Na/H, K/H, and Cl/HCO_3 exchange have been described in a number of different types of epithelial cells, and the exchange of SCFA anions for HCO_3 has been proposed for hindgut and rumen epithelium. Nonelectrogenic cotransport of Na-K-2Cl and cotransport of Na, Cl, and imino or amino acids has been demonstrated in the midgut epithelium of fish. Many of these mechanisms for carrier-mediated transport of electrolytes are found in basolateral membranes as well.

The mechanisms responsible for HCO_3, SCFA, and ammonia transport have proved most difficult to study. The Na/H and Cl/HCO$_3$ exchange mechanisms are coupled in some cells by the intracellular hydration of CO$_2$ to carbonic acid, which dissociates into H and HCO$_3$ (see Fig. 9.15). The HCO$_3$ found in the lumen also may result from its transport from blood, production in the lumen, or the absorption of H or secretion of OH. Disappearance of HCO$_3$ from the lumen may be due to direct absorption or absorption as CO$_2$ and water, following its titration by H. Studies of SCFA and ammonia absorption are complicated by their presence in both un-ionized and ionic forms and, in the case of SCFA, by their metabolism by epithelial cells. Water is transported by passive diffusion and principally by the paracellular pathway through the junction between cells. Its direction of transport is determined by osmotic gradients, which largely depend on the transepithelial distribution and transport of electrolytes. Electrolytes and water are absorbed principally by epithelial cells on the intestinal villi and are secreted by cells in the crypts.

Salivary secretion

Variations in salivary secretion among different species are discussed in a number of reviews (Burgen 1961, 1967; Ellison 1967; Leeson 1967; Phillipson 1970; Cook et al. 1994). The composition and volume of salivary secretions varies among species, between different pairs of glands in a given species, and with the rate of secretion. Parotid glands usually secrete a serous fluid. Secretions of the submaxillary (submandibular) and sublingual glands tend to contain large amounts of mucus. The parotid gland is relatively large in ungulates, kangaroos, beavers, manatees and fruit-eating bats, but it is approximately the same size as the submaxillary gland in the rat and smaller than the submaxillary gland in insectivorous bats. The submaxillary gland of the giant anteater is extremely large and is provided with a storage bladder. Schneyer and Schneyer (1967) compared the rates of intensely stimulated flow per gram of salivary tissue in humans, rats, dogs, cats, and sheep. The sublingual glands of these species secreted at a similar rate per unit of gland weight, as did the parotid and submaxillary glands of humans and rats. However, the submaxillary glands of the dog, cat, and sheep secreted at rates 7.5, 4.5, and 1.5 times greater than that of humans, respectively, and the parotids of the dog and sheep secreted at rates seven and four times the rate of humans. Therefore, the marked species differences in volume flow of saliva are due to differences in secretory rate per gram of tissue, as well as to variations in gland size.

Fig. 9.9 shows the organization of the submaxillary salivary gland of the rat. The acinus contains secretory cells, yielding mucous, serous, or mixed secretions. Myoepithelial cells are often closely associated with the cells of the acini and are presumed to aid in the release of their contents. In some species, the secretory

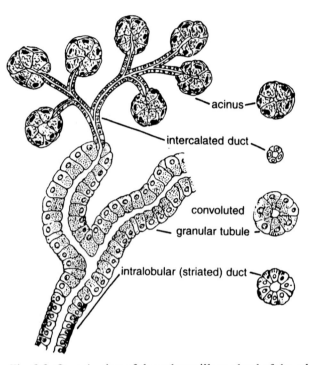

Fig. 9.9. Organization of the submaxillary gland of the adult male rat. The granular tubule is interposed between the intercalated duct and intralobular (striated) duct. (From Leeson 1967.)

portion of the gland is in the form of acinotubules, which are said to be tubular or "alveolar" in ruminants. The secretory portion of the gland connects to the intercalated duct, which is lined with small cuboidal cells containing few secretory granules. Myoepithelial cells are present in some species, and the length, or even the presence, of this section of duct varies. It is relatively long in ruminants.

The intercalated duct is confluent with a duct that is lined with columnar cells, which is called the *striated duct* because of the striations in its basal portion. This section of duct is rich in mitochondria and is believed to be very active in the exchange of ions between blood and the lumen. Its degree of development varies considerably among the glands in a given species, but it is extremely well developed in monotremes, marsupials, and ruminants. The last section of duct, called the *extratubular*, or *excretory*, *duct*, is lined with a double-layered, or stratified, squamous epithelium. The ducts of salivary glands are richly supplied with blood. Burgen and Seeman (1958) concluded that the capillary flow runs countercurrent to the flow of duct saliva, providing an acinoductal portal system, which may act to recycle electrolytes that are absorbed from the duct contents back to the acini.

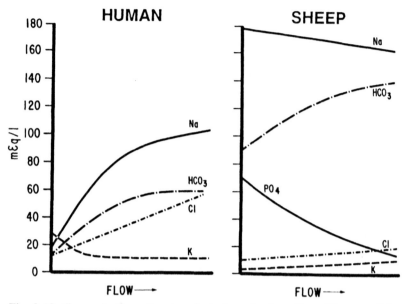

Fig. 9.10. Concentration of major electrolytes in the saliva of humans (Thaysen, Thorn, and Schwartz 1954) and sheep (Argenzio 1984) as the salivary flow rate is increased.

The major inorganic ions in the saliva of mammals are Na, K, Cl, and HCO_3, although PO_4 concentrations are high in ruminants and kangaroos. All of these ions seem to be derived from the plasma with the exception of HCO_3, which appears to be largely supplied by secretory cells. Saliva from the sheep parotid and cat sublingual glands is isotonic at all flow rates. The parotid and submaxillary saliva of humans, dogs, and cats, as well as the parotid saliva of the rat and red kangaroo (Beal 1984), are hypotonic at low rates of secretion. However, at higher rates of flow, they approach isotonicity, suggesting that an originally isotonic secretion is rendered hypotonic if allowed sufficient time for ion resorption by the tubules. The ionic composition of saliva varies with both species and the rate of salivary flow. At high rates of flow, the parotid and submaxillary secretions released into the mouth of humans, dogs, cats, and rats have an osmolality and Na/K ratio similar to blood plasma, but a higher ratio of HCO_3 to Cl and higher pH (Fig. 9.10). At slower rates of flow, there is a reduction in the osmolality, Na/K ratio and the concentrations of Cl and HCO_3. The pH of parotid saliva also is reduced in dogs and humans. This type of secretion does not necessarily hold for other salivary glands in these species. For example the sublingual gland of the dog and cat secretes Cl in concentrations higher than those of plasma, and a pH of 8.9 has been recorded for mixed saliva of the rat.

As previously mentioned, the parotid glands of herbivores tend to be considerably larger than those of other species. It is estimated that sheep secrete 8–16 liters and cattle 98–190 liters of saliva each day (Phillipson 1970). The parotid gland accounts for a major portion of this and its secretory mechanisms have been extensively examined in sheep (Blair-West et al. 1965; Blair-West et al. 1970). Saliva was spontaneously secreted in the absence of stimulation or innervation. At the lowest rate of flow, the parotid saliva was isotonic, with a high Na/K ratio, high concentrations of HCO_3 and PO_4, a high pH, and relatively low levels of Cl (Fig. 9.10). Stimulation of flow rate increased the Na/K ratio and the concentration of HCO_3, with only a slight increase in Cl levels, but neither the osmolality nor pH appeared to change. The salivary glands of camelids (Engelhardt and Höller 1982) and kangaroos (Beal 1984, 1986) seem to function in a similar manner. This pattern of salivary secretion provides a continuous flow of saliva to buffer the large quantities of SCFA produced in the forestomach. However, it results in the loss of large quantities of electrolytes and water into the digestive tract. Part of the Na in the initial salivary secretions was absorbed by the ducts and replaced by K (Blair-West et al. 1970). The degree of replacement correlated with the degree of Na deficiency and was stimulated by administration of aldosterone.

The rate of salivary secretion is increased by the release of acetylcholine and decreased by the release of norepinephrine by the nerves that innervate the gland. It was once believed that the acinar cell secreted a fluid similar to that of plasma and that this was adjusted in the salivary ducts of most species. However, recent studies show a much more complex arrangement (Cook et al. 1994). Acinar cells secrete Na and K at levels similar to those in the plasma, but they actively secrete Cl and HCO_3 in proportions that can be chloride-rich, as in the rat and rabbit mandibular gland, or bicarbonate-rich, as seen in the parotid gland of sheep, goats, kangaroos, and some primates. The salivary ducts of all species absorb Na and secrete K, but the rate of Na absorption can be high (rat mandibular gland), low (rat parotid gland), or absent unless stimulated by aldosterone (sheep parotid). The rate of K absorption can also vary with species. The excretion patterns of HCO_3 suggest that it is secreted by the ducts of some glands, absorbed by the ducts of others, and either secreted or absorbed, depending on autonomic nerve activity, by the ducts in yet other salivary glands.

Pancreatic secretion

As with saliva, the electrolyte composition of the pancreatic fluid released into the intestine varies with species and, in most species, with the rate of flow. Intermittent feeders such as humans, dogs, and cats secrete chiefly during the digestive phase following a meal. Daily secretions of up to 2 liters have been reported in humans (Davenport 1982). Continuous feeders such as the sheep and horse secrete pancreatic fluid at a relatively continuous rate. The daily secretions per 100 kg body mass have been reported to range between 0.6 and 1.1 liters in sheep, 3 and 5 liters in cattle and up to 12 liters in ponies (see Table 9.3). The concentrations of Na and K in pancreatic fluid are similar to those found in plasma, and the levels of Cl and HCO_3 in equine pancreatic fluid remained relatively constant at all rates of flow. However, in most species the concentration of HCO_3 and the pH are high at the most rapid rates of pancreatic flow and low at slow rates of flow.

The mechanisms responsible for pancreatic secretion have been recently reviewed by Yule and Williams (1994) and Agent and Case (1994). The exocrine pancreas consists of acinar cells, which can constitute more than 80 percent of the pancreatic volume and ducts that comprise 2 percent of the volume in rats and 14 percent in humans. Acinar cells function principally for the synthesis of digestive enzymes. Their secretion is stimulated by acetylcholine, cholecystokinin, secretin, and a number of other hormones (see Chapter 10). The duct cells secrete Na, by paracellular diffusion, and actively secrete HCO_3. They take up CO_2 from the blood and hydrate it with the aid of carbonic anhydrase to H_2CO_3, which dissociates to H and HCO_3. The H is returned to the blood, probably by Na/H exchange, and HCO_3 is secreted into the lumen in exchange for Cl. The rate of HCO_3 secretion depends on the availability of luminal Cl. This, in turn, depends on the opening of a Cl channel in the apical membrane, which is stimulated by the hormone secretin.

Biliary secretion

The electrolytes and water in bile are derived from the enterohepatic canaliculi and bile ducts and are adjusted in some species during storage in the gall bladder (Hofmann 1994). Biliary secretions have a Na/K ratio similar to plasma, and a high HCO_3/Cl ratio and pH at rapid rates of biliary flow. In many species, the HCO_3 and pH levels decrease with a reduction in flow, in a manner similar to that noted with salivary and pancreatic secretions. As in the pancreas, HCO_3 is believed to be secreted by the hepatic duct in exchange for Cl, with its rate of secretion controlled by the rate of Cl secretion into the lumen. Bile is stored in the gallbladder of most species for release during the digestive phase following a meal. The gallbladder of some species can concentrate bile twentyfold to thirtyfold and reduce its pH during interdigestive periods by the absorption of electrolytes and water and the secretion of H. This appears to be accomplished by Na–H exchange and cotransport of Na, K, and Cl by the apical membrane (Reuss 1991). However, little absorptive ability is apparent in other species such as the pig and ruminants, and bile is continuously secreted into the duodenum of the horse and other species that lack a gallbladder.

Gastric secretion and absorption

The four different regions of epithelium that may be found in the stomach of vertebrates are described in Chapter 2. The three glandular regions are principally secretory and, with the exception of a few electrolytes, relatively impermeable to the absorption of water-soluble substances. The secretory and absorptive characteristics of stratified squamous epithelium appear to vary among species.

Proper gastric glandular mucosa

The proper gastric region contains glands that secrete HCl and pepsinogen. One cell is responsible for secretion of both the acid and the enzyme by the stomach of fish (Materazzi and Menghi 1975; Bondi et al. 1982). The same is said to be true for adult amphibians, reptiles, and birds, although other studies suggest that HCl and pepsinogen can be secreted by separate areas of the amphibian esophagus and stomach. The proper gastric glands of mammals contain both parietal cells, which secrete HCl, and neck chief cells, which secrete pepsinogen.

The apical membrane of parietal cells transports H and Cl into the lumen of the stomach (Fig. 9.11). Transport of Cl into the stomach is electrogenic, but H is transported in nonelectrogenic exchange for K in the lumen. Intracellular hydration of CO_2, catalyzed by carbonic anhydrase, results in equal quantities of H for secretion into the lumen and HCO_3 for release into the blood of mammals. The dual-purpose oxynticopeptic cells of amphibians and birds contained little or no carbonic anhydrase, whereas the surface mucosal cells of frogs, birds, and mammals contained relatively high levels of this enzyme (Gay, Schraer, and Shanabrook 1981). This suggested that the surface cells may be the source of the acid secreted by lower vertebrates. However, surface epithelial cells in the proper gastric mucosa of the frog, guinea pig, and dog have been shown to secrete HCO_3 (Flemström 1977; Garner and Flemström 1978; Kauffman, Reeve, and Grossman 1980), apparently in exchange for Cl absorption (Flemström, Heylings, and Garner 1982). Secretion of HCO_3 by surface epithelium is believed to establish a microlayer of neutral pH, which protects against the back-diffusion of H into these cells. Both the glandular and surface mucosa are relatively impermeable to other water-soluble substances.

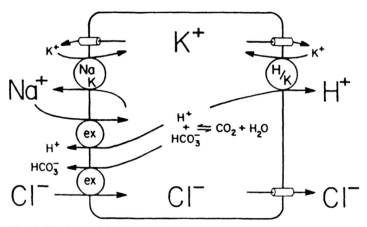

Fig. 9.11. A model proposed for secretion of HCl by gastric parietal cells. The apical membrane (*right side of figure*) contains a H–K exchange pump and conductive pathways for passive diffusion of K and Cl. The basolateral membrane (*left side*) contains a Na–K pump, K-conductive pathway, and mechanisms for Na–H and Cl–HCO_3 exchange. K and Cl are actively transported across the apical membrane by separate pumps, which transport H in exchange for K and Cl by itself. Intracellular production of H_2CO_3 by hydration of $CO_2$2 may produce the H that is secreted into the lumen and the HCO_3 that passively diffuses into the blood. The Cl enters the cell from the blood in exchange for HCO_3. (From Reenstra, Bettencourt, and Forte 1987.)

Cardiac glandular mucosa

As noted in Chapter 2, a region of cardiac glandular mucosa is found near the gastroesophageal junction of reptiles, some adult amphibians, and most mammals. Its secretions are difficult to analyze in most species because of the small area occupied by this tissue and the lack of a clear demarcation between proper gastric and cardiac glandular regions. For example, the proper gastric glands of the human stomach simply decrease in number in a gradual blend with the cardiac glandular region. However, the stomach of some species such as the pig, camelids, and most macropod marsupials contains extensive regions of cardiac mucosa. Pouches prepared from the cardiac glandular region of the swine stomach secreted a slightly alkaline fluid containing high concentrations of HCO_3 (Höller 1970a,b). Infusion of solutions containing HCl or NaCl resulted in absorption of Cl and a rise in HCO_3 levels, which indicated HCO_3–Cl exchange. Pigs 30 to 40 kg in body weight secreted at least 500 to 600 ml of this fluid per day, and the secretory rate increased at night or during forced starvation. Gastrin, a hormone that stimulates the secretion of HCl by parietal cells, inhibited the secretion of HCO_3 by the cardiac glandular mucosa. Secretion of HCO_3 may protect both the cardiac mucosal and stratified squamous regions from damage by HCl and pepsin. It also would provide a more favorable pH for the growth of bacteria in this region.

Studies of the cardiac glandular regions of the llama stomach have given conflicting results. Eckerlin and Stevens (1973) concluded that the glandular pouches in the first two compartments absorbed Cl in exchange for HCO_3 secretion, in a manner similar to that of the pig stomach. However, Rübsamen and Engelhardt (1978) found that although the glandular pouches absorbed Cl, there was little net appearance of HCO_3, and the cardiac glandular region of the third compartment of the llama stomach appeared to secrete Cl (Engelhardt, Ali, and Wipper 1979).

Pyloric glandular mucosa

The pyloric glandular region of the stomach secretes HCO_3 and mucus, which protects its cells from digestion by HCl and pepsin. However, there appears to be little information on mechanisms of transport.

Stratified squamous epithelium

Although nonglandular, stratified squamous epithelium occupies at least some portion of the stomach of species belonging to a wide range of mammalian orders, most of the information on its secretory and absorptive characteristics comes from studies of the ruminant forestomach. This tissue, which is embryologically derived from the gastric rather than the esophageal tissue of ruminants, is responsible for the absorption of about half of the 1.2–1.5 mol of Na secreted in the saliva of sheep each day (Dobson 1959) and for most of the SCFA produced by its microbes. In vitro studies of short-circuited rumen epithelium from the cow, goat, and sheep showed that it rapidly absorbs Na and Cl and secretes small amounts of K (Stevens 1964; Ferreira, Harrison, and Keynes 1964; Harrison, Keynes, and Zurich 1968). Ion replacement experiments indicated that in the absence of SCFA most of the Na was absorbed either in cotransport with Cl or in exchange with H, and a significant portion of the Cl was absorbed in exchange for HCO_3 (Chien and Stevens 1972). Subsequent studies indicated that most of the Na and Cl are absorbed from the rumen and omasum by Na–H and Cl–HCO_3 exchange and that Na absorption was increased by the replacement of Cl with SCFA (Martens and Gäbel 1988; Gäbel and Martens 1991). Absorption of Na and Cl from the forestomach replaces the large quantities secreted into the saliva of these animals, and secretion of HCO_3 aids in the buffering of SCFA produced by microbial fermentation.

As mentioned in Chapter 8, the stratified squamous epithelium of the ruminant forestomach rapidly absorbs the SCFA produced by microbial fermentation. However, the mechanisms of absorption have been difficult to determine. SCFA are present in two forms, as undissociated acids and as anions. Because of the low pK (4.75 to 4.81) of these organic acids, the undissociated form constitutes only

1 to 6 percent of the SCFA within the range of pH normally found in the forestomach. However, cell membranes tend to be permeable to the passive diffusion of the lipid-soluble undissociated organic acids, and the degree of lipid solubility increases by a factor of approximately 2.8 with each additional CH_2 group (acetate < propionate < butyrate). The anions, which constitute the major percentage of the SCFA in digesta, would need to be absorbed either by carrier-mediated transport across the cell membranes or by passive diffusion between the cells.

Early studies showed that although SCFA were rapidly absorbed at the relatively neutral pH of sheep rumen contents, their rate of absorption increased with both an increase in the chain-length of the fatty acid and a decrease in the pH of rumen contents (Danielli et al. 1945). This suggested that they were passively absorbed in their undissociated form. Ash and Dobson (1963) found that absorption of acetate from the sheep rumen was accompanied by a decrease in P_{CO_2} and an increase in the HCO_3 concentration of luminal contents (Table 9.2). They concluded that hydration of CO_2 to H_2CO_3 in the rumen generated a continuous source of protons for absorption of undissociated acetate and HCO_3 for release into luminal contents. However, this would account for only half of the acetate absorbed, suggesting that the remainder was absorbed as anions.

In vitro studies showed that the increase in transport with chain length could be attributed entirely to a parallel increase in their rates of metabolism by bovine rumen epithelium and that a tenfold increase in the concentration of undissociated acid in the lumen bath only doubled the rate of SCFA transport (Stevens and Stettler, 1966a,b). SCFA transport was unaffected by the application of an electrical potential across the tissue, which appeared to rule out passive diffusion of anions through paracellular pathways. When tissues were short-circuited and bathed on both surfaces with solutions of identical composition and pH, they secreted small amounts of acetate into the lumen bath (Stevens and Stettler 1967), which appeared to rule out active absorption as well. Furthermore, rumen epithelium secreted a variety of other, nonphysiological weak organic acids in the same direction as acetate and absorbed a variety of weak organic bases in the absence of transepithelial electrochemical gradients (Stevens, Dobson, and Mammano 1969).

Hogben et al. (1959) had noted the similar transport of drugs that were weak organic acids or bases in opposite (but reverse) directions across the small intestinal epithelium of the rat. They suggested that this was due to the presence of a microlayer at the apical surface of these cells, which was maintained at a pH lower than that of the bulk lumen contents. This would increase the concentration of undissociated acid and reduce the concentration of undissociated base at the lumen surface, favoring their passive diffusion in opposite directions across the cell. Impermeability of these tissues to the passive back-diffusion of anions

Table 9.2. *In vivo absorption or net appearance of short-chain fatty acids (SCFA, Na, HCO₃, and water by the isolated reticulorumen of sheep and large intestine of the pony, pig, human, goat, and dog*

Species	Segment	Bathing solution[a]					Daily absorption or appearance per kg body weight				
		pH	SCFA (mmole/L)	Sodium (mmole/L)	Bicarbonate (mmole/L)	Chloride (mmol/eL)	SCFA (mmole)	Sodium (mmole)	Bicarbonate (mmole)	Water (ml)	Source
Sheep	Reticulo-rumen	7.0	134	149	9.1	29	30	15	−15	40	Dobson (1959)
Pony	Ventral colon	6.1	100	140	20	20	31	7	−12	18	Argenzio et al. (1977)
Pig	Proximal colon, distal colon, and rectum	6.4	107	122	15	15	55	26	−14	170	Argenzio and Whipp (1979)
Human	Cecum, colon, and rectum	7.4	90	120	20	40	40 10	20 3	−9 −5	110 24	Ruppin et al. (1980)
Goat	Colon and rectum	6.0	70	100	20	30	31	29	NS	190	Argenzio, Miller, and Engel-hardt (1975)
Dog	Colon and rectum	6.4	90	122	27	15	8	8	NS	58	Herschel et al. (1981)

[a] Bathing solutions were isotonic to plasma of each species but varied in initial composition. In sheep, pony, pig, and human studies the SCFA was only acetate. The solution used in the goat study contained 60 mM acetate and 10 mM propionate. The solution used to perfuse the dog colon was an equimolar mixture of acetate, propionate, and butyrate. Positive values designate net absorption and negative values net appearance within the lumen. Appearance of HCO₃ in the goat and dog colon was not significant (NS). A steady-state perfusion technique was used on all species except the sheep and pony, in which a static system was used. (Modified from Stevens, Argenzio, and Clemens 1980).
NS, Not significant.

would result in a net accumulation of these organic electrolytes in the opposing bathing solution. A microlayer, which is maintained at a pH lower than that of the bulk lumen contents, subsequently was found at the apical surface of midgut cells (Lucas et al. 1975). However, the net transport of SCFA from the serosal to lumen bath of rumen epithelium would require a microlayer at the luminal surface with a pH higher than that of its bathing solution.

Stevens et al. (1969) showed that if the luminal- and serosal-facing (basolateral) membranes differed in their *relative* permeability to the dissociated and undissociated forms of organic electrolytes, their net transport could be driven by the pH (H ion) gradient between the epithelial cell contents and the bathing solutions. Jackson et al. (1974) proposed an alternate model for transport of organic electrolytes across the intestinal epithelium in which the middle compartment consisted of the extracellular space maintained at a pH higher than that of the bathing solutions. However, this also fails to account for the direction of net SCFA transport across rumen epithelium in the absence of a transepithelial electrochemical gradient. Studies of SCFA transport across hindgut epithelium, which provide additional clues to their mechanism of transport, are included in the following discussion.

The stratified squamous epithelium of the pig stomach transported SCFA at a rate similar to that of rumen epithelium, but that of the pony stomach appeared to be relatively impermeable to these organic acids (see Table 8.4). There appears to be no information on the other transport characteristics of these tissues or of the stratified squamous epithelium in the stomach of other mammals.

The ruminant forestomach absorbs the ammonia produced by microbial degradation of urea and other nitrogenous compounds that also are present in nonionized (NH_3) and ionized (NH_4) forms. With a pK of approximately 9, it is present principally as NH_4 ions at the normal pH of rumen contents. It was assumed to be absorbed as the more lipid-soluble NH_3, but the concentration of ammonia at the luminal surface, where much of the microbial hydrolysis of urea may occur, is unknown, and there is evidence for absorption of NH_4 from the rat ileum (Koch and Hall 1992) and the hindgut of locusts (see Chapter 11).

Intestine

Transport of electrolytes and water across intestinal mucosa varies among segments of the intestine and between cells in the crypts and villi. Some ions, and most of the water, are transported across intestinal epithelium by paracellular diffusion, but this varies with ionic species ($P_K > P_{Na} > P_{Cl}$) and decreases progressively between the duodenum and rectum. Most of the secretion and absorption of electrolytes is dependent on transport across epithelial cell membranes. Crypt cells secrete electrolytes. Cells on the villi of the small intestine and on the surface of the hindgut serve both functions. Water can be absorbed from the hindgut against a substantial transepithelial osmotic gradient, but this has been attributed to an osmotic gradient between the lumen and intracellular space, produced by electrolyte transport through the apical and lateral membranes of the epithelial cells (Curran 1968; Diamond 1977; Lundgren 1984).

Midgut

Information on the secretion and absorption of electrolytes by the midgut epithelium was reviewed by Powell (1986) and, more recently, by Chang and Rao (1994). The midgut secretes Na, K, Cl, HCO_3, and water. The first three ions appear to be secreted by the cotransport of Na-K-2Cl across the basolateral membrane. Chloride then diffuses through a Cl channel in the apical membrane (Sullivan and Field 1991), and the transepithelial potential produced by Cl secretion results in the passive secretion of Na and water. A mechanism for K secretion is shown in Fig. 9.12, *A*. There is good evidence for HCO_3 secretion in exchange for Cl by ileal cells in a number of species. There also is evidence for cotransport of Na and HCO_3 across the basolateral membrane and for passive diffusion of HCO_3 through an anion-selective channel in the apical membrane of cells in the rabbit ileum (Minhas, Sullivan, and Field 1990). Most of the Cl and HCO_3 is believed to be secreted by crypt epithelium (Welsh et al. 1982; Minhas and Field 1994).

The midgut absorbs substantial amounts of electrolytes. The major mechanisms responsible for Na, Cl, and K absorption from different regions of the human intestine are shown in Fig. 9.13. Much of the sodium is absorbed from the proximal midgut in cotransport with organic solutes such as glucose, amino acids, and B vitamins across the apical membrane (Fig. 9.13, *A*). Studies of humans and a variety of laboratory animals show that Na absorption is stimulated by increasing the glucose concentration of luminal contents up to levels of 50 mM and that Na and glucose are transported by a 2:1 coupling. Addition of glucose can result in a fourfold increase in the absorption of Na from the jejunum. Galactose and fructose had a lesser effect. Sodium and amino acids were absorbed in a 1:1 ratio by the separate mechanisms for amino acid transport. Early suggestions that oligopeptides may be absorbed by Na-dependent mechanisms have not been confirmed by studies using brush border vesicles.

Most of the Na that is not absorbed across the apical membranes of midgut cells in cotransport with organic solutes is transported by Na/H exchange (Fig. 9.13, *A*). This is not coupled to $Cl-HCO_3$ in the proximal midgut. Therefore, the secretion of H results in conversion of HCO_3 to CO_2 and water, which are readily absorbed. However, in the distal midgut Na–H exchange is coupled with $Cl–HCO_3$ exchange. An interrelationship between Na and Cl absorption also has been described in the rabbit ileum (Nellans, Frizzell, and Schultz 1973); flounder

Mucosa   **Serosa**

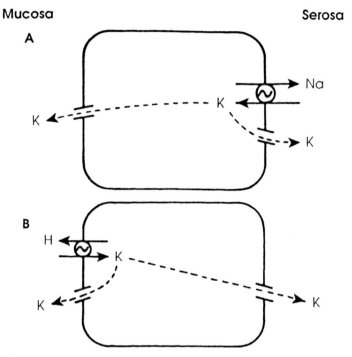

Fig. 9.12. Models for transcellular K transport by the intestinal cells. **A,** K secretion in which the Na-K-ATPase creates an electrochemical gradient for K
across the apical cell membrane with subsequent K diffusion through conduc-
tance channels. The direction of net transport in this model could be governed
by relative K conductances at the two cell membranes. **B,** K absorption in which
a K-H-ATPase mediates entry across the apical membrane to achieve intracellular concentrations above electrochemical equilibrium. The basolateral and apical
membrane K conductances determine the direction of net K movement across
the epithelium. (From Smith and McCabe 1984.)

intestine (Frizzell et al. 1979; Rao et al. 1981); and the colon of rabbits (Sellin
and De Soignie 1984), monkeys (Powell et al. 1982), and humans (Hawker,
Mashiter, and Turnberg 1978). Although the low levels of short-circuit current
suggested nonelectrogenic cotransport of NaCl, Turnberg et al. (1970) showed that
it could be explained by simultaneous exchange of Na for H and Cl for HCO_3,
coupled in varying degrees by the hydration of intracellular CO_2. This has been
supported by studies of transport across brush border membrane vesicles (Murer,
Hopfer, and Kinne 1976; Powell and Fan 1984; Knickelbein et al. 1985).

Chloride is absorbed from the midgut by passive diffusion down its electrochemical gradient and by Cl–HCO_3 exchange in its distal segment (Fig. 9.13, *B*).
It also is absorbed in cotransport with Na and K across the apical membrane of
the flounder midgut (Fig. 9.14). Cotransport with Na and imino or amino acid

A. Na⁺ Absorption

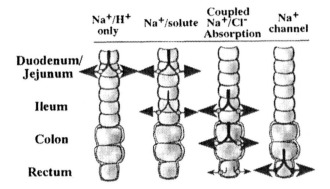

B. Cl⁻ Absorption

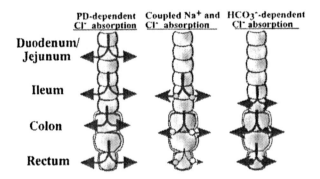

C. Potassium transport

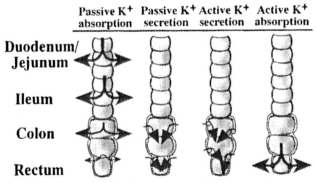

Fig. 9.13. Pathways for the transport of ions across human intestinal epithelium. The thickness of arrow heads represents the relative degree of transport. (From Chang and Rao 1994.)

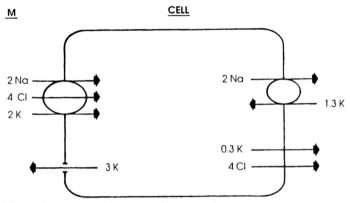

Fig. 9.14. A model that would account for Na, K, and Cl transport by flounder intestine; the Na-K-Cl cotransport mechanism. (From Frizzell et al. 1984.)

also has been reported for midgut epithelium in fish (Bogé, Rigal, and Pérès 1983) and rabbit (Miyamoto et al. 1989; Munck and Munck 1990; Munck 1993) (Fig. 9.8, *H*). Potassium absorption can be attributed to passive diffusion, and the disappearance of HCO_3 can be attributed to its conversion to CO_2 and water by Na–H exchange. However, there is evidence for K–H exchange (Fig. 9.12, *B*) in the rat jejunum (Lucas 1976), rabbit ileum (Smith, Cascairo, and Sullivan 1985), and amphibian midgut (Imon and White 1984) and electrogenic absorption of HCO_3 was reported in the jejunum of the congo eel (White 1985).

Hindgut

Information on electrolyte transport by mammalian hindgut epithelium has been reviewed by Wrong, Edmonds, and Chadwick (1981); Engelhardt and Rechkemmer (1983); Argenzio (1991); and Binder and Sandle (1994). The hindgut absorbs Na, K, Cl, HCO_3, SCFA, and ammonia. Secretion and absorption varies between segments (Fig. 9.13), among species, and with an animal's need to conserve Na and water. The proximal colon of humans and rabbits actively secretes K. It also is secreted by the distal colon of sodium-depleted rats (Fordtran, Hayslett, and Binder 1984). This appears to be the result of diffusion through channels in the apical membrane following its transport across the basolateral membrane by Na-K-ATPase exchange and possibly the cotransport of Na-K-2Cl. It is believed that the secretion of HCO_3 depends on its basolateral uptake by Cl–HCO_3 exchange or $NaHCO_3$ cotransport, and its secretion across the apical membrane on Cl–HCO_3 exchange.

The hindgut of adult mammals does not appear to absorb glucose or amino acids by any carrier-mediated process. However, the structure and function of the colon in neonatal mammals is similar to those of the midgut. The colon of neonatal rats and pigs is lined with villi and a dense population of microvilli, which absorbed Na by cotransport with glucose and amino acids and in exchange for H (Bentley and Smith 1975; Potter 1989). However, shortly after birth the villi disappeared and the microvilli became sparse, and the colon lost its ability to absorb glucose and amino acids.

Sodium is absorbed from the proximal colon of humans by nonelectrogenic Na–H exchange across the apical membrane, which is coupled to Cl–HCO$_3$ exchange, as it is in the distal midgut (Fig. 9.13). The same is true for the distal colon of normal rats. However, Engelhardt (1995) reviewed other studies, which showed that the relationship between Na and SCFA absorption can vary with both species and segments of the large intestine. The distal colon of humans, rabbits, pigs, sheep, ponies, and sodium-depleted rats absorbs Na by an electrogenic process (Fig. 9.13, *A*), which involves its diffusion across the apical membrane via sodium-permeable channels and Na–K exchange by the basolateral membranes. The distal colon can also absorb K by a mechanism for K–H exchange in the apical membrane (Fig. 9.12, *B*) similar to the one responsible for H secretion by the parietal cells of the stomach. Absorption of Na by Na–H exchange and electrogenic transport, the switch to electrogenic transport in the distal colon of rats, and the secretion of K are simulated by the release of aldosterone in sodium-depleted animals. Therefore, Na depletion increases both the recovery of Na and water and the excretion of K. Excretion of hard and soft fecal pellets by rabbits (see Chapter 6) also has been attributed to a circadian increase and decrease in plasma aldosterone and Na absorption (Clauss 1983). Glucocorticoids stimulate Na–H exchange, but they appear to have no effect on either the electrogenic absorption of Na or K secretion.

Some major differences between the absorption of Na from the hindgut of mammals and birds were reviewed by Skadhauge (1993). On low-sodium diets, Na is absorbed electrogenically from the hindgut of chickens by cotransport with glucose and amino acids by the cecal epithelium and diffusion through sodium-selective channels in the apical membranes of colonic and coprodeal cells. A high Na intake suppressed the rate of Na absorption from the coprodeum of chickens and other seed-eating birds, and transport via Na channels in the colon was replaced by cotransport of Na with glucose and amino acid (Lind et al. 1980a,b; Skadhauge et al. 1983; Munck 1989; Arnason and Skadhauge 1991). This adaptation to Na–nutrient cotransport with high-sodium diets was not seen in the domestic duck, which can excrete Na via salt glands, or in the desert-dwelling Australian emu. However, it appears that the hindgut of some birds can switch their mechanisms for Na absorption from those seen in the midgut and neonate hindgut of mammals to the more efficient mechanism seen in the distal colon of mammals on demand for Na absorption. The ability of reptiles or amphibians to make this adaptation does not appear to have been explored.

Chloride is absorbed from the hindgut by passive diffusion down its electrochemical gradient and by Cl–HCO_3 exchange, which is coupled to Na–H exchange in the proximal colon (Fig. 9.13, *B*). Studies of apical vesicles from the rat distal colon indicated the additional presence of Cl–OH exchange in the colon of the rabbit (Duffey 1984) and rat (Foster, Dudeja, and Brasitus 1990). Absorption of HCO_3 can be attributed to its titration by H secreted in exchange for Na or released by the production of SCFA.

SCFA are absorbed from the hindgut in a manner similar to that seen in the rumen. Perfusion studies yielded two different patterns of SCFA absorption (Table 9.2). Absorption of SCFA from the pony, pig, and human colon was accompanied by a decrease in the Pco_2 and an increase in the HCO_3 levels of luminal contents, similar to that seen in the rumen. This was accompanied by the absorption of Na at a rate much slower than that of the SCFA. The goat and dog colons absorbed SCFA and Na at similar rates with no net appearance of HCO_3. Replacement of Na with choline reduced the rate of SCFA absorption from the colon of rats (Umesaki et al. 1979), sheep (Rübsamen and Engelhardt 1981), and pigs (Roe and Stevens, unpublished). Replacement of SCFA with Cl resulted in a reciprocal decrease in the rate of Na absorption from the colon of the goat (Argenzio, Miller, and Engelhardt 1975) and pig (Crump, Argenzio, and Whipp 1980). These results led to the conclusion that SCFA are absorbed following their conversion to undissociated acids by the H generated from hydration of CO_2 in the lumen and secretion into the lumen by Na–H exchange. Although SCFA and Na absorption were coupled in the proximal colon of the guinea pig, this was not true for the distal colon (Rechkemmer 1988), where it appears to be coupled with K-H exchange (Stingelin, Wolffram, and Sharrer 1986; Ohmer 1993; Engelhardt et al. 1993).

The absence of a net increase in luminal HCO_3 in the goat and dog colons could be attributed to a more rapid Na–H exchange. The differences in patterns of hindgut absorption and secretion reflect the results of perfusing the entire colon. However, Argenzio (1991) pointed out that a less efficient absorption of water by Na–H exchange and a net increase in lumen HCO_3 levels would allow the proximal colon to retain the water and provide the HCO_3 required for efficient microbial fermentation. Electrogenic absorption of Na from a greater percentage of the colon of carnivores and foregut-fermenting herbivores would provide a more effective system for the recovery of both Na and water.

In vitro studies also show many similarities in the transport of SCFA by rumen epithelium and hindgut mucosa. SCFA were absorbed at a similar rate by rumen epithelium and the hindgut mucosa of the pig, pony, and dog and were metabolized to varying degrees by all of these tissues (see Table 8.4). They are also metabolized by colonic epithelia in a manner similar to that of rumen epithelium (Roediger 1991). Inhibition of Na–H exchange with amiloride reduced SCFA transport across the cecal and proximal colon epithelia of the guinea pig (Engelhardt et al. 1993), showing a similar interrelationship between Na and SCFA absorption. SCFA transport was similarly unaffected by the application of an electrical gradient across the colonic epithelium of the pony (Argenzio et al. 1977) and guinea pig (Rechkemmer and Engelhardt 1993). Epithelium from the cecum and proximal colon of the rabbit (Hatch 1987; Sellin and De Soignie 1990; Sellin 1992) and guinea pig (Rechkemmer and Engelhardt 1993) also secreted SCFA into the lumen bath in the absence of an electrochemical gradient. A marked decrease in the lumen bath pH had little effect on transport across the colonic epithelium of ponies (Argenzio et al. 1977) and no effect on SCFA absorption from the proximal colon of the guinea pig (Rechkemmer 1987) or the large intestine of humans (McNeil, Cummings, and James 1978). This could be attributed to a microlayer at the luminal surface maintained at a pH of 7.0 (Rechkemmer et al. 1986) and 6.5 (McNeil, Ling, and Wager 1987), respectively. There also is evidence for a greater degree of lipid solubility in the apical versus the basolateral membrane (Brasitus and Dudeja 1988: Luciano, Konitz, and Reale 1989).

SCFA transport has been examined more recently by the use of membrane vesicles, prepared from epithelial cells of the tilapia midgut (Titus and Ahearn 1988, 1991), rat distal colon (Mascola, Rajendran, and Binder 1991; Reynolds, Rajendran, and Binder 1992), and human proximal colon (Harig et al. 1990). These studies indicated that a major portion of the SCFA was transported by carrier-mediated anion exchange across both the apical and basolateral membranes. The carrier in the apical membrane exchanged SCFA for HCO_3. The carrier in the basolateral membrane exchanged SCFA for either HCO_3 or Cl, but showed a lower affinity for the SCFA and transported them at lower rates. Vesicles from the basolateral membranes of rat jejunal epithelium showed a similar mechanism for carrier-mediated transport of lactic acid (Cheeseman, Shariff, and O'Neill 1994). SCFA transport across these vesicle membranes was unaffected by the presence of Na in the solutions and affected relatively little by changes in the pH, indicating that only a small percentage was transported as undissociated acid.

Absorption of SCFA anions by carrier-mediated exchange would account for their rapid uptake from digesta maintained at relatively neutral pH and the poor correlation between SCFA transport and luminal pH. The low permeability to undissociated acids is more difficult to explain. However, carrier-mediated transport would be saturated at low levels of SCFA. Titus and Ahearn (1991) estimated that passive diffusion could account for 60 percent of the acetate absorbed by the tilapia intestine from solutions containing 15 mM acetate, but less at higher concentrations, and Hume, Karasov, and Darken (1993) concluded that acetate was absorbed from the hindgut of the prairie vole *Microtus ochrogaster* principally by passive diffusion when present in the concentrations normally found in the hindgut of these animals. Engelhardt (1995) concluded that differences in the effects of lumen pH and SCFA chain length on the rates of SCFA absorption by different segments of hindgut and among different species could be due to variations in the pH of the microlayer at the lumen surface and the lipid solubility of the apical and basolateral membranes.

Fig. 9.15 summarizes the three models that have been proposed for the transport of SCFA across rumen and hindgut epithelia. SCFA may be absorbed by passive diffusion of the undissociated acids and carrier-mediated transport of the anions. Transport of undissociated acids could be attributed to passive diffusion down their concentration gradients, which would be influenced by the pH of both a microlayer at the luminal surface of the epithelial cells and of the cell contents. Carrier-mediated absorption of anions would be especially important when the lumen contains low levels of SCFA. Regardless of the mechanisms involved, the absorption of SCFA and Na are interdependent in both the rumen and hindgut, and their combined absorption is responsible for most of the osmotic absorption of water from these organs.

The difficulties in determining the mechanisms responsible for ammonium absorption, its presence in two forms; and its production from urea by bacteria at the apical surface of epithelial cells, were discussed earlier. Although it was assumed to be absorbed in its more lipid-soluble undissociated form, ammonium absorption was reduced by inhibitors of K absorption and coupled with NaCl absorption in the ileum of rats (Koch and Hall 1992). Apical membranes of insect rectal cells also appear capable of transporting NH_4 in exchange for Na (see Fig. 11.2).

Enterosystemic circulation

The digestive tract expends much of its energy on the resorption of electrolytes and water secreted into the gut. The gastrointestinal tract of humans contains only 2 percent of the total body water (Table 9.1). However, Soergel and Hofmann (1972) estimated that 5.9 liters of fluid were secreted by the salivary glands, stomach, pancreas, and small intestine of human subjects who had been starved for 24 hours (Table 9.3). This was equivalent to approximately 40 percent of their ex-

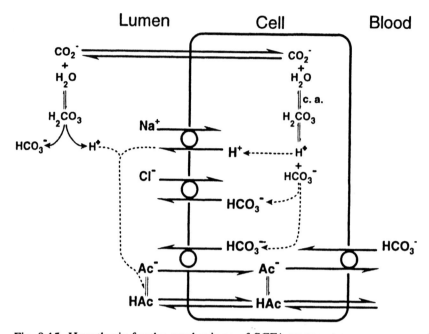

Fig. 9.15. Hypothesis for the mechanisms of SCFA transport across rumen and hindgut epithelium. Hydrogen ions produced by microbial fermentation in the lumen and secreted by carrier-mediated Na–H exchange in the lumen-facing (apical) membrane provide for the protonation of SCFA anions (*Ac*) to their undissociated form (*HAc*), which passively diffuse across cell membranes. Intracellular hydration of CO_2 derived from metabolism of SCFA and other substrates and catalyzed by carbonic anhydrase, produces H for carrier-mediated Na–H exchange and HCO_3 for exchange with both Cl and SCFA anions. The dependence of both Na and SCFA anion absorption upon CO_2 hydration would account for their interdependent absorption. SCFA are transported across the basolateral membrane by diffusion of the undissociated form and by a carrier-mediated exchange of SCFA anions with blood HCO_3. The relatively low pH of epithelial cell contents and differences in the net rate of transport of the two forms of SCFA across the apical and basolateral membranes could account for the net transport of SCFA from blood to lumen in the absence of an electrochemical gradient under experimental conditions. Na-H exchange is replaced with K-H exchange in the distal colon, and the relative contributions of carrier-mediated anion transport and passive diffusion of undissociated acid can vary with segment of gut and species. (Modifications and combination of models from Stevens, Dobson, and Kammano 1969; Stevens, Argenzio, and Roberts 1986; and Titus and Ahearn 1992.)

Table 9.3. *Daily secretion and absorption of fluid by the gastrointestinal tract of human, sheep, and pony*

Parameter	L/100 kg body weight		
	Human	Sheep	Pony
Salivary secretion	1.3	17	12
Gastric secretion	1.3	16	—
Pancreatic secretion	1.3	0.6–1.1	10–12
Biliary secretion	0.5	1.1–4.0	4
Small intestinal secretion	3.3	—	—
Ileocecal flow	1.6	8.6	12
Large intestinal secretion	—	—	7.5
Fecal excretion	0.1	0.7	0.9

Values for humans are estimates per 100 kg body weight for an individual starved 24 hours prior to measurements (Soergel and Hofmann 1972). Values for the sheep and pony are from studies conducted under a variety of experimental conditions. Those for the sheep are mean values from various sources for salivary (Denton 1957; Kay 1960), abomasal (Hill 1965), pancreatic (McGee 1961; Taylor 1962), and biliary (Harrison 1962) secretion. Estimates for ileocecal flow excretion in sheep are from Kay and Pfeffer (1970). Estimates for salivary, pancreatic, and biliary secretions of the pony are from Alexander and Hickson (1970). Those for ileocecal flow, large intestinal secretion, and fecal excretion are from Argenzio et al. (1974.)

tracellular fluid volume (ECFV) or 12 percent of the total body water. These secretions contained approximately 30 percent of the Na and Cl and 45 percent of the HCO_3 normally present in their extracellular fluid. The volume increased to 7 liters of fluid in fed subjects (Powell 1986). Clinical signs of dehydration can be seen with an uncompensated loss of 15 percent of the ECFV, or 20 percent of the total body water, and a 30–35 percent loss in body water can be fatal. Of the ions and water secreted into the digestive tract, 98 percent were reabsorbed. The midgut is a major site of electrolyte and water absorption in carnivores and omnivores. However, the hindgut is the final site, and the most important site in animals that are dehydrated as a result of malfunction of the midgut or a low water intake.

The gastrointestinal tract of herbivores contains a much higher percentage of the body water–29 percent in sheep (Table 9.1). The salivary, abomasal, pancreatic, and biliary secretions of sheep (Table 9.3) were equivalent to approximately 2.5 times the ECFV. Of the fluid entering the digestive tract, 98 percent was recovered prior to fecal excretion. About 20 percent of this can be recovered by reabsorption of salivary electrolytes from the forestomach (Dobson 1959), and the remainder was absorbed from the midgut and hindgut. Salivary, pancreatic, and biliary secretions of the horse were equivalent to two times the ECFV, and this can be raised to 2.5 times the ECFV by secretions into the large intestine (Table 9.3). The large intestine also absorbed about half of the fluid secreted into the equine digestive tract (Argenzio et al. 1974). Therefore, the digestive tract of herbivores must absorb much larger volumes of electrolytes and water, and it is a major site for this in hindgut-fermenting species.

Secretion and absorption by the equine hindgut can be greatly influenced by the feeding schedule. Fig. 9.16 shows cyclic changes in the levels of SCFA, protein, and nonprotein nitrogen in the large intestine of ponies fed a high-concentrate diet at twelve-hour intervals. Each compartment underwent cyclic changes in volume with time after feeding. This was most marked in the ventral colon, where there was a fivefold increase in volume during the first eight hours following a meal. These changes in volume were associated with cyclic increases and decreases in protein–nitrogen and SCFA concentrations. This showed that the initial arrival of digesta from the meal instigated the multiplication of bacteria and production of SCFA. During the next six hours, SCFA were produced more rapidly than they could be absorbed, resulting in the secretion of water into the lumen. However, during the final four hours of the cycle, the rate of absorption of SCFA exceeded their rate of production, with a net absorption of water. Despite these large fluctuations in the net production and absorption of SCFA, their concentration remained relatively stable (see Fig. 8.4) because of the compensatory effects of the secretion and absorption of water.

Additional studies were conducted to compare the effects of a twelve-hour versus a two-hour feeding schedule on plasma volume (Clarke, Roberts, and Argenzio 1990). Ponies fed at twelve-hour intervals showed a 15 percent reduction in plasma volume within one hour after feeding. This was regained within two hours and then was followed by a smaller reduction six hours after feeding. The initial reduction in plasma volume was attributed to salivary and pancreatic secretions. The second reduction was attributed to hindgut secretion in response to microbial fermentation. The decrease in plasma volume was associated with an increase in plasma levels of renin and aldosterone (Fig. 9.17). Renin, which is released from the kidney in response to hypovolemia, generates the release of angiotensin. Angiotensin reduces urinary excretion and stimulates both thirst and the more gradual release of aldosterone. Aldosterone stimulates the absorption of Na and water by both the kidney and hindgut. The plasma levels of renin and aldosterone gradually returned to prefeeding levels twelve hours after the meal. Ponies fed at two-hour intervals demonstrated neither the reduction in plasma volume nor the increase in plasma renin and aldosterone levels, and urinary excretion of Na remained unchanged.

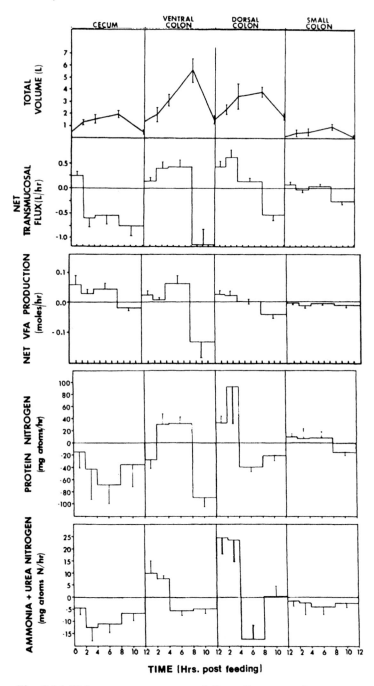

Fig. 9.16. Volume, net transmucosal flux of water, and net appearance and disappearance of VFA (SCFA), protein, and urea plus ammonia nitrogen in the pony large intestine as a function of time. All values, other than volume, are corrected for exchange between segments resulting from digesta flow. (Modified from Argenzio et al. 1974; Argenzio, Southworth, and Stevens 1974; Wootton and Argenzio 1975).

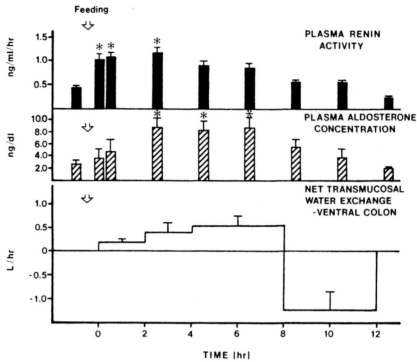

Fig. 9.17. Relationship between colonic water exchange, plasma renin activity, and aldosterone levels in ponies fed at twelve-hour intervals. The net secretion of water during the first eight hours following a meal is accompanied by a rapid increase in plasma renin activity and a slower increase in plasma aldosterone levels. The plasma levels of both of these agents return to prefeeding levels 12 hours after the meal. (From Clarke, Roberts, and Argenzio 1990.)

The hindgut of birds must conserve electrolytes and water from both the digestive secretions and cloacal urine. Björnhag and Sperber (1977) estimated that 20–24 percent of the urine entering the cloaca was transported to the ceca of turkeys and that 87–97 percent of the cecal nitrogen content of turkeys originated from urine. A low nitrogen content in the food increased the reflux of urine into the ceca of chickens (Björnhag 1989). Paxton (1976) estimated that the ceca of Japanese quail could absorb 80–90 percent of the Na derived from the rectum and ileum. Birds conserve water by resorption of urinary and digesta electrolytes from their hindgut and by excretion of nitrogenous wastes and electrolytes in osmotically inactive uric acid–urate complexes in their feces. The emu, which can survive in an extremely arid environment and utilizes its midgut as the principal site for microbial digestion, has relatively short ceca and a short colon (see Fig. 3.14). In normal hydration, most of the Na and Cl was in the fluid fraction of the feces (Dawson, Herd, and Skadhauge 1985). However, dehydrated birds showed a tenfold increase in the rate of Na absorption from the hindgut and a marked increase in the osmotic activity of the feces, with most of these electrolytes in the solid phase. The ostrich, which has a similar ability to survive in arid climates, has a very long colon (see Fig. 3.14) and does not appear to reflux its urinary excretions from the cloaca (Skadhauge et al. 1984). However, electrolyte and water conservation is aided by nasal salt glands and the ability to concentrate the urine to twice the osmotic activity of the emu.

Most mammals have a greater capacity to concentrate urine and tend to have a longer hindgut than other vertebrates. Herbivores have the greatest need for the efficient recovery of the electrolytes and water from their digestive tract, and this may be one of the major advantages of foregut fermentation. It allows camels and other desert ruminants to rapidly ingest large volumes of water without damage to the osmotic equilibrium of their body fluids. The black Bedouin goat can graze for several days without access to water by decreasing the rate of reticuloruminal outflow (Shkolnik, Maltz, and Choshniak 1980). When they return to an oasis, they can drink a volume of water equivalent to 40 percent of their dehydrated body weight with a twofold increase in reticuloruminal volume. However, the outflow from the reticulorumen is delayed by the hypotonicity of its contents. Dehydrated cattle also can replace an 18 percent loss in body weight in one drinking period (Silanikove 1989). Administration of the antidiuretic hormone vasopressin, which reduces renal excretion, closed the esophageal groove of dehydrated cattle (Mikhail et al. 1988). Therefore the forestomach serves as a reservoir, slowly releasing water for absorption by the remainder of the digestive tract. Therefore, use of the forestomach as a major site for microbial fermentation allows for the direct recovery of some of the salivary secretions and the storage of water. It also allows the midgut to participate in the recovery of the large quantities of electrolytes and water secreted and released in the process of foregut fermentation.

Excessive sequestration of electrolytes and water in any segment of the gut that results from blockage, or excessive loss resulting from emesis or diarrhea can be life-threatening to any animal. Any condition that results in oversecretion of electrolytes or in a reduction of carbohydrate digestion in the small intestine can produce diarrhea. Human patients suffering from cholera may lose 1 liter of fluid per hour in their feces because of the stimulation of small intestinal secretion by cholera toxin. As mentioned in Chapter 7, the small intestine of neonates and adults of some species is deficient in lactase. In these animals, ingestion of lactose-containing milk or milk products results in diarrhea because of the fermentation of lactose to SCFA and lactic acid at rates too rapid for their efficient absorption.

Transmissible gastroenteritis (TGE), a viral disease of pigs that is characterized by atrophy of small intestinal villi, produces a diarrhea that can result in 100 percent mortality in pigs infected the first week after birth but only a 2 percent mortality rate in pigs three or more weeks of age. Argenzio et al. (1984) found that pigs infected with TGE showed a marked increase in the amount of fluid secreted into their proximal small intestine and a twofold increase in the volume of fluid presented to the hindgut. This resulted in a profuse diarrhea in three-day-old pigs, which were unable to ferment carbohydrate and absorb the fluid. However, unabsorbed carbohydrate was almost completely converted to SCFA in the hindgut of the older animals, with a sixfold increase in fluid absorption from their hindgut, compared with controls of the same age.

Summary

Secretion of electrolytes and water provides the fluid and optimal pH necessary for digestion. Fluids secreted by the salivary glands, pancreas, and biliary system tend to contain high levels of HCO_3, but the volume, composition, and pH can be adjusted in most species by glandular ducts or during the storage of bile in the gallbladder of some animals. Rapid rates of flow add large volumes of highly buffered solutions to the digestive tract. Slower rates of salivary and pancreatic secretion and intermittent release of bile from the gallbladder tend to conserve Na, HCO_3, and water in animals that are discontinuous feeders. Gastric secretion of HCl requires the protection of gut epithelium by titration of H with HCO_3. The large quantities of SCFA produced in the forestomach of some mammalian herbivores and the hindgut of terrestrial vertebrates must also be neutralized by titration and absorption. Most of the electrolytes and water secreted into the digestive tract must be reabsorbed to preserve the electrolyte–water and acid–base balance of body fluids.

Secretion and absorption of electrolytes involve a limited number of mechanisms that appear to be present in all vertebrates but that are arranged in various combinations and permutations in the epithelial cells lining the digestive tract and the ducts of its ancillary glands. Beginning with the study of frog skin, comparative studies have contributed a great deal to our present understanding of the underlying mechanisms. They also provide excellent examples of the adaptations of animals to their diet and feeding practice. The major differences among vertebrates seem to be in the segments of gut involved and in the relative degree of secretion or absorption that takes place. The digestive system of carnivores, which ingest meals high in protein and energy, at infrequent intervals is adapted to the episodic secretion and absorption of electrolytes and water. The herbivore digestive system is adapted to a more continuous secretion of much larger quantities of electrolytes and water and a greater need for their resorption from the gut. The ultimate need for efficiency of recovery of electrolytes and water is seen in desert herbivores. It should come as little surprise that the infrequent feeding and/or high-concentrate diets to herbivores can result in malfunction of their digestive system. The digestive system of carnivores is equally ill-suited to the feeding of low-concentrate, high-fiber diets. Although the diet and feeding strategies of omnivores tends to fall between these two extremes, the effects of sudden changes in diet and/or feeding frequency need equal consideration.

10

Neuroendocrine control

The motor, secretory, digestive, and absorptive activities of the digestive system are coordinated by neuroendocrine control. Functions that involve the most cranial and caudal ends of the digestive tract, such as mastication, salivary secretion, deglutition, and defecation, are principally or entirely controlled by the nervous system. Those involving the remainder of the digestive system are coordinated by a combination of neural and endocrine controls. Nerve stimulation tends to produce rapid and transient effects. Hormones that circulate in the blood or paracrine secretions that act on neighboring cells tend to result in a slower and more prolonged response. However, the distinction between nervous and endocrine control is complicated by the large number of neurotransmitters, neuromodulators, hormones, and paracrine agents involved and the production of some agents by both neurons and endocrine cells. A subject that once could be comfortably explained by the action of two neurotransmitters and four or five hormones now occupies 20 percent of the chapters in the latest edition of *Physiology of the Gastrointestinal Tract* (1994). Species comparisons are further complicated by differences in the sites of neuroendocrine secretions and the receptors on target organs and cells that respond to these agents. Therefore, this chapter will describe only the major characteristics of neuroendocrine control and differences among classes of vertebrates.

The understanding of neurohumoral control mechanisms has rapidly advanced as a result of the chemical characterization of many of the agents involved and the use of immunoassay techniques to locate their sites of production and action. However, many of these agents are present in multiple molecular forms, and Van Noorden and Polak (1979) pointed out that economy of nature often adapts one molecule to many uses that depend on both its structure and target organ. Furthermore, the site on a molecule that determines its antigenic response may have evolved separately from the amino acid sequence that determines its biological activity. Therefore, immunoreactivity does not necessarily indicate the presence

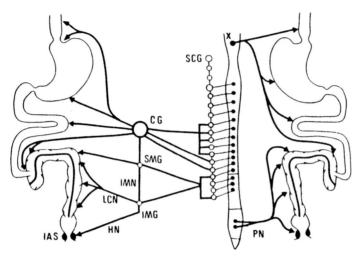

Fig. 10.1. Schema of the extrinsic efferent innervation of the gut. The sympathetic innervation is represented to the left of the figure; the parasympathetic innervation is represented to the right. This representation is a synthesis of various data and may present variations according to different species. *SCG,* Superior cervical ganglion; *CG,* celiac ganglion; *SMG,* superior mesenteric ganglion; *IMG,* inferior mesenteric ganglion; *IMN,* intermesenteric nerve; *LCN,* lumbar colonic nerves; *HN,* hypogastric nerves; *X,* vagus dorsal motor nucleus and vagus nerve; *PN,* pelvic nerves; *IAS,* internal anal sphincter. (From Roman and Gonella 1981.)

of a given neurotransmitter or hormone, nor does the absence of immunoreactivity necessarily indicate its absence.

Neural control of the digestive system

Anatomy

The digestive system is innervated by afferent sensory nerves and visceral efferent nerves from the central nervous system (CNS) to the digestive tract and its ancillary glands, and an intrinsic nervous system. Sensory nerve receptors can distinguish changes in the chemical composition, pH, or osmolality of digesta or in the degree of gut wall distension and can transmit this information to the enteric and central nervous systems. The intrinsic (enteric) nervous system of the digestive tract consists of the neurons that form the myenteric and submucosal plexuses in its wall and innervate its motor and secretory cells. The visceral efferent, or autonomic nervous system, provides CNS modulation of the intrinsic nervous system of the gut and the salivary glands, pancreas, and biliary system.

The autonomic (visceral efferent) innervation of the human gastrointestinal tract is illustrated in Fig. 10.1. It consists of parasympathetic and sympathetic nerves that innervate the digestive tract and its ancillary glands. The parasympathetic division is often referred to as the *craniosacral system* because of the location of its preganglionic nerve cell bodies in the brain and sacral spinal cord. Cranial nerves innervate the salivary glands and cranial esophagus and join the vagal nerves to the pancreas, gallbladder, distal esophagus, stomach, small intestine, and proximal segment of the large intestine. Vagal nerves synapse with neurons in the enteric nervous system of the digestive tract. The sacral segment of the spinal cord provides similar parasympathetic innervation to the more distal segments of the large intestine via the pelvic nerves.

The cell bodies of preganglionic neurons of the sympathetic division are located in the thoracolumbar region of the spinal cord and synapse with neurons in the paravertebral, cervical, celiac, or mesenteric ganglia. These ganglia consist of masses of cell bodies, which provide postganglionic sympathetic axons that can terminate on blood vessels, muscles, and secretory cells, but the majority of these axons terminate in the myenteric plexus.

All preganglionic neurons of the autonomic nervous system are cholinergic, releasing acetylcholine, which is excitatory to postsynaptic neurons or effector cells. Postganglionic sympathetic neurons to the digestive system are adrenergic, releasing norepinephrine, which is generally an inhibitory agent. Although the neurons of the autonomic nervous system are either cholinergic or adrenergic, their stimulation can produce a variety of effects. Stimulation of craniosacral parasympathetic nerves can produce cholinergic excitation, noncholinergic excitation, or nonadrenergic inhibition (Burnstock 1972; Mishra and Raviprakash 1981; Komori and Ohashi 1982; Meldrum and Burnstock 1985; Burnstock 1986). Stimulation of sympathetic neurons can inhibit muscular contraction and glandular secretion, but their major inhibitory effect appears to be on the release of acetylcholine from cholinergic neurons. Furthermore, their stimulation can excite salivary secretion and the contraction of muscles in sphincters and the muscularis mucosa. This led to the discovery that nervous control of the digestive tract is mediated to a large extent by a variety of neurotransmitting and modulating agents released by the enteric nervous system.

The functional anatomy of the human enteric nervous system (ENS) was reviewed by Gershon, Kirchgessner, and Wade (1994). The ENS differs from other divisions of the peripheral nervous system in that it is modified for the local regulation of behavior, independently of the brain or spinal cord. A large proportion of the vagal nerve axons are afferent (sensory). Vagal efferent innervation is highest in the stomach and decreases distally along the intestine, and its principal target is the myenteric plexus. Whereas the vagal efferent innervation consists of thousands of neurons, the ENS of the small intestine alone contains millions of neurons. It was first suggested that the motor axons innervated specialized "com-

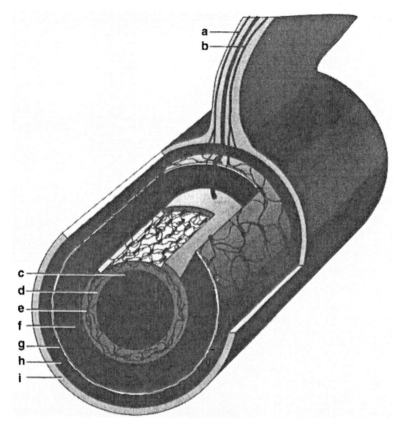

Fig. 10.2. Cross-sectional and cutaway diagram of the wall of the small intestine showing the extrinsic parasympathetic (*a*) and sympathetic (*b*) nerve fibers entering via the mesentery and the submucosal (*e*) and myenteric (*g*) plexuses of the enteric nervous system. Cross section shows the successive layers of mucosa (*c*), submucosa (*d*), circular muscle (*f*), longitudinal muscle (*h*), and serosa (*i*). (Modified from Gershon and Erde 1981.)

mand" neurons in the myenteric plexus, but it is now believed that they exert their effect via a widespread modulatory influence, at least in the stomach.

Fig. 10.2 illustrates a cross section of the small intestine and its innervation. The ENS is arranged in a myenteric (Auerbach's) and a submucosal (Meissner's) ganglionated plexus over most of its length. The myenteric plexus, which is the larger of the two, is located between the longitudinal and circular muscle layers. It innervates intestinal muscle, and some of its axons project to the pancreas and gallbladder. The submucosal plexus probably consists of two interconnecting plexuses, Meissner's plexus (nearest the mucosa) and the plexus of Shabadasch (nearest the circular muscle). It innervates mucosal epithelial cells. Sympathetic

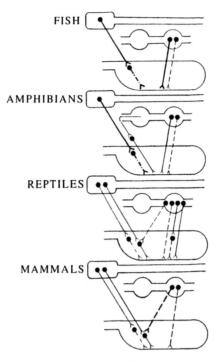

Fig. 10.3. Diagrammatic representation of the autonomic cholinergic excitatory (*solid line*), adrenergic (*broken line*) and nonadrenergic inhibitory (*dashes and dots*) nerves to the stomach of vertebrates. The vagal parasympathetic outflow is purely inhibitory to the fish and amphibian stomachs and is opposed by excitatory cholinergic sympathetic fibers. In reptiles and mammals the cholinergic excitatory nerves have been switched to the parasympathetic outflow, and sympathetic fibers become adrenergic and inhibitory. Adrenergic modulation of intramural ganglion cell activity is rudimentary in reptiles and strongly developed in mammals. The diagram depicting the innervation of the reptile stomach is largely conjectural. Intramural neurons that are independent of the extrinsic nerve supply are not included in the diagrams. (From Burnstock 1969.)

neurons accompany the blood vessels. Vagal parasympathetic neurons enter separately and terminate in the myenteric plexus.

Christensen, et al. (1984) found a distinct difference between the anatomy of the myenteric plexus of the distal colon of rodents (rat and guinea pig) and rabbits) compared to that of the cat, dog, cynomolgus monkey, American oppossum (*Didelphis virginiana*), and the brushtail possum *Trichsuorus vulpecula*. These differences may account for some of the differences in the transit of digesta through the distal colon of these animals.

Descriptions of the autonomic nervous system of fish (Campbell 1970), reptiles (Berger and Burnstock 1979), and birds (Bennett 1974) are not as complete

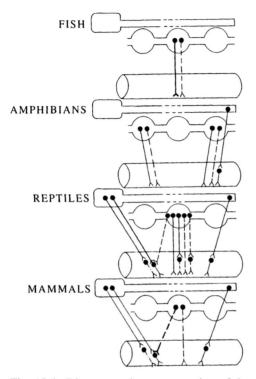

Fig. 10.4. Diagrammatic representation of the autonomic cholinergic excitatory (*solid line*), adrenergic (*broken line*), and nonadrenergic inhibitory (*dashes and dots*) nerves to the intestine of vertebrates. The diagram illustrating the reptile condition is partly conjectural, being based largely on the results of fluorescent histochemical studies. The vagus nerve does not extend far enough down the gut in lower vertebrates to influence most of the intestine, although in mammals it extends at least as far as the ileocolonic junction. Adrenergic modulation of ganglion activity appears first in the reptiles and is strongly developed in mammals. It is not known whether adrenergic nerve terminals envelop postganglionic sacral parasympathetic neuron cell bodies. A separate sacral parasympathetic outflow in amphibians is depicted, although this is debatable. Intramural neurons that are independent of the extrinsic nerve supply are not included, although in the large intestine there is evidence that intramural, nonadrenergic inhibitory neurons, as well as cholinergic neurons are involved in local reflex pathways. (From Burnstock 1969.)

as those for mammals. However, Burnstock (1969) concluded that there were some marked differences in the autonomic nervous system of the different classes of vertebrates (Figs. 10.3 and 10.4). Vagal innervation of the digestive tract of fish did not extend beyond the stomach, and stimulation of the vagal nerves produced a nonadrenergic, noncholinergic inhibitory response, whereas stimulation of the spinal autonomic nerves to the gastrointestinal tract produced both a cholinergic excitatory and adrenergic inhibitory response. Autonomic innervation of the gastrointestinal tract of adult amphibians appeared to be similar to that of fish, except for evidence of a sacral cholinergic nerve supply. The autonomic nervous system of reptiles appeared to be similar to that of mammals, with the complete exchange of cholinergic excitatory function from the sympathetic to the parasympathetic outflow; retention of some vagal nonadrenergic, noncholinergic inhibitory innervation; and sacral parasympathetic innervation to the hindgut. The autonomic nervous system of birds also appears to be similar to that of mammals (Bennett 1974; Tindall 1979; Hodgkiss 1984).

The salivary glands of mammals receive their parasympathetic stimulation via the glossopharyngeal and facial nerves and their sympathetic stimulation of blood vessels and secretory cells via the superior cervical ganglia. The salivary glands of birds (Bennett 1974) and reptiles (Berger and Burnstock 1979) appear to be provided with similar innervation. Vagal nerve fibers innervate postganglionic cholinergic neurons in the pancreas of mammals. This provides a cephalic phase of pancreatic secretion, stimulated by the sight, smell, or ingestion of food. Cholinergic innervation of pancreatic cells has been observed in the snake *Elaphe quadrivirgata* (Watari 1968), but there appeared to be no evidence of vagal stimulation of the fish pancreas (Barrington 1957). A cephalic phase also stimulates the secretion of HCl and pepsinogen in mammals, both directly and through release of the hormone gastrin from pyloric mucosa. Geese (Karpov 1919), ducks (Walter 1939), and barn owls (Smith and Richmond 1972), are reported to have a cephalic phase of gastric secretion, but there is disagreement over its presence in chickens (Burhol 1982; Duke and Bedbury 1985). Vagal stimulation resulted in the secretion of small volumes of gastric juice, high in HCl and pepsinogen, by the tortoise *Testuda graeca* and lizard *Trachydosaurus rugosus* (Wright, Florey, and Sanders 1957), but Barrington (1957) found no evidence of a cephalic phase for gastric secretion in fish.

Distension of the stomach with food results in the further release of gastrin via local nerve reflexes. The gastric phase of gastric secretion appears to be common to all vertebrates studied. Evidence for autonomic control of electrolyte transport by the intestinal mucosa was reviewed by Powell and Tapper (1979). They concluded that the cholinergic response of the small and large intestine was secretion of Cl and that the adrenergic response was absorption of Na and Cl, with perhaps an associated inhibition of HCO_3 secretion.

Neurotransmitters and neuromodulators

The discovery of nonadrenergic and noncholinergic neurons led to the discovery of a large number of purines, amines, peptides, and other agents that served as neurotransmitters and neuromodulators. Burks (1994) and Dockray (1994) reviewed recent developments in the search for these agents and their mechanism of action. Acetylcholine is believed to be the principal neural regulator of gastrointestinal motility and to have a significant effect on mucosal transport of electrolytes. Contractions of the digestive tract are increased by an increase in cholinergic activity in the enteric nervous system and are reduced by a decrease in cholinergic activity in most regions of the tract. However, the release of acetylcholine from cholinergic nerves is governed by a number of other agents. Stimulation of some receptors on these nerve cells by 5-hydroxytryptamine (5-HT) releases acetylcholine, but other receptors for 5-HT, norepinephrine, and dopamine inhibit its release. Adenine and adenine nucleotides (ATP, ADP, AMP) can stimulate either contraction or relaxation of gastrointestinal muscle by inhibiting the release of acetylcholine and norepinephrine or by depolarization of cell membranes.

Some neurotransmitters act directly on target cells. Histamine mediates contraction of gastrointestinal muscle by direct action on smooth muscle cells. Nitric oxide, a fat-soluble gas generated by oxidation of arginine in enteric neurons, is now believed to be a major mediator of the muscular relaxation attributed to nonadrenergic, noncholinergic nerves. However, unlike other neurotransmitters, it is not stored in nerve cells, and it produces its action by penetrating the effector cell to act on guanylyl cyclase.

The major neuropeptides that affect the mammalian digestive system are substance P, vasoactive intestinal peptide (VIP), opioids, gastrin-releasing peptide (bombesin), and neuropeptide Y (NPY) (Dockray 1994). Substance P was the first peptide identified in both the enteric and central nervous systems (von Euler and Gaddum 1931). It is a potent vasodilator and a strong stimulus to salivary secretion (Van Noorden and Polak 1979). It can also stimulate the secretion or inhibit the absorption of electrolytes by the intestine (Powell and Tapper 1979) and can act as a transmitter, together with acetylcholine, for excitation of intestinal motor neurons. Vasoactive intestinal peptide (VIP) is believed to be one of the most important nonadrenergic, noncholinergic transmitters in peripheral nerves. It relaxes many types of vascular and nonvascular smooth muscle and stimulates the intestinal and pancreatic secretion of electrolytes. The opioids such as enkephalin that are found in enteric nerve cells can influence gut motility by a direct (generally excitatory) action on effector cells or by suppressing the release of other transmitters by enteric neurons. NPY inhibits muscular contraction and stimulates secretion of Na and Cl. Its inhibitory effect on muscle may be due to the release of norepinephrine.

Bombesin was first isolated from amphibian skin and found to affect gut motility and secretion in a wide range of vertebrates. A bombesinlike peptide, which showed potent gastrin-releasing activity, was isolated from intestinal tissue of the pig and named *gastrin-releasing peptide (GRP)* (McDonald et al. 1979). It is now known to belong to a family of bombesinlike peptides found in the enteric neurons of vertebrates. Gastrin-releasing peptide stimulates gastric secretion by the release of gastrin from the stomach and stimulates pancreatic secretion by release of secretin in mammals.

Somatostatin, cholecystokinin, and neurotensin are neurotransmitters/hormones that have been identified in nerve cells of the peripheral and central nervous system and in endocrine cells of the gastrointestinal tract or pancreas of a wide range of vertebrates. The principal action of somatostatin in nerves appears to be the inhibition of acetylcholine release (Van Noorden and Polak 1979; Walsh 1994). The roles of somatostatin, cholecystokinin, and neurotensin as hormones are discussed in the next section on endocrine control.

With the exception of nitric oxide, most of the agents that have been identified as neurotransmitters or neuromodulators in mammals have been found in the nervous system of other classes of vertebrates. Substance P-like immunoreactivity has been identified in the gut of birds, reptiles, amphibians, and teleosts (Buchan, Polak, and Pearse 1980; Rawdon and Andrew 1981; Brodin et al. 1981; Holmgren, Vaillant, and Dimaline 1982; Buchan, Lance, and Polak 1983; Rombout and Reinecke 1984). VIP-like activity was found in the nerve cells of all vertebrates except cyclostomes (Falkmer et al. 1980; Fouchereau-Peron et al. 1980; Vaillant, Dimaline, and Dockray 1980; Buchan et al. 1981; Fontaine-Peruse, Chanconie, Polak, and Le Douarin 1981; Rawdon and Andrew 1981; Reinecke, et al. 1981; Holmgren et al. 1982; Buchan et al. 1983; Holmgren and Nilsson 1983; el Salhy 1984; Rombout and Reinecke 1984; Rawden 1984). Enkephalins have been found in birds (Epstein, Lindberg, and Dahl 1980), as well as mammals.

Bombesinlike immunoreactivity has been identified in the gut of mammals, birds, reptiles, amphibians, teleosts, and cartilaginous fish (Timson et al. 1979; Vaillant, Dockray, and Walsh 1979; Buchan, Polak, and Pearse 1980; Rawdon and Andrew 1981; Holmgren, Vaillant, and Dimaline 1982; Buchan, Lance, and Polak 1983; Holmgren and Nilsson 1983). Administration of bombesin stimulated both secretion of acid (Holstein and Humphrey 1980) and motility (Holmgren and Jonsson 1988) in the Atlantic codfish *Gadus morhua*.

Studies of neurotransmitting and neuromodulating agents are complicated by the fact that some neurons produce more than one of these agents, and one may modulate the release of another. This includes neurons that release ATP and norepinephrine, ATP and substance P, and acetylcholine and VIP. Species comparisons are further complicated by the fact that many of these agents are released from endocrine cells as well. For example, substance P, bombesin-GRP, and VIP are present in the gut endocrine cells of many species. Bombesinlike activity was found in endocrine cells of the stomach of birds (Table 10.1) and in the stomach

Table 10.1. *Distribution of cells showing immunoreactivities for gut peptides and serotonin in the gastrointestinal tract of the chicken*

	SOM	APP	PYY[a]	GLUC	SEC	VIP	GAS	CCK	NT	BN	SP	ENK[b]	MOT	5-HT[c]
Proventriculus	●	●							●	●●				●
Gizzard				●●									●	
Pylorus	●●						●●			●	●●		●●	●●
Duodenum	●	●	●	●	●	●	●●●	●[d]	●●		●	●		●
Upper ileum	●	●	●	●	●	●		●	●		●	●		●
Lower ileum		●		●		●		●	●			●		●
Cecum						●					●			●●
Rectum						●			●●					●

Abbreviations: *SOM*, Somatostatin; *APP*, avian pancreatic polypeptide; *PYY*, polypeptide YY; *GLUC*, glucagon; *SEC*, secretin; *VIP*, vasoactive intestinal peptide; *GAS*, gastrin; *CCK*, cholecystokinin; *NT*, neurotensin; *BN*, bombesin; *SP*, substance P; *ENK*, leu-enkephalin; *MOT*, motilin; *5-HT*, serotonin.

[a]El-Salhy et al. (1982b), recently hatched chicks.
[b]Alumets, Hakanson, and Sundler (1978), chickens.
[c]Unpublished observations, chicks at hatching.
[d]Larrson and Rehfeld (1977), chickens. All other data from Rawdon and Andrew (1981), chicks at hatching.
Source: Rawdon (1984).

and intestine of amphibians (see Fig. 10.6). GRP was found in endocrine cells of the proventriculus of birds (Table 10.1) and amphibians (see Fig. 10.6), and VIP was present in the gut epithelial cells of many nonmammalian vertebrates (Reinecke et al. 1981; Rawdon 1984). Therefore, some of these agents can serve as neurotransmitters or modulators and/or endocrine agents in various species.

Endocrine control of the digestive system

Hormones are agents that are secreted in small amounts by endocrine cells and act on the receptors of target cells. The classical hormones are secreted into the blood and act on distant target organs of the digestive system, but some act in a paracrine fashion on neighboring cells. As with the neurotransmitters and modulators, studies conducted in recent years have added a great deal of new information on sites of production and mechanisms of action of hormones. Many of these agents are produced in both endocrine and nerve cells and can act as either hormones, neuromodulators, or neurotransmitters in the same species or different classes of vertebrates. Walsh (1994) reviewed information on the principal hormones that act on the digestive system of mammals. These consisted of gastrin, cholecystokinin, secretin, gastric-inhibitory polypeptide, enteroglucagons, pancreatic polypeptide, peptide YY, somatostatin, motilin, and neurotensin. Fig. 10.5 illustrates sites of secretion and actions of some of these hormones on the digestive system of mammals. The distribution of endocrine cells that contain some of these agents in amphibians, reptiles, and birds is illustrated in Fig. 10.6 and Tables 10.1 and 10.2.

Gastrin

Gastrin is the physiological stimulator for the gastric phase of acid secretion and an important regulator of gastric cell proliferation. It is synthesized by and stored in the G cells of the pyloric glandular and duodenal mucosa of mammals. The structural characteristics of gastrin have been determined for a number of species. As many as six different gastrinlike molecules have been isolated from a single species, and many amino acid substitutions are found in the gastrins of different animals. However, the C-terminal pentapeptide is common to all forms. Gastrin can be released by vagal stimulation or the presence of peptides, amino acids, and Ca salts in gastric contents. Its release is inhibited by gastric acidification and the local release of somatostatin.

In addition to its effect on gastric secretion, gastrin stimulates the growth of gastric mucosa and the contraction of circular muscle of the stomach and lower esophageal sphincter. Systemic administration of pentagastrin, a synthetic peptide with gastrin activity, inhibited the motility of the reticulorumen of sheep (McLeay

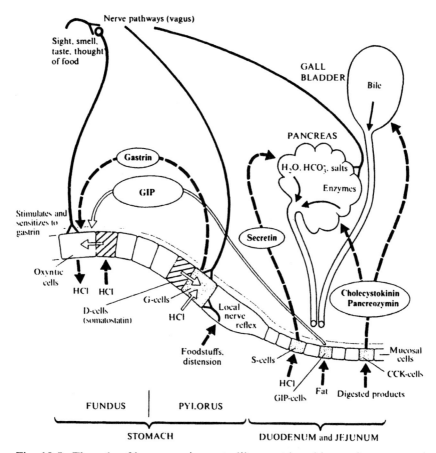

Fig. 10.5. The role of hormones in controlling gastric acid secretion, pancreatic secretion of salts and enzymes, and contraction of the gallbladder. Gastrin, from the pylorus, initiates secretion of hydrochloric acid by the oxyntic cells in the fundus. The duodenojejunal hormones, secretin and cholecystokinin-pancreozymin, initiate the secretion of pancreatic juice and enzymes, respectively. Gastric-inhibitory polypeptide (GIP) inhibits gastric acid secretion. *Open arrows* indicate inhibition; *closed arrows* indicate stimulation. (From Bentley 1982.)

and Titchen 1970, 1975) and cattle (Ruckebusch 1971). Subsequent studies in sheep showed this to be a direct effect on the brain (Grovum and Chapman 1982; Nicholson 1982). Intracerebroventricular administration of pentagastrin, tetragastrin, or gastrin-17 inhibited rumination in sheep (Honde and Bueno 1984). There is evidence that gastrin also stimulates secretion and inhibits absorption of electrolytes by the intestine (Powell and Tapper 1979).

Gastrin appears to be absent from the intestine of lampreys (Holmquist 1979), but it was present in the intestine of the stomachless carp *Cyprinus carpio* (Noaillac-Depeyre and Hollande 1981). Gastrinlike activity was found in the stomach of the coho salmon *Oncorhynchus kisutch* (Vigna 1979) and rainbow trout *Salmo*

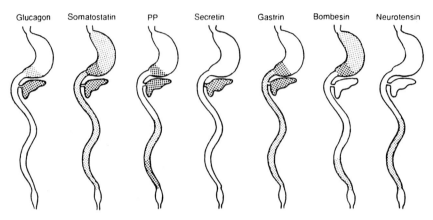

Glucagon   Somatostatin   PP   Secretin   Gastrin   Bombesin   Neurotensin

Fig. 10.6. Diagram showing the populations of endocrine cells containing different peptide immunoreactivities in the digestive system of the frog *Rana catesbeiana*. (From Fujita et al. 1981.)

gairdneri (Holmgren, Vaillant, and Dimaline 1982); the stomach and intestine of the perch *Perca fluviatilis* and catfish *Ameiurus nebulosus*; and in the rectum of the dogfish (Holmgren and Nilsson 1983). It was found in the stomach, intestine, and pancreas of the frog (Fig. 10.5); the stomach of the salamander *Salamandra salamandra* crocodile *Crocodylus niloticus* and alligator *Alligator mississippiensis* (Dimaline et al. 1982; Buchan, Lance, and Polak 1983); and in the stomach and intestine of the caiman (Table 10.2). It was also present in the stomach and duodenum of the chicken (Table 10.1), duck (Larsson and Rehfeld 1977), and turkey (Dockray 1979b). Pentagastrin was a stronger stimulus for secretion of pepsinogen than HCl by the cells that produce both of these in the chicken (Burhol 1982).

Cholecystokinin

Cholecystokinin (CCK) was first described as a hormone that was produced in the duodenal and jejunal mucosa of mammals, was released by the presence of fat in the duodenum, and was responsible for contraction of the gallbladder and relaxation of the choledochal sphincter. This was followed by the discovery that the presence of HCl and the digestive products of fat and protein in the duodenum released a hormone called *pancreozymin (PZ)*, which stimulated the secretion of pancreatic enzymes. Pancreozymin proved to be identical to CCK, with the result that it was subsequently labeled CCK-PZ and then, simply, CCK.

Cholecystokinin stimulates contraction of the gallbladder and pancreatic secretion of enzymes, inhibits gastric emptying and acid secretion, and acts as a

Table 10.2. *Distribution and frequency of gastrointestinal endocrine cells in the caiman*

	Stomach		Intestine					
	Fundic	Pyloric	Initial region	Middle of proximal half	Middle of whole length	Middle of distal half	Ampulla	Cloaca
5-HT	+	−	++	++	++	++	++	−
Somatostatin	+	++	+	+	+	+	±	−
Gastrin	−	+++	++	++	+	−	−	−
Motilin	±	++	++	++	+	±	−	−
Neurotensin	−	−	+	+	+	+	++	++
BPP	−	−	±	±	±	−	−	+
APP	−	−	−	±	−	−	−	−
Pancreatic glucagon	±	−	±	±	±	−	−	−
Enteroglucagon	−−	−	++	++	++	+	±	−
Glicentin	−	−	±	±	±	−	−	−
Secretin	−	−	±	−	−	−	−	−
CCK-33	−	−	±	−	−	−	−	−
GRP and bombesin	−	−	−	−	−	−	−	−
GIP	−	−	−	−	−	−	−	−

− Absent, ± rare (not detected in every animal), + few (detected in every animal but not in every section), ++ moderate, +++ numerous.
From Yamada et al. 1987.

281

short-term satiety factor. It also appears to play an important role in peristalsis and is found in both endocrine cells and nerves. Its activity on the gallbladder appears to be purely hormonal, but its other activities appear to involve either hormonal and neural or strictly neural pathways (Liddle 1993).

The structure of CCK can vary with species, but the C-terminal octapeptide possesses all of the hormone activity, and the C-terminal pentapeptide is identical to that of gastrin. Cholecystokininlike immunoreactivity has been demonstrated in the intestine of birds, reptiles, amphibians, teleosts, chondrichthyins, and cyclostomes (Barrington and Dockray 1970, 1972; Nilsson 1970, 1973; Östberg et al. 1976; Dockray 1979a; Vigna 1979; Dimaline et al. 1982; Holmgren, Vaillant, and Dimaline 1982; Buchan, Lance, and Polak 1983; Dimaline 1983; Holmgren and Nilsson 1983; Vigna et al. 1985; Jonsson, Holmgren, and Holstein 1987). It also was found in the stomach of the chicken (Vigna 1984), crocodile (Dimaline et al. 1982), and cod (Jonsson, Holmgren, and Holstein 1987).

The actions of CCK appear to vary among the different classes of vertebrates. Infusion of mammalian CCK into chickens stimulated gastric acid secretion and release of pancreatic enzymes (Burhol 1982). These may have been pharmacological rather than physiological effects caused by the large doses used. However, Martinez et al. (1993) found that whereas mammalian CCK-octapeptide (CCK-8) and CCK-tetrapeptide (CCK-4) inhibited gastric motility in the chicken, duodenal motility was stimulated by CCK-8 and inhibited by CCK-4 and chicken gastrin. This suggested differences between the CCK-mediating receptors of birds and mammals. Intracerebral or intravenous administration of CCK-8 to plaice, *Platessa platessa*, inhibited motor activity in the cranial stomach and increased peristaltic activity in the pyloric stomach and anterior midgut (Shparkovskii 1988). The intracerebral effects were blocked by vagotomy, and the intravenous effects were blocked with atropine, suggesting that these activities were mediated by neural pathways.

Cerulein is a decapeptide initially found in frog skin, with a structure identical to that of the C-terminal decapeptide of CCK, except for a single amino acid substitution. Ceruleinlike activity has been identified in the stomach of teleosts and amphibians (Larsson and Rehfeld 1977, 1978; Dimaline 1983).

Secretin

Secretin was the first hormone to be discovered, when Bayless and Starling (1902) showed that acid placed in a denervated loop of the upper intestine of the anesthetized dog resulted in secretion of pancreatic fluid. It is released by acidification of luminal contents in the proximal small intestine, and it stimulates the release of HCO_3-containing fluid from the pancreas. It also stimulates bile salt-independent biliary secretion in humans and a number of other species. Although secretin has only a weak stimulatory effect on pancreatic enzyme secretion, and CCK is only a weak stimulus for pancreatic and biliary secretion of fluids, the two hormones have a synergistic effect on both types of secretion when released together. Secretin also stimulated the secretion of pepsinogen by the dog and cat stomachs. It was reported to inhibit gastric secretion of HCl in dogs and rats but appeared to have no effect in humans, and it stimulated the secretion of HCl and pepsin in birds (Burhol 1982). Secretin inhibited gastric emptying in dogs and humans and appears to have a physiological role in its regulation. It also stimulated Cl secretion and inhibited Na and HCO_3 absorption in the rat small intestine.

Secretin is a peptide containing twenty-seven amino acid residues, many of which are common to vasoactive intestinal polypeptide (VIP), gastric-inhibitory polypeptide (GIP), and glucagon. VIP shares nine of its twenty-eight amino acids with secretin. Secretinlike activity has been identified in the gut of birds, reptiles, teleosts, and cyclostomes (Barrington and Dockray 1970; Dockray 1974, 1975, 1978; Nilsson 1970, 1973; Nilsson et al. 1980; Rawdon and Andrew 1981; Buchan, Lance, and Polak 1983). It was also found in the pancreas of frogs (Fig. 10.6) but was absent from the gut of the salamander (Buchan, Polak, and Pearse 1980).

Gastric-inhibitory polypeptide

Gastric-inhibitory polypeptide (GIP) is released from the middle and lower small intestinal mucosa by the presence of glucose, fat digestion products, HCl, and amino acids in the lumen. This hormone inhibits gastric acid secretion, and it stimulates Cl secretion and reduces Na and water absorption by the intestine. The rise in plasma GIP concentrations after a meal supports its role as a true hormone. The peptide isolated from pig intestinal mucosa contains forty-three amino acids, and ten of the first twenty-seven are identical to those of porcine secretin (Bacarese-Hamilton, Adrian, and Bloom 1984). GIP has been considered a prime candidate for the role of enterogastrone, a hormone proposed to be responsible for the inhibitory effect of fat ingestion on gastric acid secretion.

GIP-like immunoreactivity has been reported in the pancreas and intestinal mucosa of the dogfish (el-Salhy 1984), but was not demonstrated in the gut of the salamander (Buchan, Polak, and Pearse 1980) or reptiles (Buchan et al. 1982; Buchan, Lance, and Polak 1983). However, porcine GIP was shown to stimulate HCO_3 secretion by the duodenum of the bullfrog *Rana catesbeiana* (Flemström and Garner 1980).

Enteroglucagons

Mammalian intestinal cells produce a number of glucagons that stimulate the secretion of insulin by pancreatic islet cells, and a proglucagon called *GLP*, which

inhibits gastric acid secretion. Glucagons were also found in the stomach and intestine of chickens (Table 10.1), the caiman (Table 10.2), and the frog (Fig. 10.6).

Pancreatic polypeptide

Pancreatic polypeptide (PP) belongs to a family of peptides that includes peptide
YY (PYY) and neuropeptide Y (NPY). Pancreatic polypeptide is present in pancreatic islet and F cells and is released by cholinergic stimulation. It inhibits pancreatic exocrine secretion of HCO_3 and protein, and increases gastric and intestinal motility in mammals. It has been isolated from the pancreas of many mammals
(Floyd et al. 1977), and sheep, pigs, and dogs have a PP identical in molecular
structure. Pancreatic polypeptide was first isolated from the pancreas of chickens
by Kimmel, Hayden, and Pollock (1975), where it was shown to be a potent stimulus of acid and pepsin secretion (Kimmel, Pollock, and Hayden 1978). It was
subsequently isolated from the pancreas of several mammals. However, twenty of
the thirty-six amino acids in mammalian PP differ from those of the chicken hormone (Walsh 1981), and mammalian PP has little or no effect on gastric secretion of either the chicken or mammals.

PP-like immunoreactivity was demonstrated in the pancreas of ten additional
species of birds, as well as in reptiles and amphibians (Langslow, Kimmel, and
Pollock 1973). It also was found in the intestine of the chicken (Table 10.1),
caiman (Table 10.2), the intestine and pyloric stomach of frogs (Fig. 10.5), the
gut and pancreas of teleosts, and the intestine of cyclostomes (Van Noorden and
Polak 1979). However, it was not found in the gut of the salamander (Buchan,
Polak, and Pearse 1980) or alligator (Buchan, Lance, and Polak 1983).

Peptide YY

Peptide YY (PYY) was found in endocrine cells in the ileum and colon of the rat,
pig, and human and in the midgut of chickens (Table 10.1). It inhibits gastric and
pancreatic secretion and motility of the gastrointestinal tract.

Somatostatin

Somatostatin is found in peripheral and CNS nerves and in gastrointestinal and
pancreatic endocrine cells. It can inhibit gastric secretion of acid and pepsinogen,
pancreatic secretion of enzymes, and interdigestive electrical and motor activity
of the stomach and small intestine. It also can block the secretion and stimulate
the absorption of electrolytes in the small intestine. It is present in blood but appears to act principally as a paracrine inhibitor on the release of acetylcholine,

gastrin, cholecystokinin, secretin, GIP, VIP, PP, PYY and enteroglucagon. So-matostatinlike immunoreactivity has been identified in the gut of mammals, birds, reptiles, amphibians, teleosts, chondrichthyeans, and cyclostomes (Falkmer and Östberg 1977; Alumets, Sundler, and Hakanson 1977; Langer et al. 1979; Seino, Porte, and Smith 1979; Buchan, Polak, and Pearse 1980; Rawdon and Andrew 1981; Holmgren and Nilsson 1983). Somatostatin is widely distributed in the gastric and intestinal mucosa of the chicken (Table 10.1), caiman (Table 10.2), and both the gastrointestinal tract and pancreas of the frog (Fig. 10.6).

Motilin

Motilin, which is found in the intestinal mucosa of mammals, appears to stimulate the interdigestive migrating myoelectric complexes in the stomach and small intestine of the dog and human (Lee et al. 1983; Nakaya et al. 1983; Usellini et al. 1984; Borody, Byrnes, and Titchen 1984) but does not stimulate those of the pig (Borody, Byrnes, and Titchen 1984). Although the plasma levels of motilin increase during the fasting state in dogs and decrease after ingestion of a meal, the stimulus is unknown. Motilinlike immunoreactivity has been identified in the stomach and intestine of the chicken (Table 10.1) and in the intestine of Japanese quail *Coturnix c. japonica*. It also was found in the caiman stomach and intestine (Table 10.2) but not in the lizard *Anolis carolinesis* or the Pacific hagfish *Eptatretus stouti* (Seino et al. 1979).

Neurotensin

Neurotensin is produced by gut endocrine cells and neurons in the brain. It has been considered as a hormone mediator for a number of functions stimulated by ingestion of fat, including inhibition of gastric secretion, stimulation of pancreatic and intestinal fluid secretion, and changes in intestinal motility from the fasting to the fed pattern. However, no physiological role has been proven for neurotensin at normal blood concentrations. Neurotensinlike immunoreactivity has been reported in the stomach and intestine of chickens (Table 10.1); reptiles (Table 10.2); amphibians (Fig. 10.6); and freshwater, stomachless teleosts (Sundler et al. 1977a,b; Reinecke et al. 1980).

Melatonin

Although the pineal gland has long been known to be a source of melatonin, its synthesis by gut tissues is a relatively recent discovery. Melatonin and its synthesizing enzymes have been detected in the esophagus, glandular stomach, duodenum, and colon of rats and pigeons, but their presence in the digestive tract of fish remains unresolved (Lee, Hong, and Pang 1991). Pinealectomy has no effect on melatonin levels in the gut, and the villi are thought to be the site of synthesis in the mammalian midgut (Raikhlin and Kvetnoy 1976). Like pineal melatonin, gut melatonin levels were severalfold higher in the dark than in the light phase of the circadian cycle.

Melatonin had several direct gastrointestinal effects, including inhibition of proliferation of jejunal epithelium, reduction in the transport of Na by the colonic epithelium of Na-deficient rats, and inhibition of serotonin-induced contraction of gut muscle (Lee and Pang 1993). These paracrine and/or hormonal actions of melatonin on both gastric and intestinal tissues suggest the presence of melatonin receptors at multiple sites along the gut. Findings of significant immunoreactivity supported this suggestion in the gut of the mouse, duck, chicken, and human. In the duck, densities of melatonin binding sites were greatest in the midgut, followed by the colon, cecum, and esophagus (Lee and Pang 1993). How gut melatonin functions in the integrated control of gut activities under various nutritional regimens and during different seasons of the year remains to be established.

Cytokines

The cytokines represent another group of autocrine and paracrine substances that affect the gastrointestinal epithelial cells. They are comprised of a large group of polypeptides or glycoproteins that mediate intercellular signaling and include lymphokines, monokines, interleukins, interferons, and growth factors. The cytokines are principally involved in communication among the lymphoid cells of the mucosal immune system. However, recent evidence shows that the epithelial cells produce and respond to these cytokines, and that they are major regulators of epithelial cell growth, development, and restitution following injury. Elson and Beagly (1994) have reviewed the role of these mediators in immunity and other functions of the digestive tract.

Summary

The autonomic nervous system of fish appears to differ from that of reptiles, birds, and mammals in both the absence of a sacral parasympathetic nerve supply to the intestinal tract and the action of its cranial and spinal nerves. The sacral parasympathetic nerve supply may have first appeared in the amphibians, but reptiles show the first evidence of craniosacral parasympathetic and thoracolumbar sympathetic nervous systems that function in a role similar to those of mammals.

Neuroendocrine control of the digestive system is effected by a wide range of agents, which are secreted by nerves and endocrine cells and act directly on the effector cells or the release of one another. Many of these agents act as both neurotransmitters/modulators and hormones in either the same species or different classes of vertebrates. The receptors on target organs or cells also appear to vary among classes of vertebrates.

As with many previous topics, comparative data on neuroendocrine control in nonmammalian vertebrates is limited to relatively few species. Interpretation is often further hampered by a lack of information on the molecular structure of the agents produced in a given species and their physiological effects in the same animal. However, comparative studies have yielded a great deal of information on the agents that are involved in neuroendocrine control and have generated a number of theories on how they may have evolved. These are discussed in Chapter 11.

11

Evolution of the digestive system

This chapter summarizes the diversity of adaptations of the vertebrate digestive system to the environment, the diet and other physiological characteristics of animals, and it speculates on how these may have evolved. The fossil remains of the digestive system are mainly limited to teeth and jaws, which relate principally to the diet and feeding practices. Therefore, much of the evidence must be derived from studies of extant species. However, speculations on the evolution of this system can provide clues as to how animals have adapted to their environment and the mechanisms that made this possible.

Although the subject of invertebrates is beyond the scope of this book, brief mention of their digestive systems provides some necessary perspective. Barnes (1974) pointed out that 95 percent of all described species in the Animal Kingdom are invertebrates, and some are more closely related to vertebrates than to other invertebrate groups. Therefore, a taxonomist less biased than Man might provide a better division of animals into arthropods (75% of all species) and nonarthropods (25%). It is not surprising that most of the basic characteristics of the vertebrate digestive system are found in various invertebrate species. The reader is referred to Barnes (1974) and Andrew (1959) for broad discussions of invertebrate digestive systems, Wigglesworth (1984) for similar discussion of insects, and Barnard and Prosser (1973) and Vonk and Western (1984) for comparison of invertebrates and vertebrates.

Protozoa and parazoa (sponges) do not have a digestive tract. Some protozoa absorb nutrients across their body cell membrane, but many ingest food by phagocytosis at a specific site or various points on the membrane. Ingested material is taken up into food vacuoles, which undergo prolonged passage through the cell, with digestion of their contents, absorption of nutrients across the vacuolar membrane, and subsequent evacuation of waste products from the cell. Many of the digestive enzymes found in vertebrates have been isolated from various species of protozoa (Vonk and Western 1984). In some, such as *Paramecium*, intravac-

288

uolar digestion is associated with a sequential acidification and alkalinization in a manner similar to that seen in the digestive system of most vertebrates.

Most other invertebrates have a mouth and either a blind digestive cavity (gastrovascular cavity) or a digestive tract that terminates in an anus. The digestive tract of most advanced invertebrates can be divided into a headgut (oral cavity and throat), foregut (esophagus and stomach), and intestine. The intestine may be further subdivided into a midgut, which serves primarily for digestion and absorption, and a hindgut, which retains undigested material for the reabsorption of water and electrolytes secreted into the digestive tract.

As one might expect, the ability to conserve water is particularly necessary for terrestrial species. Movement of food and transport of digestive secretions is aided by cilia in many invertebrates. However, the mechanical breakdown of food requires muscular activity, and as the size of food particles and the diameter of the digestive tract increase, cilia lose their effectiveness. In many of the more advanced invertebrates, the motor activity required for ingestion, physical breakdown, and transit of digesta is accomplished by muscular contraction. Both cilia and muscles are used for this purpose in many invertebrates, but cilia are absent from the digestive tract of others, such as the nematodes and insects.

The evolution of vertebrates is illustrated in Fig. 11.1. The earliest vertebrates, which are believed to have been freshwater fish most closely related to the present-day cyclostomes, evolved into the cartilaginous fish and teleosts. Amphibians are believed to have evolved from predacious lobe-finned fishes (the Crossopterygii), and the earliest amphibians demonstrate the same well-developed, sharp, pointed teeth in their lateral jaw (Romer 1966). Some were quite large, up to 1 m in length. The reptiles are believed to have evolved from the amphibian anthrocosaurs, over 300 million years ago (mya), resulting in the dinosaurs, which became the dominant reptiles from the mid-Triassic to the end of the Cretaceous period, and the chelonians, crocodiles, snakes, and lizards that are present today. Birds are believed to have evolved from reptiles and resemble them in many ways other than in their feathers, flight, and endothermy. Bakker (1987) has suggested that they evolved from the dinosaurs, but other evidence suggests an earlier origin.

Mammals first appeared in the late Triassic period, at least 200 mya (Lillegraven, Kraus, and Brown 1979), evolving from therapsid reptiles into the Prototheria or monotremes, and, via the Pantotheria, into the metatherian (marsupial) and eutherian (placental) mammals. However, they did not compete successfully with the dinosaurs until the latter's demise at the end of the Cretaceous period (70 mya). Analysis of dental and jaw morphology of Mesozoic mammals suggests that the earliest mammals were small (20–30 g) carnivores that fed mainly on insects, other arthropods, and small vertebrates. Endothermy would have allowed nocturnal feeding and avoidance of competition from and predation by reptiles

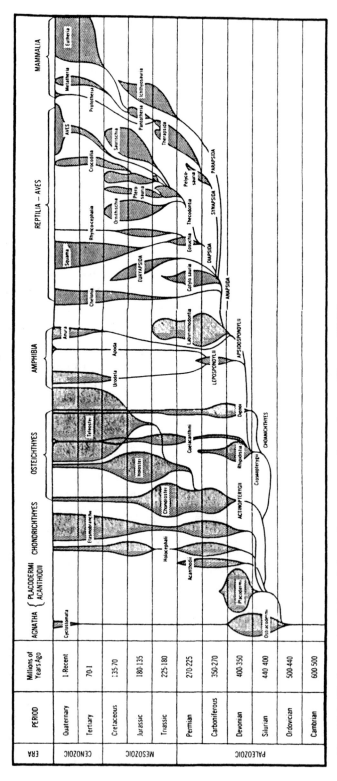

Fig. 11.1. Phylogenetic origins of the various groups of vertebrates. (Torrey 1971.)

(Crompton 1980). Forms such as the shrews, moles, and hedgehogs, which have retained similar feeding habits and many primitive or conservative characters, are regarded as the most direct modern descendants of the primitive mammals (Romer 1966).

Headgut

Efficient digestion of food requires that it be presented in particles small enough for ingestion and hydrolysis by enzymes. The headgut of invertebrates shows a wide range of structural variation that can be related directly to their diet. Some advanced invertebrates are filter-feeders, with complex arrangements of their mouth parts that selectively sort out the small particles of food for ingestion. Some predigest their food by regurgitation of enzymes into their prey (scorpions, mites). Some echinoderms completely evert their stomach to engulf their prey. Some species ingest their entire prey in one piece, and store it in a stomach for enzymatic and physical reduction in size. However, most of the advanced invertebrates reduce their food into smaller particles by cutting, tearing, or grinding it with mouth parts designed for these purposes. Once particles have been reduced to appropriate size, they are ingested and passed through the digestive tract to undergo the serial events associated with digestion, absorption, and the excretion of waste products.

The headgut of vertebrates also shows a great degree of divergence and convergence with respect to diet. This is reflected in numerous adaptations of mouthparts for the procurement of food and in the development of oral glands for its preparation and deglutition. The major adaptations involve the jaws and teeth. Fish, which constitute almost half of the living vertebrate species, show the widest diversity. Although jaws are absent in the cyclostomes, they are present in all other vertebrates. The teeth of fish vary in both their location (jaw, tongue, pharynx, or other surfaces of the orobranchial cavity) and function. In most species they are used for grasping, cutting, or tearing. The ability to reduce food to small particles is limited to those that are filter-feeders, grazers, or those that grind their food with pharyngeal teeth. Potential loss of small food particles through the gills can account for the gill rakers in many species and the adaptation of the pharynx for mastication in others. Many of these characteristics are seen in larval amphibians as well, but adult amphibians are carnivorous, with weak dentition that serves only for the grasping of food while it is being swallowed.

The mouth parts of most reptiles are used for grasping, cutting, and tearing their food. This is accomplished with a beak in the chelonians and fangs or jaw teeth in other reptiles. A few species have teeth that can crush their food, but the scissorslike action of the jaw prevents the grinding action associated with mastication by mammals. The mouth parts of birds show a wide range of diversification with the diet of the species. This is particularly true for the bill, which can serve for cutting, tearing, crushing, filter-feeding, or other purposes. However, as with other nonmammalian vertebrates, the jaws are not constructed for efficient grinding of food.

One of the major advances in the evolution of mammals was the development of an extremely efficient masticatory apparatus. Although a few species have lost their teeth or adapted them for other purposes, most mammals have a set of teeth that includes incisors for cutting, canines or fang teeth for grasping and tearing, and premolars and molars with uneven occluding surfaces. The latter are used for crushing and, because of the articulation and musculature of the jaw, for grinding of food. Placement of food between the shearing, crushing, and grinding surfaces is aided by muscular cheeks and a mobile tongue.

Foregut

The foregut of advanced invertebrates often includes an esophagus and a stomach. The stomach is used for preliminary digestion of food in most species. Although the foregut of vertebrates shows less diversity than the headgut, it also demonstrates a wide range of adaptations to diet. With the exception of cyclostomes, the foregut and remainder of the digestive tract are enveloped to varying degrees with layers of longitudinal and circular muscle. The esophagus aids in the mechanical breakdown of food in some egg-eating reptiles and the storage of food in some reptiles and most birds. It can absorb NaCl and reduce the salinity and osmolality of sea water reaching the stomach of euryhalene and marine fish. However, its major function is to transport ingesta from the mouth to the midgut of stomachless fish, to the crop and proventriculus of birds, or to the stomach of other vertebrates. The presence of striated muscle in the esophagus of fish and at least a major portion of the esophagus of all mammalian species aids in the rapid ingestion of food. Complete replacement of this with more readily distensible smooth muscle in adult amphibians, reptiles and birds, and its partial replacement in many mammals may have allowed their esophagus to serve as an additional storage area.

The absence of a stomach in cyclostomes appears to be a primitive characteristic, but its absence in some of the more advanced species of fish is probably a regressive characteristic or secondary loss, which is often associated with the presence of pharyngeal teeth. Pharyngeal mastication allows the reduction of food into small particles without their loss through the gills, and it may have obviated the need for further storage and digestion in a stomach. However, a stomach serves for the storage, maceration, and initial digestion of food in all advanced vertebrates other than birds, in which these functions are carried out separately by the crop, proventriculus, and gizzard. With the exception of some larval amphibians and the monotremes, the stomach contains a proper glandular region, which secretes mucus, HCl, and pepsinogen. It also contains a pyloric glandular region, which secretes mucus and HCO_3. This is joined by an additional area of cardiac glandular mucosa, which also secretes mucus and HCO_3 in some amphibians and the reptiles.

The stomach of most mammals shows the same general distribution of cardiac, proper gastric, and pyloric glandular mucosa as that of reptiles. However, the stomach of some species differs from those of all other vertebrates with respect to its epithelial lining and its relative size and complexity. Ten of the twenty mammalian orders contain species in which at least part of the stomach is lined with stratified squamous epithelium. Stratified squamous epithelium lines the entire stomach of the monotremes and armadillo; most of the stomach of many anteaters; approximately half of the horse, rhino, and tapir stomach; and large segments of the forestomach of the macropod marsupials, sloths, whales, ruminants, hippos, and hyrax; and some rodents. It also occupies a small area of the colobus and langur monkey forestomach and the stomach of the domestic pig. The cardiac glandular region is similarly expanded in the forestomach of many herbivorous species. This led Oppel (1887) and Bensley (1902–1903) to conclude that it represented regression of the highly specialized proper gastric glandular mucosa to less complex cardiac mucosa and then nonglandular stratified squamous epithelium.

The widespread presence of stratified squamous epithelium in the stomach of animals considered to represent both primitive and advanced species suggests that it performs an important function. Its presence in anteaters of various orders (Monotremata, Pholidota, Edentata) and numerous herbivores suggests that it may protect the lining from mechanical abrasion. Although it is relatively impervious to SCFA in the horse, it absorbs SCFA from the stomach of the domestic pig and plays an important role in the absorption of Na, Cl, and SCFA and in the secretion of HCO_3 by the ruminant forestomach. Therefore, the presence of this tissue, plus the buffering capacity of expanded areas of cardiac glandular mucosa, may provide protection against the rapid absorption of undissociated SCFA. Expansion of the proper gastric glandular mucosa may be limited by both the energy cost for production of HCl and the ability of the duodenum to neutralize it. Therefore, the presence of stratified squamous epithelium and cardiac mucosa in the forestomach of mammalian herbivores may represent the most parsimonious alternatives available for stomach expansion.

Eight mammalian orders include species with a voluminous stomach, which is either drawn up into haustrations by longitudinal bands of muscle or partially divided into permanent compartments or diverticula. The expanded stomach appears to serve principally for the storage of food in the vampire, nectarivorous, and frugivorous bats. However, it also serves as an additional or primary site for microbial digestion in many mammals.

Salivary, pancreatic, and biliary secretion

Digestion is accomplished entirely by intracellular enzymes in the protozoa and sponges. However, the coelenterates and higher forms of invertebrates demonstrate an increasing dependence on extracellular digestion. Cells lining the gastrovascular cavity may secrete mucus or digestive enzymes, absorb nutrients or serve for food storage. These functions, which are carried out by multipurpose cells in the lower forms of invertebrates, become the property of specialized cells in more advanced species. For example, the intestinal ceca of trematodes contain both absorptive and secretory cells. Cells that are specialized for secretion of mucus, enzymes, and other substances are located in ceca or glands along the digestive tract of many advanced species. Salivary glands are highly developed in many mollusks and arthropods. Some annelids such as the earthworms have a well-developed system of extracellular digestion, with digestive glands and a general arrangement of alimentary organs and tissue similar to those of vertebrates. The accessory digestive glands of mollusks are especially interesting because various species demonstrate a broad spectrum of functions normally associated with the salivary glands, pancreas, and liver of vertebrates.

Phylum Mollusca, which includes the clams, oysters, mussels, snails, slugs, squid, and octopus, contains the largest number of species of any phylum other than the Arthropoda. It includes carnivores, omnivores, herbivores, scavengers, parasites, and animals that live in marine, freshwater, and terrestrial environments. Digestion is at least partly extracellular in all species, and enzymes may be secreted by salivary glands, esophageal pouches, portions of the stomach, intestinal digestive glands, or a combination of these.

Digestive diverticula in the stomach of bivalves (clams, oysters, mussels) contain phagocytic, vacuolated cells, which serve largely for absorption and intracellular digestion. These cells eventually undergo fragmentation to form spheres containing vacuoles of undigested material, excretory products, and residual enzymes. These spheres are released into the gut lumen and may serve as an important source of the extracellular enzymes. Although the latter may apply to the gastropods, it appears less likely in bivalves because the efflux from the diverticula is not remixed with stomach contents (Barrington 1962). Therefore, dissolution of wandering phagocytes has been suggested as a more probable source of their weak gastric protease and lipase activities. However, some species have few phagocytes, and at least one bivalve genus, a small clam (*Nucula*), lacks the mechanism for a two-way flow of digesta in the diverticula ducts. Therefore, its cells secrete and excrete but do not absorb.

This specialization of the digestive glands proceeds further in the gastropods to the point where in the snail, *Helix*, digestive glands secrete a large variety of extracellular enzymes and appear to have relatively little absorptive capacity. These glands have been referred to as the *hepatopancreas* in a number of species and have been called the *pancreas* and *liver* in cephalopods (squid and octopus).

The development of hepatic functions similar to those of vertebrates appears as another prominent feature in many species of the phylum Mollusca. The liver develops as a gland of many cell types, which can function for absorption, intracellular digestion, secretion, excretion, and storage. In many species, food particles are seen in the vacuoles of its cells. In the squid, however, food must be absorbed and must reach these cells via the blood (Campbell and Burnstock 1968). The squid "pancreatic" duct empties into the "liver" duct, and the combined secretions flow from a common duct, which can be directed into the stomach or cecum. A sphincter of striated muscle, between the hepatic and common duct, prevents reflux of lumen contents into the liver. The liver produces carboxypeptidase, aminopeptidase, and dipeptidase. The latter two, plus lipase, also are produced by the "pancreas."

Phylum Arthropoda includes crustaceans, arachnids, and insects. Digestion is almost exclusively extracellular, except for the final stages in the brush border and contents of intestinal cells, and it is confined principally to the midgut. Salivary glands, which are highly developed in many insects, usually empty into the buccal cavity. They may secrete mucus, enzymes, anticoagulants, agglutinins, venomous spreading agents, or silk in various species. The "hepatic ceca" of the horseshoe crab consist of two large glands, which function for both digestion and absorption. The crustacean midgut has a pair of ceca, modified to form a group of ducts with blind secretory tubules. These tubules, which provide a site for absorption, digestive enzyme secretion and storage of glycogen, fat, and calcium, have been labeled the *hepatopancreas*.

Van Weel (1974) concluded that the terms *hepatopancreas*, *pancreas*, and *liver* were inappropriate when applied to either mollusks or crustaceans and that the structures should be simply referred to as *midgut glands*. Bidder (1976) agreed, with respect to the squid, and proposed that the terms *digestive glands* and *digestive gland appendages* should be used for the respective description of the liver and pancreas. Gibson and Barker (1979), however, concluded that the digestive glands of decapod crustaceans are ". . . rightly and properly named" the *hepatopancreas*. The presence of agents that serve the emulsifying function of bile salts in the invertebrates was discussed by Haslewood (1967).

Among vertebrates, the salivary secretions of reptiles, birds, and mammals aid in the deglutition of food and can aid in its digestion by some species. Kochva (1978) commented on the evolution of the oral glands in reptiles. Junqueira and de Moraes (1965) compared the major salivary glands of vertebrates and concluded that these, unlike the ". . . ready-built pancreas, developed gradually during vertebrate phylogenesis to their maximal complexity in mammals." Tissue that secretes endogenous enzymes is distributed along the intestinal tract of cyclostomes and some of the more advanced species of fish, but this becomes organized within a compact pancreas in most fish and all of the higher classes of vertebrate. Despite differences in structure and composition, and the readoption of endocytosis and intracellular digestion by some mammalian neonates, most of the endogenous digestive enzymes are seen in all classes of vertebrates.

Hagey (1992) concluded that the bile acids found in alligators and crocodiles developed into the bile acids presently found in the chelonians, lizards, tuatara, birds, and most mammals. However, the manatee, elephant, hyrax, rhinoceros, and horse all show a separate development of bile alcohols, which indicates a separate pathway for these related protoungulate and ungulate species that was not shared by the artiodactyls or other mammals.

Intestine

Phagocytosis and intracellular digestion within food vacuoles is absent and the intestine is divided into a midgut and hindgut in many advanced invertebrates. Kermak (cited by Barrington 1962) demonstrated that gut epithelial cells of *Arenicola marina*, a marine-burrowing worm that ingests much sand to obtain organic matter, phagocytize food particles and then transfer them to amoebocytes for digestion in a manner similar to that of many primitive metazoa, echinoderms, and other invertebrates with similar feeding habits. However, *Turbellaria lapidaria*, an annelid more selective in its feeding habits, has intestinal cells that lack food vacuoles and demonstrate a brush border of microvilli on their luminal surface. Absorptive cells of the insect midgut also have microvilli, which increase the surface area available for absorption and contain enzymes that complete the digestion of carbohydrates and protein. Most of the mechanisms for carrier-mediated transport in the gut epithelium of vertebrates are found in advanced species of invertebrates. For example, the gut of the mollusk *Alplysia californica* transported Na electrogenically in a manner similar to vertebrates (Gerencser 1988), and the apical membrane of the hindgut rectal pad of locusts demonstrated Na-dependent absorption of amino acids and Na–H exchange (Fig. 11.2).

The invertebrate gut also contains transport systems that have not been described in vertebrates. The gut of the mollusk *Alplysia* appears to absorb Cl electrogeni-

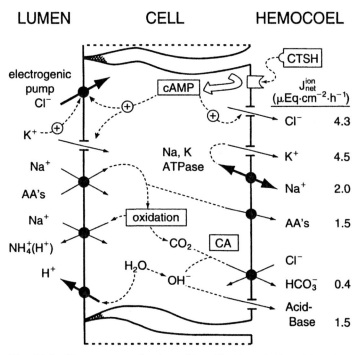

Fig. 11.2. Transport mechanisms identified in apical and basolateral membranes of locust rectal pad epithelium. Arrows through solid circles indicate carrier-mediated transport. Thick arrows indicate major ion pumps. Sodium is transported across the apical membrane in cotransport with amino acids and in exchange for intracellular H and intracellular NH_4 that is produced by metabolism of amino acids. (From Phillips et al. 1988.)

cally (Gerencser 1988) and brush-border membranes of the hepatopancreas of the freshwater prawn *Macrobrachium rosenbergii*, and marine lobster *Homarus americanus*, demonstrated an electrogenic exchange of two Na for one H (Ahern, Franco, and Clay 1990).

Midgut

Most of the basic structures and functions of the vertebrate midgut appear to be highly conserved from the invertebrates and maintained through the subsequent evolution of the vertebrates. The relative surface area and length of the midgut vary among classes and species, and the midgut of herbivores can show continuous motor activity. Given the same rate of digesta transit per unit length, a longer midgut should promote more complete digestion and absorption, and the extremely long midgut of the Florida red-bellied turtle and emu appear to serve as the major site for microbial fermentation, as well. However, the rate of digesta transit also is affected by retropulsion, observed in the midgut of some birds and mammals, and the higher frequency of peristaltic rushes in herbivorous mammals.

Hindgut

Advanced species of terrestrial invertebrates have a hindgut that aids in the conservation of electrolytes and water. The hindgut of insects generally contains high concentrations of K and Cl and low levels of Na, and the absorption of these ions is under the control of neuropeptides and hormones (Phillips et al. 1986). The hindgut is highly developed in herbivores, and the rectal pad of some insects can absorb water against an extremely high osmotic gradient. Studies of electrolyte transport mechanisms in the intestine of the desert locust, recently reviewed by Phillips et al. (1994), are of particular interest in reference to ammonia transport. The Malpighian tubules of these insects secrete Na, K, Cl, ammonia-urate, proline, and water into the junction between the midgut and hindgut. The hindgut reabsorbs Na, K, H, Cl, proline, and water and secretes H and ammonia. The proximal hindgut secretes ammonia by ion trapping. However, the rectal pad of the hindgut absorbs proline and secretes NH_4 in exchange for Na (Fig. 11.2). Therefore, it appears that ammonia is transported by both diffusion of NH_3 and by carrier-mediated $Na–NH_4$ exchange in the hindgut of these insects.

The hindgut of vertebrates shows numerous adaptations to both the environment and diet. Only a few fish such as the kyphosids have a distinct hindgut separated from the midgut by a sphincter or valve. However, a distinct hindgut is present in adult amphibians, reptiles, birds, and most mammals. It terminates in a cloaca, along with the urinary tract in the adult amphibians, reptiles, and birds, but exits the body separately in most mammals. The hindgut of adult amphibians is relatively short and appears to lack a valvular separation from the midgut, except in frogs. However, a sphincter or valve separates the midgut and hindgut of most higher vertebrates. The reptilian hindgut tends to be longer than that of adult amphibians, particularly in omnivores and herbivores, and the hindgut of some herbivores includes a cecum. The avian hindgut often contains paired ceca, which tend to be larger in omnivores and largest in herbivores. However, the colon is short in most avian species, probably because of the weight restrictions of flight.

A distinct hindgut is absent in some cetaceans and terrestrial mammals, and it is short and may even lack a valvular separation from the midgut in other species. However, the hindgut of most mammals tends to be longer than that of other vertebrates. In many species it includes a cecum, which can be very voluminous and highly developed in the smaller herbivores. Another characteristic limited to the hindgut of some mammals is the presence of haustrations in the cecum and over varying lengths of the colon, which serve to delay digesta transit.

The hindgut appears to have first evolved in response to the need of terrestrial invertebrates and vertebrates to conserve electrolytes, water, and nitrogen secreted and excreted into the digestive tract. Although the midgut recovers substantial quantities of water, this is largely dependent on the availability of monosaccharides and amino acids for coabsorption with Na. At some stage in evolution the hindgut adopted mechanisms for the efficient absorption of Na and water, governed by the release of aldosterone in response to the body's need. The advantages of this system are demonstrated in the adjustment of the chicken colon from cotransport of Na and nutrients to the electrogenic absorption of Na in birds placed on a Na-deficient diet.

Although most fish show little evidence of a distinct hindgut, the terrestrial vertebrates required a more effective means for the conservation of water. Termination of both the renal and digestive systems of adult amphibians, reptiles, and birds in a cloaca allowed the retention and resorption of urinary electrolytes and water in the hindgut, as well. At some stage in the evolution of adult amphibians or reptiles, retention of urine and digesta was prolonged by the development of a terminal pacemaker, which generated antiperistaltic waves of contraction for the reflux of digesta the length of the hindgut, and a sphincter or valve that prevented its reflux into the midgut. The presence of such a system has been demonstrated in reptiles and its importance to the conservation of electrolytes and water has been amply demonstrated in many birds.

The urinary and digestive systems of mammalian embryos are similarly joined at a cloaca. This persists in the monotremes, marsupials, and a few other species, and a pacemaker may be present at the termination of the opossum hindgut. However, in most mammals the urinary and digestive tracts form separate exits prior to birth and the pacemaker is located in a more proximal segment of a longer colon. This allowed the proximal hindgut of some mammals to serve as the principal site for digesta retention and its more distal segments to serve as the principal site for absorption of electrolytes and water from digesta. It also may account for the tendency of mammals to have a longer colon than other vertebrates.

Retention of digesta in the hindgut of terrestrial vertebrates resulted in the growth of large populations of indigenous bacteria, which convert endogenous and dietary carbohydrates to SCFA and endogenous nitrogenous compounds to ammonia and microbial protein. Absorption of SCFA aided in the absorption of Na and water and provided an additional source of energy. Absorption of ammonia aided in the conservation of the nitrogen contained in the endogenous enzymes secreted into the upper digestive tract and uric acid or urea recycled into the digestive tract. With the further expansion of the hindgut in some species, this became an increasingly important source of nutrients and led to the evolution of hindgut-fermenting herbivores.

Neuroendocrine control

The various functions of the digestive system are controlled and integrated by the action and interaction of a large number of agents secreted by nerves and endocrine cells. Neurotransmitterlike substances are found in invertebrates, and a nervous system is present in sponges and is well developed in annelids, mollusks, and arthropods (Fig. 11.3). The various functions of the vertebrate digestive system are modulated by the central nervous system (CNS) through the autonomic nervous system. In fish and amphibians, the cranial parasympathetic nerves of the autonomic nervous system innervate the stomach, and the spinal sympathetic nerves innervate both the stomach and intestine. In reptiles and more advanced vertebrates, the cranial nerves innervate most of the gastrointestinal tract. Spinal nerves continue to innervate the entire tract, but a separate sacral nerve supply, which first appeared in the amphibians, innervates part of the hindgut.

As noted in Chapter 10, stimulation of preganglionic cranial nerves in fish and amphibians resulted in noncholinergic, nonadrinergic inhibition, and stimulation of the spinal preganglionic nerves produced either excitation or inhibition. However, in reptiles and more advanced vertebrates, stimulation of parasympathetic nerves can also produce cholinergic excitation, and stimulation of sympathetic nerves is principally inhibitory. Burnstock (1969) concluded that the reversal of the roles of the parasympathetic cranial nerves at the level of reptiles may be associated with conversion of the respiratory system from gills to lungs. These various responses to extrinsic nerve stimulation are due to the intrinsic nervous system of the gastrointestinal tract, which determines its response to CNS stimulation and can control many of its functions without CNS supervision. This is the result of a variety of purines, amines, peptides, and other agents that are secreted by these neurons, and either modulate the release of neurotransmitters by cholinergic or adrenergic neurons or have a direct effect on muscle, secretory, or absorptive cells.

Control over the motor, secretory, and absorptive functions of the digestive system is shared by a variety of peptides secreted by endocrine cells. Many of these peptides are found in both nerve and endocrine cells and may serve as neurotransmitters or modulators and as hormones. Similar peptides provide intercellular communication in higher plants and unicellular organisms (Fig. 11.3). Therefore, some of these agents appear to have been present before these systems evolved. Peptides that serve as neurotransmitters or neurohormones are present in coelenterates, annelids, and insects (Barrington 1982; Krieger 1983).

Some families of hormones appear to have evolved from similar ancestral peptides, and some may have served initially as neurotransmitting or modulating agents. This led to the theory that the hormones evolved originally from neuroectodermal tissue (Pearse 1969). However, Barrington (1982) stressed the need

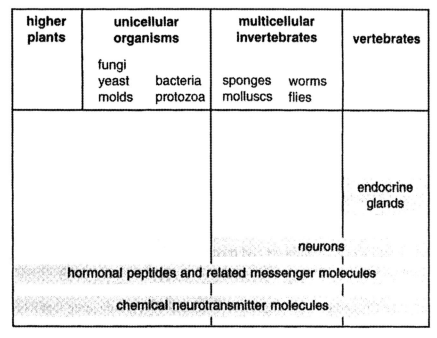

higher plants	unicellular organisms		multicellular invertebrates		vertebrates
	fungi yeast molds	bacteria protozoa	sponges molluscs	worms flies	

endocrine glands

neurons

hormonal peptides and related messenger molecules

chemical neurotransmitter molecules

Fig. 11.3. Evolutionary origins of the biochemical elements of the endocrine system and the nervous system. (Modified from LeRoith, Shiloach, and Roth 1982.)

for caution in interpreting fragmentary evidence collected from a few species and pointed out that many of the adaptations are determined by the evolution of receptors and modulation of the programing of receptor cells, rather than by changes in the molecular structure of the peptides.

Among the peptides of the secretin family, VIP has been considered to be the ancestral form because of its presence in nervous tissue of prochordates and its wide range of activity in mammals. However, secretinlike activity also has been reported in prochordates and mollusks. Pancreatic polypeptidelike immunoreactivity has been identified in the intestinal epithelium of cyclostomes and in the gut and pancreas of teleosts and birds, and a similar substance was found in the gut of prochordates and the nervous system of an earthworm, a mollusk, and insects. Substance P has been identified in the nervous system of a coelenterate (*Hydra*), prochordates, and all of the major classes of vertebrates.

Vigna (1983, 1986) reviewed information on the evolution of endocrine regulation of gastric acid secretion, pancreatic secretion, and contraction of the gallbladder in vertebrates. The CCK/gastrin family of peptides is believed to have been derived from a single ancestral molecule, rather than through parallel evolution. Cholecystokininlike peptides have been demonstrated in the gut of species belonging to all classes of vertebrates, and CCK is the only member of this hormone family found in cyclostomes. Extracts from the intestine of hagfish stimulated contraction of the guinea pig gallbladder (Vigna 1979; Vigna and Gorbman 1979). However, contraction of the hagfish gallbladder was not stimulated by either these extracts or mammalian CCK, suggesting an absence of the appropriate receptors. Vigna concluded that a cholecystokininlike agent that stimulated pancreatic secretion was present in chordates and persisted in vertebrates, but its regulation of gallbladder contraction did not appear until after the evolution of the cyclostomes.

Secretion of HCl was regulated by a CCK/gastrinlike hormone in primitive chondrichthyean and osteichthyean fish, and this continued to be the major stimulus in chondrichthyeans and primitive osteichthyeans. However, it was lost in the living osteichthyeans, and possibly replaced by bombesin. Radioimmunoassay and immunostaining studies indicated that a separate gastrinlike peptide appeared at the divergence of amphibians and reptiles. Although bioassay procedures capable of discriminating between CCK and gastrin suggested an earlier divergence between elasmobranch and teleost fish, the gallbladder of the coho salmon responded to both CCK and gastrin (Vigna and Gorbman 1977), indicating that its receptors were unable to distinguish between the two. However, Vigna (1986) also concluded that the functional evolution of these hormones involved the recruitment of new targets for old hormones, new cellular sources for old hormones, old targets for new hormones, and old targets for new hormone receptors.

Evolution of herbivores

One of the major advances in evolution was the advent of animals capable of deriving a substantial portion of their nutritional requirements from the leaves, petioles, or stems of plants. The abundance and availability of this plant material throughout the year opened the way to a much wider range of diets and ecological niches. The ability to derive nutrients from these portions of the plant requires the reduction of large quantities of plant material to small particles, and either their rapid passage through the gut for the extraction of cell contents or digestion of the structural carbohydrates that make up a high percentage of their cell walls. The first option, which was adopted by some invertebrates, some fish, and the panda, requires a highly selective diet, simple gut, and rapid transit. The second option requires cellulolytic enzymes of either endogenous or microbial origin.

Invertebrates

Digestion and the production of nutrients is aided by indigenous, symbiotic microorganisms in many invertebrate species. Buchner (1965) reviewed the widespread distribution of endosymbiosis. He pointed to the presence of algae in the cells lining the digestive tract of a wide range of invertebrates and evidence that these organisms provided their host with O_2 and carbohydrate, a site for food storage, and/or a mechanism for the excretion of CO_2, PO_4 and nitrogenous waste. In some species, such as *Paramecium bursa*, the algae even allow the animal to survive in the absence of its normal food supply, if sufficient light is provided for photosynthesis. There is a similar inclusion of bacteria within protozoa and cells lining the digestive system of other invertebrates, such as some insects.

Although the algae were noted by early workers, who only questioned their importance, the bacteria were at first believed to be specialized cellular organelles. Proof that these were bacteria that synthesized vitamins or fixed N_2 only came when means were developed for separation of the symbiont from its host. The relationship between the amoeba *Pelomyxa* and its bacterial symbionts has been examined fairly extensively. Each species appears to have a specific type or types of bacteria throughout the endoplasm and, occasionally, the ectoplasm. The ability of these protozoa to digest filter paper is believed to be the result of bacterial cellulase. In more advanced species, large numbers of bacteria or protozoa may be located at specific sites along the digestive tract. Hackstein and Stumm (1994) found that methanogenic bacteria attached to specific structures in the hindgut of millipedes, cockroaches, termites, and scarab beetles. Microbial fermentation and synthesis of nutrients have been well documented in many species, including the annelids, mollusks, echinoderms, and insects.

Cellulose digestion has been well documented in the hindgut of termites, other insects, and a few other invertebrates. However, recent reviews by Martin (1991) and Slaytor (1992) show a marked dichotomy of views on how this is accomplished by insects. Martin (1991) contends that cellulose digestion is usually mediated by protozoa and bacteria, which evolved as a result of the widespread occurrence of symbiotic relationships between insects and the bacteria and protozoa that reside in their hindguts. This is based partly on the assumption that the complete cellulase complex of endo-β-1,4-glucanase and exo-β-1,4-glucanase components is required for digestion of crystalline cellulose, and insects are unable to synthesize the exo-β-1,4-glucanase (cellobiohydrolase) component. However, Slaytor (1992) concluded that there was no evidence that an exo-β-1,4-glucanase is involved in, or needed for, the production of glucose from crystalline cellulose in termites or wood-eating cockroaches. The endo-β-1,4-glucanase components, which are active against both crystalline cellulose and carboxymethylcellulose, are found in the salivary glands, foregut, and midgut of these insects. He further contended that there is no evidence that bacteria are involved in cellulose digestion by termites or cockroaches and that the protozoa in the hindgut of lower termites produce cellulases quite distinct from those in the foregut and midgut. He concludes that although cellulase activity in the guts of other invertebrates appears to be mainly of microbial origin, the evidence is often weak, and endogenous synthesis of at least part of the activity may be involved.

Notwithstanding the debate over sources of cellulase activity in the invertebrate gut, there is no evidence of cellulase production by the digestive system of any vertebrate. Most vertebrate herbivores appear to digest the more fibrous portions of plants by the retention of large quantities of this material for periods long enough to allow microbial production of a significant fraction of the animal's nutrient requirements. This is accomplished by expansion of the hindgut as an organ of fermentation in a few species of reptiles and birds and in a high percentage of mammals, or it is accomplished by the subsequent evolution of foregut fermentation in a variety of mammalian species (Hume and Warner 1980; Stevens 1990).

Fish

Although there are numerous species of herbivorous fish, the importance of microbial fermentation remains unclear. Fish that have mouth parts for filter-feeding, teeth for grazing, or a pharyngeal mill can reduce plant material to the particle size required for efficient microbial digestion, and plant cell walls also may be partially degraded by gastric acid. Both anaerobic bacteria and SCFA have been found in the intestine of fish, and conflicting reports on the presence of cellulases may simply reflect the relative absence of cellulose in aquatic plants. Gut bacteria may contribute to the fermentation of the β-linked glucose in the storage polysaccharide (laminarin) and the β-linked uronic acids and fucoidans in the cell wall polysaccharide (alginic acid) of brown algae. Although the bacterial populations and SCFA levels found in the intestines of herbivorous fish are low in comparison with those found in herbivorous mammals, the energy requirements of ectotherms are also at least an order of magnitude lower than those of endotherms. Furthermore, the gut microbes may contribute to the conservation of nitrogen by fish on a low-protein diet or may aid in digestion of the chitin or wax esters found in the diets of many species. However, most fish appear to have a limited capacity for the storage and prolonged retention of digesta, and the nutritional significance of SCFA and other microbial end products awaits measurements of their rates of production and quantity of uptake.

Amphibians

Although little is known about microbial fermentation in herbivorous amphibian larvae, conditions for retention of digesta and multiplication of large numbers of microbes appear to have first been met by the hindgut of adult amphibians. However, adult amphibians do not need a large hindgut for recovery of Na and water. This, coupled with a dentition unsuitable for the fine reduction of plant material, could account for the absence of herbivores among surviving species.

Reptiles

In contrast to amphibians, the terrestrial reptiles required effective mechanisms for the conservation of electrolytes and water of both urinary and digestive tract origin. This resulted in further development of the hindgut. The expansion into a cecum and longer colon seen in most contemporary herbivores may have been a response to the bulk and gastrointestinal secretions associated with the ingestion of increasing amounts of plant material. The limited number of herbivorous species may be attributed to the lack of an efficient apparatus for the mastication of plant material and the effects of ambient temperature on digesta transit and microbial metabolism. It also has been attributed to the small body size and, thus, limited gut capacity of most species. However, the presence of large carnivores, such as the Komodo dragon and crocodiles, raises the question of body size as the principal limiting factor, and none of these limitations accounts for the high percentage of herbivorous dinosaurs in prehistoric times.

The dinosaurs first appeared in the late Triassic period. The earliest species were carnivores, but by the subsequent Jurassic and Cretaceous periods, the dinosaurs became the dominant terrestrial herbivores (Colbert 1980). It is now believed that the herbivorous dinosaurs included not only the largest species but also species of all sizes with a herbivore/predator ratio similar to that seen in present day, undisturbed mammalian populations (Bakker 1987).

The problem of reducing plant material to a small enough particle size may have been overcome in some herbivorous dinosaurs by the presence of a gizzard, as suggested by the presence of piles of large smooth stones found with the fossil remains of some herbivores (Bakker 1987). This finds some support in the finding of a high incidence of gastrolithis in the stomach of present-day crocodiles (Corbet 1960). However, Norman and Weishampel (1985) concluded that the ornithopods, a diverse group of dinosaurs that dominated all other contemporary herbivores in number of species per family during the late Mesozoic era, may have had a very efficient masticatory apparatus. Reconstruction of their skull indicated the presence of teeth that interlocked to form obliquely shearing blades and a jaw articulation that allowed the grinding of food by lateral rotation of the upper jaw in one group of animals and by rotation of the lower jaw in another group (Fig. 11.4, *B*).

The high ratio of body mass to surface area in large dinosaurs would tend to result in a relatively high and stable body temperature for microbial digestion and digesta transit. Farlow (1987) concluded that the low mass-specific metabolic rates

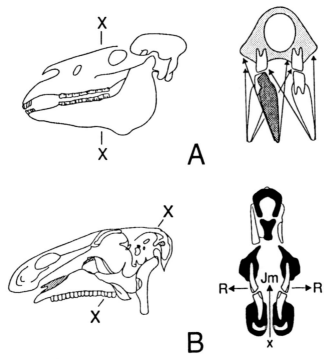

Fig. 11.4. Longitudinal and cross sections of the skulls of the horse (**A**) and ornithopod dinosaurs (**B**). **A,** Shows the well-developed premolars and molars of the horse and the lateral movements of the lower jaw that are typical of most mammals. **B,** The ornithopod dinosaurs had jaws of equal width and teeth that interlocked to form oblique, shearing surfaces. Efficient mastication of plant fiber is believed to have been accomplished either by a hinge that allowed lateral rotation of the upper jaw or, in one group, by a ball and socket joint that allowed medial and lateral rotation of the lower jaw. (From Norman and Weishampel 1985.)

that accompany large body size would reduce the total energy requirements and rate of digesta passage and that the fermentation mass would have provided enough heat for thermoregulation. However, this would not apply to the smaller herbivores, and it has been suggested that at least some species were able to control body temperature (Robertshaw 1984; Bakker 1987). This is based partly on indirect evidence such as the fact that they occupied most ecological niches and remained the most successful medium- to large-size terrestrial carnivores, omnivores, and herbivores long after the appearance of mammals (Bakker 1987).

The argument for endothermic dinosaurs finds some support from microscopic studies of their bone structure, which were recently reviewed by Chinsamy and Dodson (1995). The bones of vertebrates consist of specialized cells located in lacunae surrounded by a matrix of collagenized fibrils closely associated with the Ca-containing compound hydroxyapatite. In mammals and birds, bone is laid down in primary and secondary concentric rings to form osteons around blood vessels, and the collagen fibers are arranged in a haphazard fibrolamellar fashion. Bones stop growing at maturity and late bone is marked by laminar rings and rest lines. However, the blood vessels of reptilian bones are rarely located in primary and secondary osteons. Because of slower and more continuous growth, their bone contains a more organized lamellar arrangement of collagen fibers, and peripheral rest lines are generally replaced by annular rings during periods of slower growth. The fact that bones of dinosaurs are highly vascular and organized in a fibrolamellar fashion like that of birds and mammals led Bakker (1971) to propose that dinosaurs were endotherms.

Additional evidence of endothermy was provided by a recent examination of the oxygen isotope composition of bone phosphate at various sites in a skeleton of *Tyrannosaurus rex* (Barrik and Showers 1994). The fact that the phosphate oxygen isotope composition of bone is partially dependent on the rate of metabolism reflects variations in body temperature during their growth and turnover. Comparison of the results with studies of other ectotherms and mammals indicated that unless the seasonal temperature varied by less than 10°C, *T. rex* was a homeotherm. However, Chinsamy and Dodson (1995) point out that other studies found similar oxygen isotope ratios in the bones of other reptiles, dinosaurs, and mammals at similar latitudes. Growth rings were also found in the bones of a number of carnivorous and herbivorous dinosaurs, questioning whether the fibrolamellar arrangement was caused by periods of faster growth driven by environmental conditions. Studies of the growth rate of dinosaurs on the basis of annular rings showed differences among species, which ranged from those of present-day reptiles to those of birds and mammals. Some species appeared to stop growing in their latter years. Chinsamy and Dodson concluded that elevated growth rates and endothermy may have risen independently in different groups of dinosaurs.

The lack of preservation of soft tissues means that speculation on the digestive strategy of herbivorous dinosaurs must be based largely on what is known about present-day herbivores. The herbivorous dinosaurs may have used all of the digestive strategies seen in present-day herbivores, depending on their diet, habitat, body size, and ability to maintain a relatively high and stable body temperature. The advantages of hindgut versus foregut fermentation as a function of the quality and toxicity of the diet and the availability of water are discussed in the following section in mammals. However, Van Soest (1994) concluded that the gut capacity of the largest dinosaurs would require a foregut retention time of a week. Mean retention times of something less than four days are necessary to eliminate the more objectionable methanogens, leaving only those that convert carbon dioxide, hydrogen, and formate to methane, as they do in the digestive tracts of most extant herbivores and humans. A longer period would provide methanogenic bacteria with the opportunity to degrade acetate and other SCFA, producing carbon dioxide and methane at the levels that occur in sewers and bogs and resulting in the loss of more than 80 percent of the potential digestible energy. However, a lower body temperature and more rapid transit time, such as those seen in the forestomach of macropod marsupials, would remove some of these restrictions from the larger species, and they would not apply to those of intermediate size.

Birds

Although birds have the advantages of both endothermy and an efficient masticatory apparatus in their gizzard, they include relatively few herbivorous species. This could be due to the restrictions of flight on both the weight and distribution of gut contents, which would limit hindgut capacity and account for the short colon of most birds. The four major types of digestive strategies of avian herbivores are illustrated in Fig. 11.5. Most of the herbivores that are capable of flying, such as the grouse, use a large pair of ceca for microbial fermentation. The only known exception to this is the hoatzin, which uses its crop and distal esophagus for this purpose. This strategy, which places the fermentation chamber cranial to both the masticatory apparatus (gizzard) and center of gravity, is difficult to understand. Among the avian herbivores that do not fly, it appears that the principal sites for microbial fermentation are the ceca in the rhea, the long colon in the ostrich, and the midgut in the emu. The advantages of midgut fermentation in the emu (and Florida red-bellied turtle) are equally hard to understand, since it should conflict with the other digestive and absorptive functions in this segment of the gut.

Mammals

As mentioned earlier, the first mammals are believed to have appeared during the early Jurassic period. Their evolution was accompanied by major changes in climate and plant life. During the Jurassic and Cretaceous periods of the Mesozoic era, the climate over much of the earth was tropical or subtropical (Colbert 1980). Shallow seas and quite uniform temperatures coincided with exceptional plant growth. Gymnosperms (cycads, cycadeoids, and conifers) were the dominant terrestrial plants during the Triassic and Jurassic periods (Thomas and Spicer 1986). However, angiosperms (flowering plants) appeared in the Cretaceous period and became the predominant plants as a result of the competitive advantages of insect

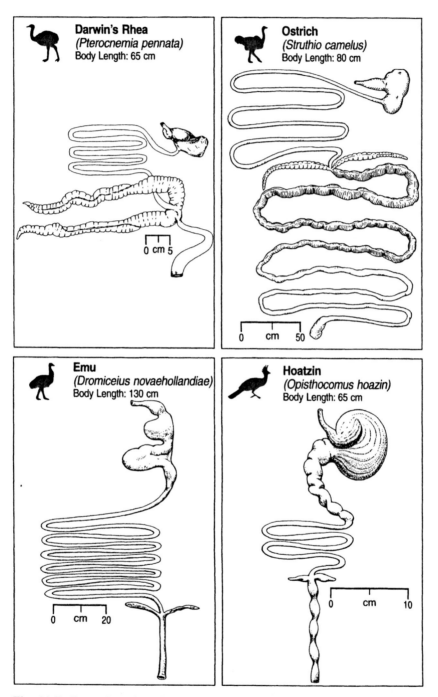

Fig. 11.5. Examples of the four major types of digestive strategies practiced by avian herbivores. The ceca are the major site for microbial digestion in the rhea and most flying species, such as grouse. The presence of a long colon used for this purpose, as seen in the ostrich, is rare. Microbial digestion in the midgut or foregut appears to be limited to the emu and hoatzin, respectively.

pollination, seed distribution by birds and mammals, and more rapid growth (Regal 1977). Thus, the initial habitat for all mammals at the start of their radiation was probably a type of "tropical" forest (Janis 1982). The rapid growth of these plants results in rapid development of a high lignin content, which would favor the evolution of browsers that could take advantage of new plant growth.

The establishment of modern continental land masses as a result of the breaking up of the supercontinent Pangaea, which began in the late Triassic period, was accomplished in the late Cretaceous period by a long period of mountain building, with the birth of the Alps and Himalayas in the Old World and of the Andes and Rockies in the New World (Lillegraven, Kraus, and Brown 1979). Growth of these and other mountain systems created definite climatic zones, with marked seasonal fluctuations in temperature. The early Tertiary period of the Cenozoic era, which began 70 mya, saw completion of these major changes, a gradual cooling of world climates, and the rapid expansion of angiosperms. Although tropical-type forests continued to extend to the Arctic Circle as late as the Eocene epoch, 54-38 mya (Wolfe 1978), the cooler and drier climates of the Miocene epoch resulted in the rapid expansion of grasses and other herbs (Harris 1980).

These profound changes in the climate and plant life of the globe, commencing about 65 mya, were accompanied by the diversification of mammals and, especially, the expansion of herbivores (see Fig. 11.8). The four major digestive strategies of mammalian herbivores are illustrated in Fig. 11.6. The earliest mammalian herbivores were probably small hindgut fermenters, which combined the advantages of using the upper digestive system for digestion by endogenous enzymes and their hindgut for microbial production of nutrients. Their small size, high rate of metabolism, and limited gut capacity would favor selective retention of the more readily fermentable plant material in a cecum and more rapid transit of less digestible components through the remainder of the intestine. Coprophagy would allow the additional recovery of microbially synthesized protein and vitamins, electrolytes, and water, particularly when combined with the periodic release and selective consumption of nutrient-rich feces (cecotrophy).

Fig. 11.6. Four major types of digestive strategies practiced by mammalian herbivores. Although the black bear is an omnivore, its gastrointestinal tract is shown as an example of the long midgut and short hindgut seen in the more herbivorous bears and the panda. The long midgut can provide for more efficient extraction of nutrients and, possibly, for microbial digestion as well for these animals. The rabbit is an example of the large number of small herbivorous mammals for whom the cecum is the major site for microbial digestion. Many of these species practice coprophagy or cecotrophy as well. For most large mammalian herbivores like the perissodactyla, elephants, and apes either their proximal colon, or the forestomach for ruminants, kangaroos, sloths, colobus, and langur monkeys is utilized for this purpose. (Rabbit, pony, and sheep drawings from Stevens 1977.)

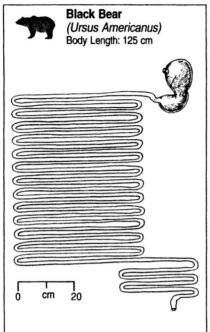

Black Bear
(Ursus Americanus)
Body Length: 125 cm

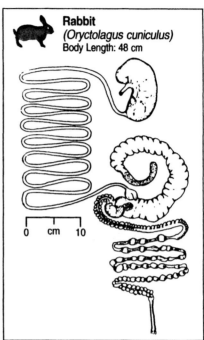

Rabbit
(Oryctolagus cuniculus)
Body Length: 48 cm

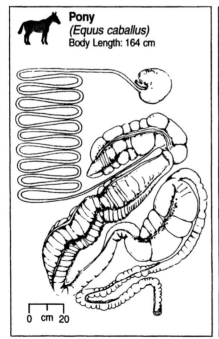

Pony
(Equus caballus)
Body Length: 164 cm

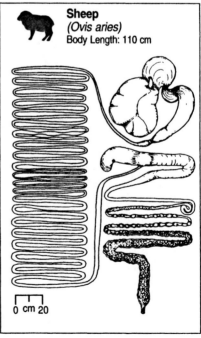

Sheep
(Ovis aries)
Body Length: 110 cm

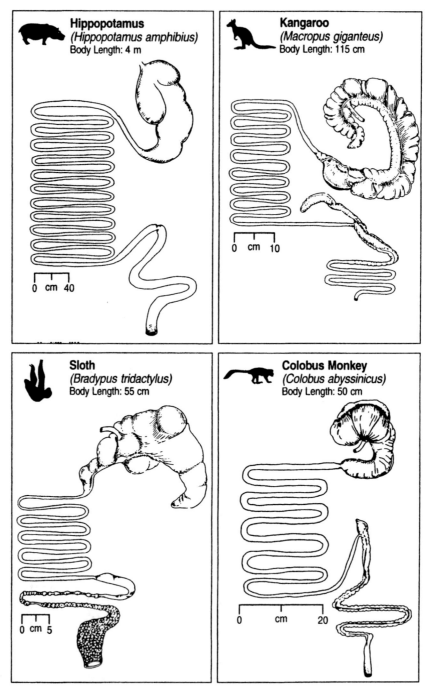

Fig. 11.7. Examples of four nonruminants with an enlarged forestomach that serves as a site for microbial fermentation; the hippopotamus, kangaroo (from Stevens 1977), sloth (from Stevens 1980), and colobus monkey (from Stevens 1983).

An increase in body size and gut capacity and a decrease in the metabolic rate per unit of body mass, allowed larger herbivores the luxury of a longer retention time for digesta. A few species such as the panda and herbivorous bears have a long midgut which allowed for extraction of plant nutrients and, possibly, for microbial fermentation as well. However, in many species the proximal colon became the major site for microbial digestion. This would favor the more efficient digestion of structural carbohydrates, and the nutritional return from low-quality, high-fiber diets would be increased by increasing the quantities ingested and passed through the gut. However, use of the proximal colon for these purposes would make it difficult to periodically release its nutrient-rich contents for coprophagy. It also required the maintenance of a fluid medium and put a greater load on the distal colon for the recovery of electrolytes and water.

The next stage in the evolution of the gut of mammalian herbivores was probably the expansion of the stomach as a supplementary site for microbial fermentation. This strategy, which is seen in some rodents and the hyrax, reduces the ability to obtain the maximum nutrient value from readily digestible starches and protein. However, it provides for the additional storage of plant material between meals, the utilization of gut microbes for detoxification, and the utilization of the protein and B vitamins synthesized by gastric microorganisms. Use of the stomach as the principal site of microbial fermentation (Fig. 11.7) appears to have evolved separately in the evolution of marsupials, edentates, primates, and artiodactyls (Hume and Warner 1980; Stevens 1980). This allowed microbial detoxification of food and the efficient recovery of microbially synthesized amino acids and B-complex vitamins in animals unable to practice cecotrophy.

The evolution of rodents, ungulates, and macropod marsupials (Fig. 11.8) gives some additional insight into how the herbivorous mammals may have evolved. The rodents, which represent the most successful order of mammals, with respect to the number of individual animals and species, include a high percentage of herbivores. The ungulates represent two orders of large herbivores with entirely different digestive strategies. A close relationship between artiodactyls and cetaceans may account for the complex stomach of whales. The marsupials represent an order that has evolved separately from other mammals but demonstrate many examples of convergence on similar digestive strategies.

Rodents

The rodents can be traced back to the late Paleocene epoch, but their number expanded during the Miocene epoch (Fig. 11.8). The Cricetidae, which includes the herbivorous voles and lemmings, underwent a major expansion with the spread of grasslands and steppes during the Miocene epoch (Vorontsov 1962). Landry (1970) contends that the phenomenal success of the Rodentia can be attributed

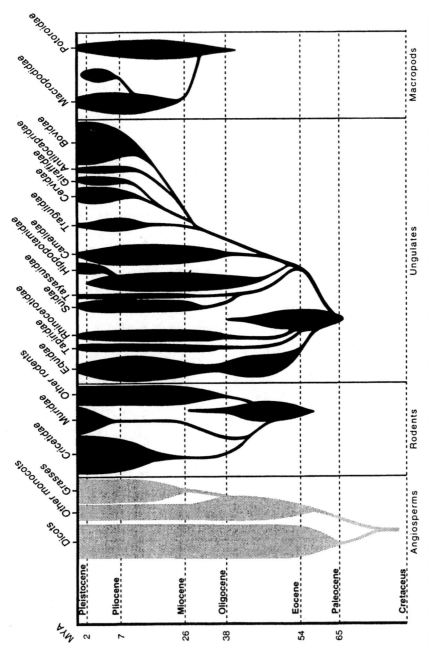

Fig. 11.8. Evolution of angiosperms, rodents, ungulates, and macropod marsupials during the Tertiary period. Width of column indicates approximate compromise between species diversity and density. Data on angiosperms from Van Soest (1994), on rodents from Romer (1966), on ungulates from Janis (1976), and on macropod marsupials from Hume (1978).

314

principally to the flexibility of their masticatory apparatus to a variety of diets. Their incisors grow continuously and occlude outside their mouth, and the articulation of their jaw allows it to move forward for occlusion of the incisors and backward for occlusion of the molars. This enables them to seize and shear prey; clip and grind plant stems, leaves, and buds; or remove bark from trees and shrubs. A few rodents are carnivores. Many are omnivores. For example, 27 percent of the stomach contents of the yellow pine chipmunk *Eutamius amoenus*, which feeds mainly on the seeds of the ponderosa pine, consisted of arthropod remains during the spring (Tevis 1953). However, a significant percentage of the rodents are herbivores than can subsist largely on the fibrous portion of plants. This is accomplished principally by microbial digestion in the cecum and aided in many species by euprophagy.

The herbivorous rodents range from some of the smallest species, such as the voles and lemmings, to the largest species, such as the porcupines, beaver, and capybara. Adaptations for herbivory include hypsodonty, continuously growing cheek teeth with a number of ridges on the occluding surfaces, and stronger masticatory muscles (Vorontsov 1962). The herbivores have a large cecum, which is often haustrated and may be lined with circular or spiral folds. The advantage of cecal retention of fluid, bacteria, and small digesta particles by herbivores of small body size is discussed in Chapter 6. The stomach of some species is enlarged, divided into two or three compartments, and partially lined with stratified squamous epithelium, providing an additional site for microbial fermentation. It also has been suggested that retention of ingesta at the high and relatively stable pH of the cranial gastric compartments may account for the prolonged digestion of starch by salivary (Carleton 1973, 1981) or microbial (Perrin and Kokkin 1986) amylase in the stomach of some of these species.

Ungulates

Fossil evidence indicates that the protoungulates, which were probably browsers in tropical forests, also appeared in the early Tertiary period and separated into the Perissodactyla and Artiodactyla in the late Paleocene epoch (Fig. 11.8). The perissodactyls (tapirs, small equids, and rhinos) predominated in the Paleocene and early Eocene epochs, but by the end of the latter (38 mya) the artiodactyls reached equal numbers. The spread of grasslands during the Miocene epoch was accompanied by a change in the cheek teeth of equids, from selenodont (low crowned) to the hypsodont (high crowned) teeth of grazers (Romer 1966). It also saw a marked increase in the number of artiodactyls and, particularly, ruminants. The larger rumen volume and more highly developed omasum of grazing species allowed the more complete fermentation of forage and less reliance on the hindgut as a supplementary source of SCFA (Gordon and Illius 1994).

The predominance of the Artiodactyla over the Perissodactyla suggests a competitive advantage of foregut fermentation over hindgut fermentation among the ungulates. Although much has been written about direct competition between artiodactyls (especially bovids) and perissodactyls (especially equids), much of their early evolution occurred separately on different continents; the equids in North America and the bovids primarily in Africa and Eurasia (Janis, Gordon, and Illius 1994). Both the Bovidae and Equidae underwent expansion at times corresponding with the spread of grasslands on their respective continents.

Illius and Gordon (1992) used a modeling approach, based on that of Mertens (see Allen and Mertens 1988), to show that living equids extract more nutrients per day than ruminants from a wide range of forage qualities, as long as the quantity is not limiting. This can be easily understood by comparing the modified plug-flow reactor gut of equids, which can increase digesta transit with an increase in dietary fiber, with the stirred-tank foregut of ruminants, which retains plant fiber until it is broken down into small particles (see Chapter 6). It also suggests that the major competitive edge of ruminants over equids comes from their ability to survive when the forage is either limited on a regular or seasonal basis or when it contains toxic substances that can be degraded by forestomach bacteria. For these reasons, Hume and Warner (1980) concluded that the special features of the ruminant forestomach that result in prolonged retention of coarse roughage may have evolved in arid climates. Transfer of the primary fermentation chamber from the proximal colon to the foregut also allows greater use of the colon for the conservation of water. The advantage of this to animals in an arid environment is demonstrated in the earliest ruminating herbivores, the camelids.

One example of the relative advantages and disadvantages of hindgut versus foregut fermentation can be seen in the different grazing strategies of equids and ruminants on the Serengeti plains of Africa (Janis 1982). In subtropical grasslands, the more nutritive components are found at the lower levels of the sward. The upper levels consist of high-fiber, low-protein grass culms, and their removal by the less selective equids makes the lower levels more accessible to ruminants. This is supported by observations on the grazing association of zebra and wildebeest by Bell (1970) and Owaga (1975). During the wet season, they share the short grass of the plains in the southeastern Serengeti. As the season dries, there is a massive migration to the northwestern region, where the rainfall is higher and the grass is longer. The zebra, which lead these migrations, take more stem and less leaf than the wildebeest (Fig. 11.9), and their ability to break down the dense strands of dry grass by grazing and trampling is of great assistance to the more selective wildebeest (Bell 1970). Owaga (1975) found that when zebra could satisfy their need for water, they withstood drought conditions better than the wilde-

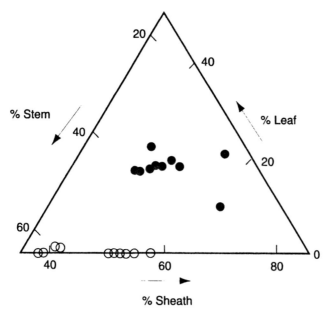

Fig. 11.9. Analysis of stomach contents of zebra (*open circles*) and wildebeest (*closed circles*) in the Serengeti during the dry season, showing the frequency of the three principal grass components. (From Bell 1970.)

beest. She attributed this to their greater tolerance for the high-fiber species and parts of grass.

Cetaceans

There is now considerable paleontological, morphological, embryological, and molecular evidence of a close relationship between the cetaceans and artiodactyls (Gingerich, Smith, and Simons 1990; Arnason, Gullberg, and Wildegren 1991; Novacek 1992; Adachi, Cao, and Hasegawa 1993; Milinkovitch, Guillermo, and Meyer 1993). From phylogenetic analysis of protein and mitochondrial DNA, Graur and Higgens (1994) concluded that the cetaceans evolved from primitive Artiodactyla about 45 to 49 mya, after the separation of suiforms and camelids and prior to the Tragulidae. This would include the cetaceans within the order Artiodactyla. It also could account for the complex cetacean forestomach, which may have been useful to herbivores inhabiting the shores of the epicontinental remnant seas of the early Eocene epoch and may continue to serve for the microbial fermentation of chitinous exoskeletons of krill in baleen whales (Mathiesen et al. 1994; Martensson et al. 1994).

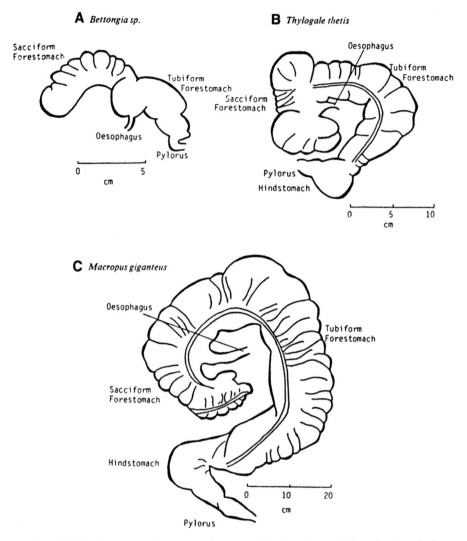

Fig. 11.10. Stomachs of macropod marsupials showing variations in the relative length of the sacciform and tubiform segments of the forestomach. (Hume and Warner 1980.)

Macropod marsupials

Grasses appeared on the Australian continent a little later than in Africa. Paleontological evidence (Bartholomai 1972) suggests that the macropod marsupials were represented in the early Oligocene epoch (Fig. 11.8) by the family Potoroidae (the rat-kangaroos), including a species of the extant genus *Bettongia*. Hume et al. (1988) suggested that the large sacciform forestomach region of present-day rat-kangaroos (Langer 1988; Fig. 11.10) may have evolved primarily as a storage

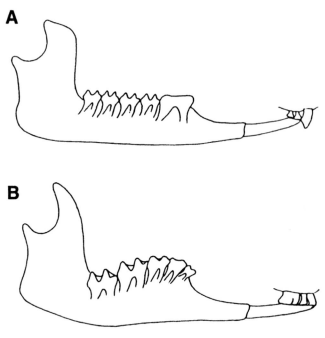

Fig. 11.11. Lateral views of the mandible and upper incisors of a macropod browser and grazer. **A,** In browsers such as *Dendrolagus* and *Wallabia,* all lophodont molars occlude simultaneously, producing a crushing action. Anterior drift of the molars is blocked by the large premolar. **B,** In grazers such as *Macropus,* the curve in the occluding surfaces of the molars results in occlusion of only the first two molars and a shearing action. However, the vestigial premolar allows anterior progression of the molars, so all molars are used during the life of the animal. (Modified from Sanson 1989.)

organ rather than as a fermentation chamber. Predator evasion by these small (0.5–3.0 kg) marsupials depends on the use of forest or other dense vegetation as refuge, but feeding areas are generally more exposed. It would be advantageous for rat-kangaroos to feed intensively over short periods for storage in an expanded stomach. Furthermore, Sanson (1989) concluded that all of the Potoroidae have a primitive dentition adapted to nonabrasive materials. They lack the molar progression and hypsodonty seen in those macropodids that are adapted to abrasive diets. Their simple dentition, combined with a relatively high ratio of metabolic requirements to absolute gut capacity, tends to prohibit the use of microbial fermentation as a major means of energy acquisition. This is consistent with the known food habits of extant rat-kangaroos, which dig for hypogenous (underground) fungi, plant roots, and rhizomes low in fiber content and supplement this with grass seeds and tree sap (Hume 1982).

The kangaroos and wallabies (family Macropodidae) were almost certainly distinct from the rat-kangaroos by the end of the Oligocene epoch and before the appearance of grasses in the fossil record (Hume 1978; Freudenberger, Wallis, and Hume 1989). The earliest members of this family were inhabitants of wet forests (Flannery 1984) and may have been similar to the Potoroidae. The stomach of the extant pademelons (*Thylogale* spp.), which are small (3–8 kg) wallabies, resembles that of the potoroids, with a large sacciform and comparatively short tubiform region (Fig. 11.10), and there is no gastric sulcus (Hume 1982). The pademelons have a lophodont (browsing grade) dentition, which maximizes the occlusal surface area and results in a crushing action suitable for soft, nonabrasive forage (Fig. 11.11, *A*). However, the browse they consume contains a wide range of toxic plant allelochemicals, and a symbiosis with detoxifying microbes may have been a critical adaptation for its exploitation.

As the Australian climate dried, the forests receded, creating ecotones (interfaces) between forests and grasslands, which are nutritionally diverse and currently support a number of medium-size wallabies (5–15 kg). Some of these species are intermediate feeders on both browse and grasses, and some retain the ancestral browsing-grade dentition, while others have derived a grazing-grade dentition, with a reduced area of contact and more effective shearing action (Fig. 11.11, *B*). The feature common to all of these forest-edge species is an enlarged tubiform forestomach, which allows the more rapid transit of less digestible but more abundant grasses, and a gastric sulcus.

Macropodid body size increased dramatically during the Pliocene and Pleistocene epochs (Flannery 1984), which corresponded with the expansion of grasslands of low nutritive value. This may indicate evolutionary pressure to minimize the ratio of metabolic requirements to gut capacity so that this new abundant food resource could be exploited. The stomach of extant grazing kangaroos, as exemplified by *Macropus giganteus* (Fig. 11.10), features the most reduced sacciform forestomach and most expanded tubiform forestomach of all the macropodids. Their colonlike forestomach and large body size would have allowed them to utilize abundant grasses of seasonally poor quality, in much the same way as the perissodactyls. Like equids, kangaroos can maintain their rate of food intake as fiber content of the forage increases, when rate of intake by ruminants of similar body size declines (Freudenberger, Wallis, and Hume 1989; Freudenberger and Hume 1992).

Summary

Speculations on the evolution of the vertebrate digestive system are derived from a picture taken at one brief instant in time and fossil evidence that is limited principally to the skeletal parts of the larger species and the effects of climatic changes on plant life. This information can be compared with a net, a large number of holes of varying size tied together with string of varying strength. However, the evidence that is available suggests that many adaptations of the digestive system of vertebrates can be attributed to the diet, environment, body size, or the ability to maintain a relatively constant body temperature.

Most adaptations of the headgut, foregut, and midgut can be attributed to the diet. Those of the headgut show the widest range of structural variation, with a great deal of functional convergence in separate classes and orders. Adaptations of the foregut are principally structural changes that can be attributed to the diet and the need for food storage. The midgut shows relatively few structural adjustments, other than differences that increase its surface area or length. Its major functional adaptations appear to consist of changes in the composition and levels of the endogenous enzymes involved in extracellular digestion. The hindgut shows numerous adaptations in its structure and function to diet, and this is particularly true for mammals. However, most of its adaptations among the lower vertebrates and many of those seen in mammals can be attributed to the need for the conservation of electrolytes, nitrogen, and water.

The earliest vertebrates appear to have been small carnivores, with a relatively simple digestive tract. A progressive increase in the capacity and complexity of the gut allowed the evolution of omnivores, which could derive additional nutrients from plants, and then the herbivores that use plants as their principal source of nutrients. These changes in gut capacity and complexity were largely limited to the hindgut of terrestrial vertebrates and served to delay digesta transit for the recovery of electrolytes and water. Therefore, the evolution of these changes was constrained by the environment and the evolution of the kidney. This can account for the reflux of digesta and cloacal urine in the hindgut of reptiles and birds. The evolution of a more efficient kidney could account for separate exits of the urinary and digestive systems of most mammals. This allowed the evolution of a longer colon and use of the proximal hindgut for retention of digesta and the distal colon for recovery of electrolytes and water.

An increase in digesta capacity and retention time of the hindgut increased the contribution of its indigenous bacteria to the nutrient requirements of their host and led to the evolution of herbivores. The relative rarity of herbivores among present-day reptiles suggests that the efficiency of microbial fermentation is limited by their masticatory apparatus and the effects of ectothermy on digesta transit and the gut microbes. However, the success of herbivorous dinosaurs indicates circumvention of these limitations. The relative rarity of herbivorous birds may be attributed to limitations of flight on gut capacity.

The success of mammalian herbivores can be largely attributed to endothermy, an efficient masticatory apparatus for the grinding of plant material and three major strategies for an increase in gut capacity. An expansion of cecal volume allowed small herbivores, with limited gut capacity, to selectively retain the small particles of digesta that are most readily fermentable by indigenous microorganisms. The efficiency of this system was increased in many species by coprophagy and, especially cecotrophy. In the larger mammalian herbivores either the proximal colon or the stomach became the major site of expansion and microbial fermentation. The relative advantages of these two strategies can be related to the characteristics of the diet and the need to recover water.

12

Conclusions and future directions

The digestive system of vertebrates has a number of common basic characteristics. Food is digested by a similar complement of endogenous enzymes secreted into the digestive tract or present in the apical membrane and cytosol of intestinal mucosal cells. Indigenous bacteria provide an additional source of nutrients. Glandular and enteric cell secretions provide mucus, electrolytes, and water to aid digestion of food and absorption of nutrients. Ingesta and digesta are mixed with secretions and moved through the digestive tract by its motor activity in a similar manner, and electrolytes, vitamins, and the end products of carbohydrate, lipid, and protein digestion are absorbed by similar mechanisms. These events are closely integrated and controlled by the nervous system and the release of hormones and paracrine agents.

Despite these common characteristics, the various parts of the digestive system show a considerable range of structural and functional variations both within and among the classes of vertebrates. The headgut shows a wide range of adaptations for the procurement and cutting, tearing, grinding, or filtering of food. The esophagus transfers food from the mouth to the stomach of most vertebrates, where it is stored and, generally, undergoes the initial processes of digestion by endogenous enzymes. However, the esophagus also stores food in birds and some fish and reptiles. The stomach is absent in some species of fish, does not secrete acid or enzymes in some larval amphibians and a few mammals, and serves as a principal site for microbial digestion in some mammalian herbivores. Some fish lack a distinct pancreas, and pancreatic and biliary secretions show numerous adaptations to the diet and feeding strategy. The mucosal surface area of the midgut may be expanded with ceca, folds, or villi, and its motor activity differs between species that are continuous versus noncontinuous feeders. The hindgut can vary from one that is short and difficult to distinguish from the midgut to a voluminous, haustrated, and compartmentalized large intestine.

Many adaptations of the digestive system are related to the diet or feeding strategy. This can account for many of the variations in headgut, stomach, and hindgut construction and the composition of endogenous enzymes. Some adaptations are related to habitat. Gill rakers and pharyngeal teeth allow fish to swallow small particles of food without loss through the gills. Absorption of Na and Cl from the esophagus and cotransport of Na-K-2Cl by the intestine represent environmental adaptations by marine fish. A more highly developed hindgut enables terrestrial vertebrates to conserve water and, with further refinements, enhances their ability to survive in arid environments. Some variations are related to other physiological characteristics of the animal, such as the concentrating ability of the kidney, requirements for flight, or the ability to control body temperature.

Previous chapters contain many examples of the contributions of studies of comparative physiology to the understanding of basic physiological mechanisms. Much of the research on the digestive system is conducted for application to human physiology, often with the assumption that the best species for these studies are humans or other primates. However, ethical, financial, and other considerations have limited most of this research to the dog, rabbit, and five species of rodents. The digestive tracts of these species are quite different from that of humans, and the human gut was not designed for either the high-concentrate diet of carnivores or the high-fiber diet of herbivores. The domesticated pig would provide a more suitable model for many of these studies.

Furthermore, the search for animals similar to humans misses the important point that many of the major contributions of comparative physiology have been derived from differences rather than from similarities among species. The simplicity of ion transport across nonglandular frog skin provided the model and procedures for the study of transport across epithelia of the digestive system. The negative electrochemical potential difference seen when the midgut of marine fish was bathed with glucose and amino acid-free solutions led to the discovery of a Na-K-2Cl cotransport system, which was later demonstrated in the crypt epithelial cells of mammals. Comparative studies demonstrated the extent to which endogenous enzymes and carrier-mediated transport systems can adapt to changes in diet. Early studies of the ruminant forestomach and the hindgut of herbivores provided much of the basic information on the contributions of indigenous organisms to the conservation of Na and water in humans and other species. Comparative studies of motor activity and neuroendocrine control have provided important clues to their characteristics and evolution.

An understanding of species variations in the diet, feeding, and digestive strategies is equally essential for the care and maintenance of domesticated, captive, and wild animals and for the preservation of endangered species. Many diseases of domesticated and captive herbivores can be traced to the intermittent feeding of high-concentrate diets to animals with a digestive system designed for continuous feeding of high-fiber diets. Problems can similarly arise from the feeding of either rapidly digestible or high-fiber diets to carnivores whose ancestors quickly ingested their prey with long intervals between meals. The survival of many species depends on a delicate balance among plants, herbivores, omnivores, and predatory carnivores. Chitin and wax esters play important roles in the food chain of vertebrates. Restriction of the migratory grazing pattern of large African herbivores can limit the availability of food and their ability to disperse parasite eggs and other pathogenic organisms in their feces. Loss of these animals can influence the survival of their predators and other species as well.

Information that has appeared since the first edition of this book allows better interpretation of a number of phenomena. We know more about the modulation of endogenous enzymes and carrier-mediated transport, the role of ambient temperature on the rate of digesta passage in ecototherms, and the digestive strategies of many species. Some researchers are probing for the rate-limiting step in the acquisition of food and its conversion to energy and nutrient output. Others are applying approaches and techniques developed in laboratory animals to the study of other species. Examination of wild species has provided important clues to their diet and their feeding and digestive strategies. This new knowledge gives us a better understanding of the relationships between animals and their food resources and habitat. It demonstrates the great variety of vertebrate digestive systems and strategies that have evolved. However, many questions remain to be answered.

Many species and groups of mammals still have received little study. The percentages of species of birds (8,600), reptiles (6,250), and amphibians (4,000) that have received any intensive study is much lower. With the exception of larval amphibians, comparative studies of fish, which account for almost half (21,500) of all vertebrate species, have received the least attention of all. Much more information is needed on the digestive strategies of these animals, and a careful selection of species in each taxon could fill many of the voids in our present knowledge.

One of the factors that limits understanding of gut adaptations is our lack of knowledge of the actual composition of the diet of many species. The diet of many fish is unknown; this is particularly true for herbivorous marine species. How do the endogenous and microbial enzymes match the storage and structural carbohydrates of marine and freshwater algae and other plants? Do indigenous bacteria aid in the digestion of chitin and wax esters in the diet of many vertebrates? The diet of many arboreal primates includes large amounts of fruit during much of the year, but the construction of their hindgut suggests that unripened fruit and leaves provide them with a high-fiber diet. Further study of the digestive strategies of arboreal primates and marsupials could provide clues to the evolution of the human digestive system.

More information is needed on the digestive system of fish, particularly the herbivores. Studies of the larval amphibians and their metamorphosis to adults could yield valuable information on the evolution of digestive functions from fish to terrestrial vertebrates. More studies are needed on the effects of body temperature on digestion by endogenous and microbial enzymes and on the absorption of nutrients in ectothermic vertebrates.

There is a great need for studies of digesta transit and retention by the digestive tract of many species. These studies should examine the mean retention time of both fluid markers and particulate markers of different sizes. Compartmental analysis of retention time is a particularly powerful tool for studying the kinetics of digesta movement. More needs to be known about the effects of ambient temperature on digesta passage rate in ectotherms. Do freshwater or marine fish and mammals improve the recovery of electrolytes and water from their distal intestine by infrequent defecation or other means? Is this more fully developed in marine species? Although the motor activity of the ruminant forestomach has received a great deal of attention, little is known about the mechanisms that control motility in the forestomach of other mammals such as the macropod marsupials, sloths, colobus and langur monkeys, or many species of rodents.

Much is unknown about electrolyte transport. The stratified squamous epithelium of the ruminant forestomach transports SCFA, ammonia, Na, H, Cl, and HCO_3 in a manner similar to that of the hindgut, without the complications of mucus or other secretions. Therefore, its usefulness for studies of SCFA and ammonia transport seem far from exhausted. The permeability and transport characteristics of stratified squamous epithelium in the stomach of other mammalian carnivores, omnivores, and herbivores also needs to be examined. Does the expanded area of cardiac glandular mucosa in some mammalian herbivores and omnivores (such as the domestic pig and some primates) serve principally for the secretion of HCO_3? If so, how is this controlled to prevent competition with HCl secretion by the proper gastric mucosa? Does the apical membrane of marine mammals cotransport Na, K, and Cl in a manner similar to that demonstrated in marine fish?

The most important functional divisions of the vertebrate hindgut appear to be determined by the location of the pacemaker(s) that initiate antiperistalsis and the segments principally responsible for secretion to sustain microbial fermentation versus absorption for the conservation of water. Yet, little is known about these divisions in most mammals and birds, and even less is known about them in reptiles and amphibians. Where are the pacemakers located in relation to the distribution of sacral parasympathetic nerves? This would be particularly interesting in animals that have both a cloaca and long colon, such as the ostrich and many marsupials. The ability of the colonic epithelium to adapt from cotransport of Na with other nutrients to transport of Na alone, seen when some birds are placed on Na-deficient diets, should be examined in other vertebrates.

Much needs to be learned about the contributions of indigenous microbes. Do the microbes indigenous to the digestive tract of fish contribute to the requirements for energy, nitrogen conservation, or to the digestion of chitin or wax esters? What is the reason for the high percentage of acetate in the foregut of the hyrax and guinea pig and in the midgut of the emu and grass-fed horse?

The symbiotic, or deleterious, effects of microbes indigenous to the stomach of simple-stomached species needs further exploration. What are the advantages and disadvantages of a high degree of microbial fermentation in the midgut, such as those seen in the red-bellied turtle, emu, and grass-fed horse, given the problems of competition with and digestion of endogenous enzymes? Are the higher levels of bacteria and SCFA a result of longer midgut retention times or regurgitation from the hindgut? Do other herbivores, such as the panda and herbivorous bears, utilize a long midgut for a similar purpose? A better understanding of the mechanisms of SCFA and ammonia absorption could provide information on the conservation of water and nitrogen, and the toxic effects of these substances on epithelial cells. Are SCFA and ammonia absorbed principally by carrier-mediated transport of ions or by passive diffusion of nonionized acid or base? If it is by carrier-mediated transport, does this account for the transport of other weak organic electrolytes as well? Are ammonium ions exchanged for Na, as appears to be the case in the insect hindgut, or are they exchanged by the same route as K, as suggested by studies of the mammalian ileum?

Are cancer and other diseases of the human large intestine associated with low-fiber, high-protein, high-fat diets due to their effects on digesta passage, pH, or increased microbial production and absorption of SCFA, ammonia, metabolites of bile acids, or other potentially toxic metabolites? Can methods be developed for a more rapid repopulation of the large intestine with a normal complement of bacteria following their disruption by diet, oral antibiotics, or other causes?

The previous edition of this book ended with a plea for greater communication and cooperation between scientists interested in basic physiological mechanisms and those interested in comparative physiology. The renewed interest in comparative physiology of the digestive system and the application of new methods to a wider range of species is encouraging. Continuation of this trend will provide a better understanding of the basic mechanisms required for the normal functions of the digestive system and how these are disrupted by abnormal diets, unnatural feeding practices, or disease. It also will provide some of the information that will be required if we hope to preserve and maintain many of these species.

References

Adachi, J., Cao, Y., and Hasegawa, M. 1993. Tempo and mode of mitochondrial DNA evolution in vertebrates at the amino acid sequence level: rapid evolution in warm-blooded vertebrates. *J. Mol. Evol.* 36: 270–281.

Adam, H. 1963. Structure and histochemistry of the alimentary canal. In *The Biology of Myxine*, ed. A. Brodal and R. Fänge, pp. 256–288. Oslo: Universitetsforlaget.

Agent, B. E., and Case, R. M. 1994. Pancreatic ducts: cellular mechanism and control of bicarbonate. In *Physiology of the Gastrointestinal Tract*. 3rd ed., Vol. 2, ed. L. R. Johnson, D. H. Alpers, J. Christensen, E. D. Jacobson, and J. H. Walsh, pp. 1473–1498. New York: Raven Press.

Agungpriyono, S., Yamamoto, Y., Kitamura, N., Yamada, J., Sigit, K., and Yamashita, T. 1992. Morphological study on the stomach of the lesser mouse deer (Tragulus *javanicus*) with special reference to the internal surface. *J. Vet. Med. Sci.* 54: 1063–1069.

Ahearn, G. A., Behnke, R. D., Zonno, V., and Storelli, C. 1992. Kinetic heterogeneity of Na-D-glucose cotransport in teleost gastrointestinal tract. *Am. J. Physiol.* 263: R1018–R1023.

Ahearn, G. A., Franco, P., and Clay, L. P. 1990. Electrogenic 2Na+/1H+ exchange in crustaceans. *J. Memb. Physiol.* 116: 215–226.

Akester, A. R., Anderson, R. S., Hill, K. J., and Osboldiston, G. W. 1967. A radiographic study of urine flow in the domestic fowl. *Br. Poult. Sci.* 8: 209–212.

Al-Hussaini, A. H. 1949a. On the functional morphology of the alimentary tract of some fish in relation to differences in their feeding habits: anatomy and histology. *Quart. J. Microsc. Sci.* 90: 109–139.

Al-Hussaini, A. H. 1949b. On the functional morphology of the alimentary tract of some fish in relation to differences in their feeding habits: cytology and physiology. *Quart. J. Microsc. Sci.* 90: 323–354.

Alexander, F. 1962. The concentration of certain electrolytes in the digestive tract of the horse and pig. *Res. Vet. Sci.* 3: 78–84.

Alexander, F. 1965. The concentration of electrolytes in the alimentary tract of the rabbit, guinea pig, dog and cat. *Res. Vet. Sci.* 6: 238–244.

Alexander, F., and Davies, M. E. 1963. Production and fermentation of lactate by bacteria in the alimentary canal of the horse and pig. *J. Comp. Pathol. Ther.* 73: 1–8.

Alexander, F., and Hickson, J. C. D. 1970. The salivary and pancreatic secretions of the horse. In *Physiology of Digestion and Metabolism in the Ruminant*, ed. A. T. Phillipson, pp. 375–389. Newcastle upon Tyne: Oriel.

Alexander, R. McN. 1991. Optimization of gut structure and diet for higher vertebrate herbivores. *Phil. Trans. R. Soc. Lond.* B333: 249–255.

Allen, M. S., and Mertens, D. R. 1988. Evaluating constraints of fibre digestion by ruminant microbes. *J. Nutr.* 118: 261–270.

Alliot, E. 1967. Absorption intestinale de 1'*N*-acétyl-glucosamine chez la petite Roussette: *Scylliorhinus canicula. C. R. Séances Soc. Biol.* 161: 2544–2546.

Allison, M. J. 1984. Microbiology of the rumen and small and large intestines. In *Dukes' Physiology of Domestic Animals*, 10th ed., ed. M. J. Swenson, pp. 304–350. Ithaca: Cornell University Press.

Allo, A. A., Oh, J. H., Longhurst, W. M., and Connolly, G. E. 1973. VFA production in the digestive systems of deer and sheep. *J. Wildl. Manage.* 37: 202–211.

Alpers, D. H. 1987. Digestion and absorption of carbohydrates and proteins. In *Physiology of the Gastrointestinal Tract*, ed. L. R. Johnson, pp. 1469–1497. New York: Raven Press.

Alpers, D. H., and Solin, M. 1970. The characterization of rat intestinal amylase. *Gastroenterology* 58: 833–842.

Alumets, J., Håkanson, R., and Sundler, F. 1978. Distribution, ontogeny and ultrastructure of pancreatic polypeptide (PP) cells in the pancreas and gut of the chicken. *Cell Tissue Res.* 194: 377–386.

Alumets, J., Sundler, F., and Håkanson, R. 1977. Distribution, ontogeny and ultrastructure of somatostatin immunoreactive cells in the pancreas and gut. *Cell Tissue Res.* 185: 465–480.

Anderson, M. D., Richardson, P. R. K., and Woodall, P. F. 1992. Functional analysis of the feeding apparatus and digestive tract anatomy of the aardwolf *Proteles cristatus. J. Zool. (Lond.)* 228: 423–434.

Andrew, W. 1959. Going to great lengths–the alimentary tract of vertebrates. In *Textbook of Comparative Histology*, pp. 227–295. New York: Oxford University Press.

Andrew, W. 1963. Mucus secretion in cell nests and surface epithelium. *Ann. N. Y. Acad. Sci.* 106: 502–517.

Angert, E. R., Clements, K. D., and Pace, N. R. 1993. The largest bacterium. *Nature (Lond.)* 362: 239–241.

Annison, E. F., Hill, K. J., and Kenworthy, K. J. 1968. Volatile fatty acids in the digestive tract of the fowl. *Br. J. Nutr.* 22: 207–216.

Antony, M. 1920. Über die sprecheldrüsen der Vögel. Zool. *Jahrb., Agt. Anat. Onlog. Tiere* 41: 547–660.

Anuras, S., and Christensen, J. 1975. Electrical slow waves of the colon do not extend into the caecum. *Rendic. Gastroenterol.* 7: 56–59.

Argenzio, R. A. 1975. Functions of the equine large intestine and their interrelationship in disease. *Cornell Vet.* 65: 303–313.

Argenzio, R. A. 1984. Secretory functions of the gastrointestinal tract. In *Dukes' Physiology of Domestic Animals*, 10th ed., ed. M. J. Swenson, pp. 290–300. Ithaca: Cornell University Press.

Argenzio, R. A. 1991. Comparative physiology of colonic electrolyte transport. In *Handbook of Physiology: The Gastrointestinal System.* Vol. 4 Intestinal Absorption and Secretion. ed. S. G. Schultz, M. Field, R. A. Frizzell, and B. A. Rauner, pp. 175–286. Bethesda: American Physiology Society.

Argenzio, R. A., and Eisemann, J. 1995. Mechanisms of acid injury in pig gastroesophageal mucosa. *Am. J. Vet. Res.* (in press).

Argenzio, R. A., Lowe, J. E., Pickard, D. W., and Stevens, C. E. 1974. Digesta passage and water exchange in the equine large intestine. *Am. J. Physiol.* 226: 1035–1042.

Argenzio, R. A., and Meuten, D. J. 1991. Short-chain fatty acids induce reversible injury of porcine colon. *Digest. Dis. Sci.* 36: 1459–1468.

Argenzio, R. A., Miller, N., and Engelhardt, W. v. 1975. Effect of volatile fatty acids on water and ion absorption from the goat colon. *Am. J. Physiol.* 229: 997–1002.

Argenzio, R. A., Moon, H. W., Kemeny, L. J., and Whipp, S. C. 1984. Colonic compensation in transmissible gastroenteritis of swine. *Gastroenterology* 86: 1501–1509.

Argenzio, R. A., and Southworth, M. 1974. Sites of organic acid production and absorption in the gastrointestinal tract of the pig. *Am. J. Physiol.* 228: 454–460.

Argenzio, R. A., Southworth, M., Lowe, J. E., and Stevens, C. E. 1977. Interrelationship of Na, HCO_3 and volatile fatty acid transport by equine large intestine. *Am. J. Physiol.* 233: E469–E478.

Argenzio, R. A., Southworth, M., and Stevens, C. E. 1974. Sites of organic acid production and absorption in the equine gastrointestinal tract. *Am. J. Physiol.* 226: 1043–1050.

Argenzio, R. A., and Stevens, C. E. 1975. Cyclic changes in ionic composition of digesta in the equine intestinal tract. *Am. J. Physiol.* 228: 1224–1230.

Argenzio, R. A., and Stevens, C. E. 1984. The large bowel–a supplementary rumen? *Proc. Nutri. Soc.* 43: 13–23.

Argenzio, R. A., and Whipp, S. 1979. Interrelationship of sodium, chloride, bicarbonate and acetate transport by the colon of the pig. *J. Physiol.* 295: 365–381.

Arnason, S. S., and Skadhauge, E. 1991. Steady-state sodium absorption and chloride secretion of colon and coprodeum, and plasma levels of osmoregulatory hormones in hens in relation to sodium intake. *J. Comp. Physiol. B* 161: 1–14.

Árnason, Ú., Gullberg, A., and Widegren, B. 1991. The complete nucleotide sequence of the mitochondrial DNA of the fin whale *Balaenoptera physalus*. *J. Mol. Evol.* 33: 566–568.

Ash, P. W., and Dobson, A. 1963. The effect of absorption on the acidity of rumen contents. *J. Physiol. (Lond.)* 169: 39–61.

Ayer, A. A. 1948. *The Anatomy of Semnopithecus entellus*. Madras: Indian Publishing House.

Babkin, B. P., Hebb, C. O., and Kreuger, L. 1941. Changes in the secretory activity of the gastric glands resulting from the application of acetic acid solutions to the gastric mucosa. *Quart. J. Exp. Physiol.* 31: 63–78.

Bacarese-Hamilton, A. J., Adrian, T. E., and Bloom, S. R. 1984. Human and porcine immunoreactive gastric inhibitory polypeptides (IR-GIP) are not identical. *FEBS Lett.* 168: 125–128.

Baillien, M., and Schoffeniels, E. 1961. Origine des potentiels bioélectriques de l'épithélium intestinal de la tortue grecque. *Biochim. Biophys. Acta* 53: 537–548.

Baker, F., Nash, H., Morrice, F., and Bruce, J. 1950. Bacterial breakdown of structural starches and starch products in the digestive tract of ruminant and non-ruminant mammals. *J. Pathol. Bacteriol.* 62: 617–638.

Bakker, R. T. 1971. Dinosaur physiology and the origin of mammals. *Evolution* 25: 632–658.

Bakker, R. T. 1987. *The Dinosaur Heresies*. New York: William Morrow and Company.

Balch, C. C., and Campling, R. C. 1965. Rate of passage of digesta through the ruminant digestive tract. In *Physiology of Digestion in the Ruminant*, ed. R. W. Dougherty, pp. 108–130. Washington: Butterworths.

Baldizan, A., Dixon, R. M., and Parra, R. 1983. Digestion in the capybara (*Hydrochoerus hydrochaeris*). *S. Afr. J. Anim. Sci.* 13: 27–28.

Baldwin, R. L., Koong, L. J., and Ulyatt, M. J. 1977. Model of ruminant digestion. *Agr-Bio Syst.*, Vol. 2, No. 4, p. 282.

Banta, C. A., Clemens, E. T., Krinsky, M. M., and Sheffy, B. E. 1979. Sites of organic acid production and pattern of digesta movement in the gastrointestinal tract of dogs. *J. Nutr.* 109: 1592–1600.

Barboza, P. S. 1993. Digestive strategies of the wombats: feed intake, fiber digestion, and digesta passage in two grazing marsupials with hindgut fermentation. *Physiol. Zool.* 66: 983–999.

Barboza, P. S., and Hume, I. D. 1992a. Digestive tract morphology and digestion in the wombats (Marsupialia: Vombatidae). *J. Comp. Physiol.* B162: 552–560.

Barboza, P. S., and Hume, I. D. 1992b. Hindgut fermentation in the wombats: two marsupial grazers. *J. Comp. Physiol.* B162: 561–566.

Barcroft, T., McAnally, R. A., and Phillipson, A. T. 1944. Absorption of volatile acid from the alimentary tract of the sheep and other animals. *J. Exp. Biol.* 20: 120–129.

Bar-Eli, A., White, H. B., and Van Vunakis, H. 1966. Proteolytic enzymes from *Mustelus canis* fundic mucosae. *Fed. Proc.* 25: 745.

Barnard, E. A. 1969a. Biological function of pancreatic ribonuclease. *Nature* (London) 221: 340–344.

Barnard, E. A. 1969b. Ribonucleases. *Annu. Rev. Biochem.* 38: 677–732.

Barnard, E. A., and Prosser, C. L. 1973. Comparative biochemistry and physiology of the digestive system. In *Comparative Animal Physiology*, 3rd ed., ed. C. L. Prosser, pp. 133–164. Philadelphia: W. B. Saunders and Co.

Barnes, E. M., and Impey, C. S. 1974. The occurance and properties of uric acid decomposing bacteria in the avian cecum. *J. Appl. Bact.* 37: 393–401.

Barnes, R. D. 1974. *Invertebrate Zoology*, 3rd ed. Philadelphia: W. B. Saunders Company.

Barnes, R. H. 1962. Nutritional implications of coprophagy. *Nutr. Rev.* 10: 289–291.

Barnes, R. H., and Fiala, G. 1958. Effects of the prevention of coprophagy in the rat. I. Growth studies. *J. Nutr.* 64: 533–540.

Barnes, R. H., Fiala, G., and Kwong, E. 1963. Decreased growth rate resulting from prevention of coprophagy. *Fed. Proc.* 22: 125–133.

Barnes, R. H., Fiala, G., McGehee, B., and Brown, A. 1957. Prevention of coprophagy in the rat. *J. Nutrition* 63: 489–498.

Barrick, R. E., and Showers, W. J. 1994. Thermophysiology of *Tyrannosaurus rex*: evidence from oxygen isotopes. *Science* 265: 222–224.

Barrington, E. J. W. 1957. The alimentary canal and digestion. In *The Physiology of Fishes*, Vol. 1, Metabolism, ed. M. E. Brown, pp. 109–161. New York: Academic Press.

Barrington, E. J. W. 1962. Digestive enzymes. *Adv. Comp. Physiol. Biochem.* 1: 1–65.

Barrington, E. J. W. 1982. Evolutionary and comparative aspects of gut and brain peptides. *Med. Bull.* 38: 227–232.

Barrington, E. J. W., and Dockray, G. J. 1970. The effect of intestinal extracts of lampreys (*Lampetra fluviatilis* and *Petromyzon marinus*) on pancreatic secretion in the rat. *Gen. Comp. Endocr.* 14: 170–177.

Barrington, E. J. W., and Dockray, G. J. 1972. Cholecystokinin-pancreozymin-like activity in the eel *Anguilla anguilla*. *Gen. Comp. Endocr.* 19: 80–87.

Bartholomai, A. 1972. Aspects of the evolution of Australian marsupials. *Proc. Roy. Soc. Qld.* 82: 5–18.

Batt, E. R., and Schachter, D. 1969. Developmental pattern of some intestinal transport mechanisms in newborn rats and mice. *Am. J. Physiol.* 216: 1064–1068.

Bauchop, T., and Martucci, R. W. 1968. Ruminant-like digestion of the langur monkey. *Science* 161: 698–700.

Baverstock, P. R., Watts, C. H. S., and Spencer, L. 1979. Water-balance of small lactating rodents. V. The total water balance picture of the mother-young unit. *Comp. Biochem. Physiol.* 63A: 247–252.

Baxter, R. M., and Meester, J. 1982. The captive behaviour of the red musk shrew, *Crocidura f. flavescens* (I. Geoffroy, 1827) (Soricidae: Crocidurinae). *Mammalia* 46: 11–27.

Bayless, W. M., and Starling, E. H. 1902. The mechanism of pancreas secretion. *J. Physiol. (Lond.)* 28: 325–353.

Beal, A. M. 1984. Electrolyte composition of parotid saliva from the sodium-repleted red kangaroo (*Macropus refus*). *J. Exp. Biol.* 111: 225–237.

Beal, A. M. 1986. Responsiveness of the parotid salivary gland of red kangaroos (*Macropus rufus*) to mineralocorticoids. *J. Exp. Biol.* 120: 309–324.

Beerten-Joly, B., Piavaux, A., and Goffart, M. 1974. Quelquesenzymes digestifs chez un prosimian, *Perodicticus potto*. *C. R. Séances Soc. Biol. (Paris)* 168: 140–143.

Bell, D. J., and Bird, T. P. 1966. Urea and volatile base in the caeca and colon of the domestic fowl: the problem of their origin. *Comp. Biochem. Physiol.* 18: 735–744.

Bell, F. R. 1958. The mechanism of regurgitation during the process of rumination in the goat. *J. Physiol. (Lond.)* 142: 503–515.

Bell, G. H., Emslie-Smith, D., and Paterson, C. R. 1980. *Textbook of Physiology*, 10th ed. New York: Churchill Livingstone.

Bell, R. H. V. 1970. The use of the herb layer by grazing ungulates in the Serengeti. In *Animal Populations in Relation to their Food Resources*, ed. A. Watson, pp. 111–124. Oxford: Blackwell.

Bellwood, D. R., and Choat, J. H. 1990. A functional analysis of grazing in parrotfishes (family Scaridae): the ecological implications. *Environ. Biol. Fishes* 28: 189–216.

Belmann, H. 1973. Vergleichend-und-funktionell-anatomische Untersuchungen am Caecum und Colon myomorpher Negetiere. *Z. wiss. Zool., Liepsig* 186: 173–294.

Bennett, A. F. 1972. A comparison of activities of metabolic enzymes in lizards and rats. *Comp. Biochem. Physiol.* 42A: 637–647.

Bennett, T. 1974. Peripheral and autonomic nervous systems. In *Avian Biology*, Vol. IV, ed. D. S. Farner, J. R. King, and K. C. Parkes, pp. 1–77. New York: Academic Press.

Bensadoun, A., and Rothfeld, A. 1972. The form of absorption of lipids in the chicken, *Gallus domesticus*. *Proc. Soc. Exp. Biol. Med.* 141: 814–817.

Bensley, R. R. 1902–03. The cardiac glands of mammals. *Am. J. Anat.* 2: 105–156.

Benson, A. A., and Lee, R. F. 1975. The role of wax in oceanic food chains. *Scient. Am.* 232: 77–86.

Benson, A. A., Lee, R. F., and Nevenzel, J. C. 1972. Wax esters: major marine metabolic energy sources. In *Current Trends in the Biochemistry of Lipids*, ed. J. Ganguly and R. M. S. Smellie, pp. 175–187. New York: Academic Press.

Bentley, P. J. 1982. *Comparative Vertebrate Endocrinology*. New York: Cambridge University Press.

Bentley, P. J., and Smith, M. W. 1975. Transport of electrolytes across helicoidal colon of the newborn pig. *J. Physiol. (Lond.)* 249: 103–117.

Berger, P. J., and Burnstock, G. 1979. Autonomic nervous system. In *Biology of the Reptilia*, Vol. 10, ed. C. Gans, R. G. Northcutt, and P. Ulinski, pp. 1–57. New York: Academic Press.

Berkeley, R. C. W. 1979. Chitin, Chitosan and their degradative enzymes. In *Microbial Polysaccharides and Polysaccharidases*, ed. R. C. W. Berkeley, G. W. Gooday, and D. C. Ellwood, pp. 205–236. New York: Academic Press.

Bhattacharya, S., and Ghose, K. C. 1971. Influence of food on the amylase system in birds. *Comp. Biochem. Physiol.* 40B: 317–320.

Bidder, A. M. 1976. New names for old: the cephalopod "mid-gut gland." *J. Zool. (Lond.)* 180: 441–443.

Binder, H. J. 1970. Amino acid absorption in the mammalian colon. *Biochim. Biophys. Acta* 219: 503–506.

Binder, H. J., and Sandle, G. I. 1994. Electrolyte transport in the mammalian colon. In *Physiology of the Gastrointestinal Tract.* 3rd ed., Vol. 2, ed. L. R. Johnson, D. H. Alpers, J. Christensen, E. D. Jacobson, and J. H. Walsh. pp. 2133–2172. New York: Raven Press.

Birney, E. C., Jenness, R., and Ayaz, K. M. 1976. Inability of bats to synthesize L-ascorbic acid. *Nature (Lond.)* 260: 626–628.

Bjorndal, K. A. 1979. Cellulose digestion and volatile fatty acid production in the green turtle *Chelonia mydas. Biochem. Physiol.* 63: 127–133.

Bjorndal, K. A. 1987. Digestive efficiency in a temperate herbivorous reptile. *Copea* 3A: 714–720.

Bjorndal, K. A. 1989. Flexibility of digestive responses in two generalist herbivores, the tortoises *Geochelone carbonaria* and *Geochelone denticulata. Oecologia* 78: 317–321.

Bjorndal, K. A. 1991. Digestive fermentation in green turtles, *Chelonia mydas*, feeding on algae. *Bull. Marine Sci.* 48: 166–171.

Bjorndal, K. A., and Bolten, A. B. 1990. Digestive processing in a herbivorous freshwater turtle: consequences of small-intestine fermentation. *Physiol. Zool.* 63: 1232–1247.

Bjorndal, K. A., and Bolten, A. B. 1992. Body size and digestive efficiency in a herbivorous freshwater turtle: advantages of small bite size. *Physiol. Zool.* 65: 1028–1039.

Björnhag, G. 1972. Separation and delay of contents in the rabbit colon. *Swed. J. Agric. Res.* 2: 125–136.

Björnhag, G. 1981. The retrograde transport of fluid in the proximal colon of rabbits. *Swed. J. Agric. Res.* 11: 63–69.

Björnhag, G. 1987. Comparative aspects of digestion in the hindgut of mammals. The colonic separation mechanism (CSM) (a review). *Dtsch. Tierarztl. Wochenschr.* 94: 33–36.

Björnhag, G. 1989. Transport of water and food particles through the avian ceca and colon. In *Function of the Avian Cecum*, ed. E. J. Braun and G. E. Duke, pp. 32–37, *J. Exp. Zool.* Suppl No. 3, New York: Alan R. Liss, Inc.

Björnhag, G. 1994. Adaptations in the large intestine allowing small animals to eat fibrous food. In *The Digestive System in Mammals: Food, Form and Function*, ed. D. J. Chivers and P. Langer, pp. 287–312. Cambridge: Cambridge University Press.

Björnhag, G., and Sjöblom, L. 1977. Demonstration of coprophagy in some rodents. *Swed. J. Agric. Res.* 7: 105–114.

Björnhag, G., and Sperber, I. 1977. Transport of various food components through the digestive tract of turkeys, geese and guinea fowl. *Swed. J. Agric. Res.* 7: 57–66.

Black, J. L., and Sharkey, M. J. 1970. Reticular groove (sulcus reticuli): an obligatory adaptation in ruminant-like herbivores? *Mammalia* 34: 294–302.

Bläckberg, L., Hernell, O., and Olivecrona, T. 1981. Hydrolysis of human milk fat globules by pancreatic lipase. *J. Clin. Invest.* 67: 1748–1752.

Blain, A. W., and Campbell, K. N. 1942. A study of digestive phenomena in snakes with the aid of the roentgen ray. *Am. J. Roentgenol. and Rad. Therapy* 48: 229–239.

Blair-West, J. R., Bott, E., Boyd, G. W., Coghlan, J. P., Denton, D. A., Goding, J. R., Weller, S., Wintour, M., and Wright, R. D. 1965. General biological aspects of salivary secretion in ruminants. In *Physiology of Digestion in the Ruminant*, ed. R. W. Dougherty, pp. 198–220. Washington, D. C.: Butterworths.

Blair-West, J. R., Coghlan, J. P., Denton, D. A., and Wright, R. D. 1970. Factors in sodium and potassium metabolism. In *Physiology of Digestion in the Ruminant*, ed. A. T. Phillipson, pp. 350–361. Newcastle upon Tyne, England: Oriel Press.

Blaxter, K. L. 1962. *The Energy Metabolism of Ruminants*. London: Hutchinson and Co.

Bogé, G., Rigal, A., and Pérès, G. 1983. Analysis of two chloride requirements for sodium-dependent amino acid and glucose transport by intestinal brush-border membrane vesicles of fish. *Biochim. Biophys. Acta* 729: 209–218.

Bolton, W. 1962. Digestion in the crop of the fowl. *Proc. Nutr. Soc.* 21: XXIV.

Bolton, W. 1965. Digestion in the crop of the fowl. *Br. Poult. Sci.* 6: 97–102.

Bondi, A. M., Menghi, G., Palatroni, P., and Materazzi, G. 1982. Ultrastructural observations on the morphology of the stomach of *Salmoirideus* and on the localization of carbonic anhydrase during development. *Anat. Anz.* 151: 473–481.

Bonnafous, R. 1973. Quelques aspects de la physiologie coliqueen relation avec la dualité de l'excretion fécale chez lelapin. *Thèse Doctorat Sciences*, Toulouse.

Bonnafous, R., and Raynaud, P. 1968. Mise en evidence d'uneactivité lysante du colon proximal sur les microorganismes dutube digestif du lapin. *Arch. Sci. Physiol.* 22: 57–64.

Bookhout, T. A. 1959. Reingestion by the snowshoe hare. *J. Mammal.* 40: 250.

Boomker, E. A. 1983. Volatile fatty acid production in the grey duiker, *Sylvicapra grimmia*. *S. Afr. J. Anim. Sci.* 13: 33–35.

Borch-Madsen, P. 1946. *Resorptionens Storrelse ved experimentelt fremkaldt Achylia gastrica*. Copenhagen: Ejvind Christensens Publishing House.

Borges, P., Dominguez-Bello, M. O., and Herrara, E. 1995. Digestive physiology of wild capybara *Hydrochaeris hydrochaeris*. *J. Compo. Physiol. B* (in press).

Borgstrom, B., and Patton, J. S. 1991. Luminal events in gastrointestinal lipid digestion. In *Handbook of Physiology Section 6: The Gastrointestinal System*, Vol. IV, ed. M. Field and R. A. Frizzell, pp. 475–504. Bethesda: American Physiological Society.

Borody, T. J., Byrnes, D. J., and Titchen, D. A. 1984. Motilin and migrating myoelectric complexes in the pig and the dog. *Quart. J. Exp. Physiol.* 69: 875–890.

Botha, G. S. 1958. A note on the comparative anatomy of the gastroesophageal junction. *Acta Anat.* 34: 52–84.

Botha, G. S. 1962. *The Gastro-Oesophageal Junction*. Boston: Little, Brown and Co.

Bottino, N. R. 1975. Lipid composition of two species of Antarctic krill: *Euphausia superba* and *E. crystal lorophias*. *Comp. Biochem. Physiol.* 50B: 479–484.

Brambell, F. W. R. 1970. *The Transmission of Passive Immunity from Mother to Young*. New York: American Elsevier.

Brannon, P. M. 1990. Adaptation of the exocrine pancreas to diet. *Annu. Rev. Nutr.* 10: 85–105.

Brasitus, T. A., and Dudeja, P. K. 1988. Effect of hypothyroidism on the lipid composition and fluidity of rat colonic apical plasma membranes. *Biochim. Biophys. Acta* 939: 189–196.

Bridie, A., Hume, I. D., and Hill, D. M. 1994. Digestive tract function and energy requirements of the rufous hare-wallaby, *Lagorchestes hirsutus*. *Aust. J. Zool.* 42: 761–774.

Britton, S. W. 1941. Form and function in the sloth. *Quart. Rev. Biol.* 16: 190–207.

Brockerhoff, H., and Hoyle, R. J. 1965. Hydrolysis of triglycerides by the pancreatic lipase of a skate. *Biochim. Biophys. Acta* 98: 435–436.

Brockerhoff, H., and Jensen, R. G. 1974. *Lipolytic Enzymes*. New York: Academic Press.

Brodin, E., Alumets, J., Hakanson, R., Leander, S., and Sundler, F. 1981. Immunoreactive substance P in the chicken gut: distribution, development and possible functional significance. *Cell Tissue Res.* 216: 455–469.

Brown, G. D. 1969. Studies on marsupial nutrition. VI. The utilization of dietary urea by the euro or hill kangaroo, *Macropus robustus* (Gould). *Aust. J. Zool.* 17: 187–194.

Brown, K. M., and Moog, F. 1967. Invertase activity in the intestine of the developing chick. *Biochim. Biophys. Acta* 132: 185–187.

Brown, R. H. 1962. The anatomy of the alimentary tract of three genera of bats. Masters Thesis, Tucson University of Arizona.

Brown, W. R. 1968. Rumination in the adult: a study of two cases. *Gastroenterology* 54: 933–939.

Browne, T. G. 1922. Some observations on the digestive system of the fowl. *J. Comp. Pathol. Ther.* 35: 12–32.

Bruoton, M. R., Davis, C. L., and Perrin, M. R. 1991. Gut microflora of vervet and samango monkeys in relation to diet. *Appl. Environ. Microbiol.* 57: 573–578.

Bryant, M. P. 1977. Microbiology of the rumen. In *Dukes' Physiology of Domestic Animals*, 9th ed., ed. M. J. Swenson, pp. 287–304. Ithaca: Cornell University Press.

Buchan, A. M. J., Ingman-Baker, J., Levy, J., and Brown, J. C. 1982. A comparison of the ability of serum and monoclonal antibodies to gastric inhibitory polypeptide to detect immunoreactive cells in the gastroenteropancreatic system of mammals and reptiles. *Histochemistry* 76: 341–349.

Buchan, A. M. J., Lance, V., and Polak, J. M. 1983. Regulatory peptides in the gastrointestinal tract of *Alligator mississipiensis*: an immunocytochemical study. *Cell Tissue Res.* 231: 439–449.

Buchan, A. M. J., Polak, J. M., Bryant, M. G., Bloom, S. R., and Pearse, A. G. E. 1981. Vasoactive intestinal polypeptide (VIP)-like immunoreactivity in anuran intestine. *Cell Tissue Res.* 216: 413–422.

Buchan, A. M. J., Polak, J. M., and Pearse, A. G. E. 1980. Gut hormones in *Salamandra salamandra*: an immunocytochemical and electron microscopic investigation. *Cell Tissue Res.* 211:311–343.

Buchner, P. 1965. *Endosymbiosis of Animals with Plant Micro-organisms*. New York: Interscience Publishers.

Buddington, R. K. 1992. Intestinal nutrient transport during ontogeny of vertebrates. *Am. J. Physiol.* 263: R503–R509.

Buddington, R. K., and Diamond, J. M. 1987. Pyloric ceca of fish: a new absorptive organ. *Am. J. Physiol.* 252: B55–G76.

Buddington, R. K., and Diamond, J. M. 1989. Ontogenetic development of intestinal nutrient transporters. *Annu. Rev. Physiol.* 51: 602–619.

Buddington, R. K., and Diamond, J. M. 1990. Ontogenic development of monosaccharide and amino acid transporters in rabbit intestine. *Am. J. Physiol.* 259: G544–G555.

Buddington, R. K., and Doroshov, S. I. 1986. Development of digestive secretions in white sturgeon juveniles (*Acipenser transmontanus*). *Comp. Biochem. Physiol.* 83A: 233–238.

Bueno, L. 1972. Action du sphincter reticulo-amasal sur le transit alimentaire chez le bovins. *Ann. Rech. Vet.* 3: 83–91.

Bueno, L., and Ruckebusch, Y. 1974. The cyclic motility of the omasum and its control in sheep. *J. Physiol. (Lond.)* 238: 295–312.

Buffenstein, R., and Yahav, S. 1991. The effect of diet on microfaunal population and function in the caecum of a subterranean naked mole-rat, *Heterocephalus glaber.* *Br. J. Nutr.* 65: 249–258.

Burgen, A. S. V. 1961. The comparative physiology of the salivary glands. In *Physiol-

ogy of the Salivary Glands, ed. A. S. V. Burgen and N. G. Emmelin, pp. 267–272. Baltimore: Williams & Wilkins.

Burgen, A. S. V. 1967. Secretory processes in salivary glands. In *Handbook of Physiology*, Section 6, *Alimentary Canal*, Vol. 2, *Secretion*, ed. C. F. Code and W. Heidel, pp. 561–579. Washington, D. C.: American Physiological Society.

Burgen, A. S. V., and Seeman, P. 1958. The role of the salivary duct system in the formation of saliva. *Can. J. Biochem. Physiol.* 36: 119–143.

Burhol, P. G. 1982. Regulation of gastric secretion in the chicken. *Scand. J. Gastroenterol.* 17: 321–323.

Burkitt, D. T. 1971. Epidemiology of cancer of the colon and rectum. *Cancer* 28: 3–13.

Burks, T. F. 1994. Neurotransmission and neurotransmitters. In *Physiology of the Gastrointestinal Tract.* 3rd ed., Vol. 2, ed. L. R. Johnson, D. H. Alpers, J. Christensen, E. D. Jacobson, and J. H. Walsh, pp. 423–482. New York: Raven Press.

Burns, W. A., Flores, P. A., Moshyedi, A., and Albacete, R. A. 1970. Clinical conditions associated with columnar lined esophagus. *Am. J. Dig. Dis.* New Series 15: 607–615.

Burnstock, G. 1969. Evolution of the autonomic innervation of visceral and cardiovascular systems in vertebrates. *Pharmacol. Rev.* 21: 247–324.

Burnstock, G. 1972. Purinergic nerves. *Pharmacol. Rev.* 24: 509–581.

Burnstock, G. 1986. The changing face of autonomic neurotransmission. *Acta Physiol. Scand.* 126: 67–91.

Cahill, M. M. 1990. Bacterial flora of fishes: a review. *Microb. Ecol.* 19: 21–41.

Caira, J. N. 1989. Gut pH in the nurse shark, *Ginglymostoma cirratum* (Bonnaterre). *Coepia* 1: 192–194.

Calligan, J. J., Costa, M., and Furness, J. B. 1985. Gastrointestinal myoelectric activity in conscious guinea pigs. *Am. J. Physiol.* 249: G92–G98.

Calloway, D. H. 1968. Gas in the alimentary canal. In *Handbook of Physiology*, Section 6, *Alimentary Canal*, Vol. 5, *Bile; Digestion; Ruminal Physiology*, ed. C. F. Code and W. Heidel, pp. 2839–2859. Washington, D. C.: American Physiological Society.

Calonge, M. L., Ilundain, A., and Bolufer, J. 1990. Glycylsarcosine transport by epithelial cells isolated from chicken proximal cecum and rectum. *Am. J. Physiol.* 258: G660–G664.

Campbell, G. 1970. Autonomic nervous systems. In *Fish Physiology*, Vol. IV, ed. W. S. Hoar and D. J. Randall, pp. 109–132. New York: Academic Press.

Campbell, G., and Burnstock, G. 1968. Comparative physiology of gastrointestinal motility. In *Handbook of Physiology*, Section 6, *Alimentary Canal*, Vol. 4, *Motility*, ed. C. F. Code and W. Heidel, pp. 2213–2266. Washington, D.C.: American Physiological Society.

Cannon, W. B. 1902. The movements of the intestines studied by means of roentgen rays. *Am. J. Physiol.* 6: 251–277.

Carl, G. R., and Brown, R. D. 1983. Protozoa in the forestomach of the collared peccary (*Tayassu tajacu*). *J. Mamm.* 64: 709.

Carleton, M. D. 1973. A survey of gross stomach morphology in New World Cricetinae (Rodenta Muroidea), with comments on functional interpretations. *Misc. Publ. Museum Zool., Univ. Michigan*, No. 46.

Carleton, M. D. 1981. A survey of gross stomach morphology in Microtinae (Rodentia: Muroidea). *Zeitschr. f. Saugetierk.* 46: 93–108.

Carr, D. H., McLeay, L. M., and Titchen, D. A. 1970. Factors affecting reflex responses of the ruminant stomach. In *Physiology of Digestion and Metabolism in the Ruminant*, ed. A. T. Phillipson, pp. 35–41. Newcastle-upon-Tyne: Oriel Press.

Chang, E. B., and Rao, M. C. 1994. Intestinal water and electrolyte transport: mechanisms of physiological and adaptive processes. In *Physiology of the gastrointestinal Tract*, 3rd ed., Vol. 2, ed. L. R. Johnson, D. H. Alpers, J. Christensen, E. D. Jacobson, and J. H. Walsh, pp. 2027–2082. New York: Raven Press.

Chao, F. C., and Tarver, H. 1953. Breakdown of urea in the rat. *Proc. Soc. Exp. Biol. Med.* 84: 406–409.

Chaudhuri, C. R., and Chatterjee, I. B. 1969. L-ascorbic acid synthesis in birds: phylogenetic trend. *Science* 164: 435–436.

Cheeseman, C. I. 1986. Expression of amino acid and peptide transport systems in rat small intestine. *Am. J. Physiol.* 251: G636–G641.

Cheeseman, C. I. 1993. GLUT2 is the transporter for fructose across the rat intestinal basolateral membrane. *Gastroenterology* 105: 1050–1056.

Cheeseman, C. I., and Harley, B. 1991. Adaptation of glucose transport across rat enterocyte basolateral membrane in response to altered dietary carbohydrate intake. *J. Physiol. (Lond.)* 437: 563–575.

Cheeseman, C. I., Shariff, S., and O'Neill, D. 1994. Evidence for a lactate-anion exchanger in the rat jejunal basolateral membrane. *Gastroenterology* 196: 559–556.

Chien, W.-J., and Stevens, C. E. 1972. Coupled active transport of Na and Cl across forestomach epithelium. *Am. J. Physiol.* 223: 997–1003.

Chilcott, M. J., and Hume, I. D. 1985. Coprophagy and selective retention of fluid digesta: their role in the nutrition of the common ringtail possum, *Pseudocheirus peregrinus*. *Aust. J. Zool.* 33: 1–15.

Chinsamy, A., and Dodson, P. 1995. Inside a dinosaur bone. *Am. Sci.* 83: 174–180.

Chivers, D. J., and Hladik, C. M. 1980. Morphology of the gastrointestinal tract of primates: comparisons with other mammals in relation to diet. *J. Morphol.* 166: 337–386.

Choat, J. H. 1991. The biology of herbivorous fishes on coral reefs. In *The Ecology of Fishes on Coral Reefs*, ed. P. F. Sale, pp. 120–155, San Diego: Academic Press.

Christensen, H. N. 1985. On the strategy of kinetic discrimination of amino acid transport systems. *J. Membrane Biol.* 84: 97–103.

Christensen, J. 1971. The controls of gastrointestinal movements: some old and new views. *New Engl. J. Med.* 285: 85–98.

Christensen, J. 1989. Colonic motility. In *Handbook of Physiology*, Section 6: *Alimentary Canal*, Vol. 1: *Motility and Circulation*, ed. S. G. Schultz, J. D. Wood, and B. B. Rauner, pp. 939–973. Washington, D.C.: American Physiological Society.

Christensen, J. 1989. Colonic motility. In *Handbook of Physiology*, Section 6: *Alimentary Canal*, Vol. 1: *Motility and Circulation*, ed. S. G. Shultz, J. D. Wood, and B. B. Rauner, pp. 939–973. Washington D. C.: American Physiological Society.

Christensen, J. 1994. The motility of the colon. In *Physiology of the Gastrointestinal Tract*, 3rd ed., Vol. 2, ed. L. R. Johnson, D. H. Alpers, J. Christensen, E. D. Jacobson, and J. H. Walsh, pp. 991–1024. New York: Raven Press.

Christensen, J., Anuras, S., and Hauser, R. L. 1974. Migrating spike bursts and electrical slow waves in the cat colon: effect of sectioning. *Gastroenterology* 66: 240–247.

Christensen, J., Caprilli, R., and Lund, G. F. 1969. Electric slow waves in circular muscle of cat colon. *Am. J. Physiol.* 217: 771–776.

Christensen, J., Stiles, M. J., Gary, A., Rick, A., and Sutherland, J. 1984. Comparative anatomy of the myenteric plexus of the distal colon in eight mammals. *Gastroenterology* 86: 706–713.

Church, D. C. 1988. *The Ruminant Animal: Digestive Physiology and Nutrition*, Englewood Cliffs, New Jersey: Prentice-Hall.

Clarke, L. L., Argenzio, R. A., and Roberts, M. C. 1990. Effect of meal feeding on plasma volume and urinary electrolyte clearance in ponies. *Am. J. Vet. Res.* 51: 571–576.

Clarks, L. L., Roberts, M. C., and Argenzio, R. A. 1990. Feeding and digestive problems in horses. *Clin. Nutr.* 6: 433–450.

Clarke, R. T. J. 1977. Protozoa in the rumen ecosystem. In *Microbial Ecology of the Gut*, ed. R. T. J. Clarke and T. Bauchop, pp. 251–275. New York: Academic Press.

Clausen, M. R., and Mortensen, P. B. 1994. Kinetic studies on metabolism of short-chain fatty acids and glucose by isolated rat colonocytes. *Gastroenterology* 106: 423–432.

Clauss, W. 1983. Circadian rhythms in Na+ transport. In *Intestinal Absorption and Secretion*, ed. E. Skadhauge and K. Heintze, pp. 273–283, Lancaster, U.K.: MTP Press.

Clemens, E. T. 1977. Sites of organic acid production and patterns of digesta movement in the gastrointestinal tract of the rock hyrax. *J. Nutr.* 107: 1954–1961.

Clemens, E. T. 1980. The digestive tract: insectivore, prosimian, and advanced primate. In *Comparative Physiology: Primitive Mammals*, ed. K. Schmidt-Nielson, L. Bolis, and C. R. Taylor, pp. 90–99. New York: Cambridge University Press.

Clemens, E. T. 1982. Comparison of polyethylene glycol and dye markers in nutrition research. *Nutr. Res.* 2: 323–334.

Clemens, E. T., and Maloiy, G. M. O. 1982. The digestive physiology of three East African herbivores: the elephant, rhinoceros and hippopotamus. *J. Zool. (Lond.)* 198: 141–156.

Clemens, E. T., and Phillips, B. 1980. Organic acid production and digesta movement in the gastrointestinal tract of the baboon and Sykes monkey. *Comp. Biochem. Physiol.* 66A: 529–532.

Clemens, E. T., and Stevens, C. E. 1979. Sites of organic acid production and patterns of digesta movement in the gastrointestinal tract of the raccoon. *J. Nutr.* 109: 1110–1116.

Clemens, E. T., and Stevens, C. E. 1980. A comparison of gastrointestinal transit time in ten species of mammal. *J. Agric. Sci. (Camb.)* 94: 735–737.

Clemens, E. T., Stevens, C. E., and Southworth, M. 1975a. Sites of organic acid production and patterns of digesta movement in the gastrointestinal tract of swine. *J. Nutr.* 105: 759–768.

Clemens, E. T., Stevens, C. E., and Southworth, M. 1975b. Sites of organic acid production and pattern of digesta movement in the gastrointestinal tract of geese. *J. Nutr.* 105: 1341–1350.

Clements, K. D. 1991. Endosymbiotic communities of two herbivorous labroid fishes, *Odax cyanomelas* and *O. pullus. Mar. Biol.* 109: 223–229.

Clements, K. D., and Bellwood, D. R. 1988. A comparison of the feeding mechanisms of two herbivorous labroid fishes, the temporate *Odax pullus* and the tropical *Scarus rubroviolaceus. Aust. J. Mar. Freshwat. Res.* 39: 87–107.

Clements, K. D., Gleeson, V. P., and Slaytor, M. 1994. Short-chain fatty acid metabolism in temperate marine herbivorous fish. *J. Comp. Physiol.* B 164: 372–377.

Clements, K. D., Sutton, D. C., and Choat, J. H. 1989. Occurrence and characteristics of unusual protistan symbiont from surgeonfishes (*Acanthuridae*) of the Great Barrier Reef, Australia. *Mar. Biol.* 102: 403–412.

Clinton, S. K., Bostwick, D. G., Olson, L. M., Manglan, H. J., and Visek, W. J. 1988. Effects of ammonium acetate and sodium cholate on N-methyl-N'-nitro-N-nitrosguanidine-induced colon carcinogenesis of rats. *Cancer Research* 48: 3035–3039.

Cloudsley-Thompson, J. L. 1972. The classification and study of animals by feeding habits. In *Biology of Nutrition*, Vol. 18, ed. R. N. T-W-Fiennes, pp. 439–470. New York: Pergamon Press.

Cochran, P. A. 1987. Optimal digestion in a batch-reactor gut: the analogy to partial prey consumption. *Oikos* 50: 268–270.

Cocimano, M. R., and Leng, R. A. 1967. Metabolism of urea in sheep. *Br. J. Nutr.* 21: 353–371.

Cohen, T., Gertler, A., and Birk, Y. 1981a. Pancreatic proteolytic enzymes from carp (*Cyprinus carpio*). I. Purification and physical properties of trypsin, chymotrypsin, elastase and carboxypeptidase B. *Comp. Biochem. Physiol.* 69B: 639–646.

Cohen, T., Gertler, A., and Birk, Y. 1981b. Pancreatic proteolytic enzymes from carp (*Cyprinus carpio*). II. Kinetic properties and inhibition studies of trypsin, chymotrypsin and elastase. *Comp. Biochem. Physiol.* 69B: 647–653.

Colbet, E. H. 1980. *Evolution of Vertebrates*, 3rd ed. New York: John Wiley and Sons.

Coleman, G. S. 1980. Rumen ciliate protozoa. *Adv. Parasitol.* 18: 121–173.

Colin, D. A. 1972. Relations entre la nature de l'alimentationet l'importance de l'activité chitinolytique du tube digestifde quelques Téleostéens marins. *C. R. Séances Soc. Biol. (Paris)* 166: 95–98.

Cook, D. I., Van Lennep, E. W., Roberts, M. L., and Young, J. A. 1994. Secretion by the major salivary glands. In *Physiology of the Gastrointestinal Tract*, 3rd ed., Vol. 2, ed. L. R. Johnson, D. H. Alpers, J. Christensen, E. D. Jacobson, and J. H. Walsh, pp. 1061–1118. New York: Raven Press.

Corbet, P. S. 1960. The food of a sample of crocodiles *Crocodilus niloticus* L. from Lake Victoria. *Proc. Zool. Soc. Lond.* 133: 561–572.

Cork, S. J., and Foley, W. J. 1991. Digestive and metabolic strategies of arboreal mammalian folivores in relation to chemical defenses in temperate and tropical forests. In *Plant Defenses Against Mammalian Herbivory*, ed. R. T. Palo and C. T. Robbins, pp. 133–166. Boca Raton, Florida: CRC Press.

Cork, S. J., and Hume, I. D. 1983. Microbial digestion in the koala (*Phascolarctos cinereus*, Marsupialia): an aboreal folivore. J. Comp. Physiol. B 152: 131–135.

Cork, S. J., and Warner, A. C. I. 1983. The passage of digesta markers through the gut of a folivorous marsupial, the koala *Phascolarctos cinereus*. *J. Comp. Physiol. B* 152: 43–51.

Cornelius, C., Dandrifosse, G., and Jeuniaux, C. 1975. Biosynthesis of chitinases by mammals of the order Carnivora. *Biochem. Syst. Ecol.* 3: 121–122.

Coulson, R. A., Coulson, T. D., and Herbert, J. D. 1990. How do digestion and assimilation rates in alligators vary with temperature? *Comp. Biochem. Physiol.* 96A: 441–449.

Crompton, A. W. 1980. Biology of the earliest mammals. In *Comparative Physiology: Primitive Mammals*, ed. K. Schmidt-Nielsen, L. Bolis, and C. R. Taylor, pp. 1–12. New York: Cambridge University Press.

Crompton, A. W., and Parker, P. 1978. Evolution of the mammalian masticatory apparatus. *Am. Sci.* 66: 192–201.

Crompton, A. W., Taylor, C. R., and Jagger, J. A. 1978. Evolution of homeothermy in mammals. *Nature (Lond.)* 272: 333–336.

Cross, D. G. 1969. Aquatic weed control using grass carp. *J. Fish Biol.* 1: 27–30.

Crowcroft, P. 1952. Refection in the common shrew. *Nature (Lond.)* 170: 627.

Crump, M. H., Argenzio, R. A., and Whipp, S. C. 1980. Effects of acetate on absorption of sodium and water from the pig colon. *Am. J. Vet. Res.* 41: 1563–1568.

Cummings, J. F., Munnell, J. F., and Vallenas, A. 1972. The mucigenous glandular mucosa in the complex stomach of two New World camelids, the llama and guanaco. *J. Morphol.* 137: 71–109.

Cummings, J. H., Pomare, E. W., Branch, W. J., Naylor, C. P. E., and Macfarlane, G. T.

1987. Short-chain fatty acids in human large intestine, portal, hepatic and venous blood. *Gut.* 28: 1221–1227.

Curran, P. 1968. Twelfth Bowditch Lecture: coupling between transport processes in intestine. *Physiologist* 11: 3–23.

Curtin, T. M., Goetsch, G. D., and Hollandbeck, R. 1963. Clinical and pathologic characterization of esophagogastric ulcers in swine. *J. Am. Vet. Med. Assoc.* 143: 854–860.

Dabrowski, K. 1979. The role of proteolytic enzymes in fish digestion. In *Cultivation of Fish Fry and its Live Food*, Vol. 4, ed. E. Jaspers, pp. 107–126. Belgium: European Mariculture Society.

Daft, F. S., McDaniel, E. G., Hermann, L. G., Pomine, M. K., and Hegner, J. R. 1963. Role of coprophagy in utilization of B vitamins synthesized by intestinal bacteria. *Fed. Proc.* 22: 129–133.

Dahlqvist, A., and Thomson, D. L. 1963. Separation and characterization of two rat-intestinal amylases. *Biochem. J.* 89: 272–277.

Dandrifosse, G. 1963. La secretion de chitinase par la muqueuse gastrique isolee. *Ann. Soc. R. Zool. Belg.* 92: 199–201.

Dandrifosse, G. 1974. Digestion in reptiles. In *Chemical Zoology*, Vol. 9, ed. M. Florkin and B. T. Sheer, pp. 249–275. New York: Academic Press.

Daniel, S. L., and Drake, H. L. 1993. Oxalate- and glyoxylate-dependent growth and acetogenesis by *Clostridium thermoaceticum. Applied and Environ. Microb.* 59: 3062–3068.

Danielli, J. R., Hitchcock, M. W. S., Marshall, R. A., and Phillipson, A. T. 1945. The mechanism of absorption from the rumen as exemplified by the behavior of acetic, propionic and butyric acids. *J. Exp. Biol.* 22: 75–84.

Danulat, E. 1986. Role of bacteria in regard to chitin degradation in the digestive tract of the cod *Gadus morhua. Mar. Biol.* 90: 335–343.

Davenport, H. W. 1982. *Physiology of the Digestive Tract*, 5th ed. Chicago: Mosby.

Davies, S. J. 1985. The role of dietary fiber in fish nutrition. In *Recent Advances in Agriculture*, Vol. 2, ed, J. F. Muir and R. J. Roberts, pp. 219–2249, Boulder, Colorado: Westview Press.

Davis, D. D. 1961. Origin of the mammalian feeding mechanism. *Am. Zoologist* 1: 229–234.

Dawson, T. J., Herd, R. M., and Skadhauge, E. 1985. Osmotic and ionic regulation during dehydration in a large bird, the emu (*Dromaius novaehollandiae*): an important role for the cloaca-rectum. *J. Exp. Physiol.* 70: 423–436.

Dawson, T. J., and Hulbert, A. J. 1970. Standard metabolism, body temperature, and surface areas of Australian marsupials. *Am. J. Physiol.* 218: 1233–1238.

deLahunta, A., and Habel, R. E. 1986. Intestines. In *Applied Veterinary Anatomy*, pp. 246–256. Philadelphia: W. B. Saunders Co.

Dellow, D. W. 1982. Studies on the nutrition of macropodine marsupials. III. The flow of digesta through the stomach and intestine of macropodines and sheep. *Aust. J. Zool.* 30: 751–765.

Dellow, D. W., and Hume, I. D. 1982. Studies on the nutrition of macropodine marsupials. IV. Digestion in the stomach and the intestine of *Macropus giganteus, Thylogale thetis* and *Macropus eugenii. Aust. J. Zool.* 30: 767–777.

Dellow, D. W., Hume, I. D., Clarke, R. T. J., and Bauchop, T. 1988. Microbial activity in the forestomach of free-living macropodid marsupials: comparisons with laboratory studies. *Aust. J. Zool.* 36: 383–395.

Dellow, D. W., Nolan, J. V., and Hume, I. D. 1983. Studies on the nutrition of macropodine marsupials. V. Microbial fermentation in the forestomach of *Thylogale thetis* and *Macropus eugenii. Aust. J. Zool.* 31: 433–443.

Demment, M. W., and Van Soest, P. J. 1985. A nutritional explanation for body-size patterns of ruminant and nonruminant herbivores. *Am. Nat.* 125: 641–672.

Denis, C., Jeuniaux, C., Gerebtzoff, M. A., and Goffart, M. 1967. La digestion stomacale chez un paresseux: L'unau *Choloepushoffmanni* Peters. *Ann. Soc. R. Zool. Belg.* 97: 9–29.

Denton, D. A. 1957. The effects of variations in blood supply on the secretion rate and composition of parotid saliva in Na^+-depleted sheep. *J. Physiol. (Lond.)* 135: 227–244.

Deswysen, A. G. 1987. Forestomach: control of digesta flow. In *Physiological and Pharmacological Aspects of the Reticulo-rumen*, ed. L. A. Ooms, A. D. Degryse, and A. S. van Miert, p. 133–154. Dordrecht, The Netherlands: Martinus Nijhoff Publishers.

Diamond, J. M. 1977. Twenty-First Bowditch Lecture. The epithelial junction: bridge, gate, and fence. *Physiologist* 20: 10–18.

Dierenfeld, E. S., Hintz, H. F., Robertson, J. B., Van Soest, P. J., and Oftedal, O. T. 1982. Utilization of bamboo by the giant panda. *J. Nutr.* 112: 636–641.

Dieterlein, F. 1959. Das Verhalten des syrischen Goldhamsters (*Mesocricetus auratus* Waterhouse). *Z. Tierpsychol.* 16: 47–103.

Dimaline, R. 1983. Is caerulein amphibian CCK? *Peptides* 4: 457–462.

Dimaline, R., Rawdon, B. B., Brandes, S., Andrew, A., and Loveridge, J. P. 1982. Biologically active gastrin/CCK-related peptides in the stomach of a reptile, *Crocodylus niloticus*: identified and characterized by immunochemical methods. *Peptides* 3: 977–984.

Dirksen, G. 1970. Acidosis. In *Physiology of Digestion and Metabolism in the Ruminant*, ed. A. T. Phillipson, pp. 612–625. Newcastle upon Tyne: Oriel Press.

Dobson, A. 1959. Active transport through the epithelium of the reticulorumen sac. *J. Physiol. (Lond.)* 146: 235–251.

Dockray, G. J. 1974. Extraction of a secretin-like factor from the intestines of Pike (*Esox lucius*). *Gen. Comp. Endocr.* 23: 340–347.

Dockray, G. J. 1975. Comparative studies on secretin. *Gen. Comp. Endocr.* 25: 203–210.

Dockray, G. J. 1978. Evolution of secretin-like hormones. In *Gut Hormones*, ed. S. R. Bloom, pp. 64–67. London: Churchill Livingstone.

Dockray, G. J. 1979a. Cholecystokinin-like peptides in avian brain and gut. *Experientia* 35: 628–630.

Dockray, G. J. 1979b. Comparative biochemistry and physiology of gut hormones. *Annu. Rev. Physiol.* 41: 83–95.

Dockray, G. J. 1994. Physiology of enteric neuropeptides. In *Physiology of the Gastrointestinal Tract*. 3rd ed., Vol. 2, ed. L. R. Johnson, D. H. Alpers, J. Christensen, E. D. Jacobson, and J. H. Walsh, pp. 211–242. New York: Raven Press.

Dominguez-Bello, M. G., Lovera, M., Suarez, P., and Michelangeli, F. 1993. Microbial digestive symbiont of the crop of the hoatzin (*Opisthocomus hoazin*): an avian foregut fermenter. *Physiol. Zool.* 66: 374–383.

Dominguez-Bello, M. G., Ruiz, M. C., and Michelangeli, F. 1993. Evolutionary significance of foregut fermentation in the hoatzin (*Opisthocomus hoazin*); Avies: Opithocomidae. *J. Comp. Physiol. B* 163: 594–601.

Donta, S. T., and Van Vunakis, H. 1970a. Chicken pepsinogens and pepsins. Their isolation and properties. *Biochemistry* 9: 2791–2797.

Donta, S. T., and Van Vunakis, H. 1970b. Immunochemical relationships of chicken pepsinogens and pepsins. *Biochemistry* 9: 2798–2802.

Doty, R. W. 1968. Neural organizations of deglutition. In *Handbook of Physiology*, Section 6, *Alimentary Canal*, Vol. 4, *Motility*, ed. C. F. Code and W. Heidel, pp. 1861–1902. Washington, D. C.: American Physiological Society.

Dougherty, R. W. 1968. Physiology of eructation in ruminants. In *Handbook of Physiology*, Section 6, *Alimentary Canal*, Vol. 5, *Bile; Digestion; Ruminal Physiology*, ed. C. F. Code and W. Heidel, pp. 2695–2698. Washington, D. C.: American Physiological Society.

Dougherty, R. W. 1977. Physiopathology of the ruminant digestive tract. In *Dukes' Physiology of Domestic Animals*, 9th ed., ed. M. J. Swenson, pp. 305–312. Ithaca: Cornell University Press.

Dubos, R. 1966. The micobiota of the gastrointestinal tract. *Gastroenterology* 51: 868–874.

Duellman, U. E., and Trueb, L. 1986. *Biology of Amphibians*. New York: McGraw-Hill.

Duffey, M. 1984. Intracellular pH and bicarbonate activities in rabbit colon. *Am. J. Physiol.* 246: C558–C561.

Duke, G. E. 1986. Alimentary canal: secretion and digestion, special digestive functions and absorption. In *Avian Physiology*, 4th ed., ed. P. D. Sturkie, pp. 289–302. New York: Springer-Verlag.

Duke, G. E., and Bedbury, H. P. 1985. Lack of a cephalic phase of gastric secretion in chickens. *Poultry Sci.* 64: 575–578.

Duke, G. E., Jegers, A. A., Loff, G., and Evanson, O. A. 1975. Gastric digestion in some raptors. *Comp. Biochem. Physiol.* 50A: 649–656.

Duke, G. E., Petrides, G. A., and Ringer, R. K. 1968. Chromium-51 in food metabolizability and passage rate studies with the ring-necked pheasant. *Poult. Sci.* 47: 1356–1364.

Duncan, D. L. 1953. The effects of vagotomy and splanchnicotomyon gastric motility in the sheep. *J. Physiol. (Lond.)* 119: 157–169.

Dyer, J. R., Beechey, B., Gorvel, J. P., Smith, R. T., Wootton, R., and Shirazi-Beechey, P. 1990. Glycyl-L-proline transport in rabbit enterocyte basolateral-membrane vesicles. *Biochem. J.* 269: 565–571.

Dziuk, H. E. 1971. Reverse flow of gastrointestinal contents in turkeys. *Fed. Proc.* 30: 610.

Dziuk, H. E. 1984. Digestion in the ruminant stomach. In *Dukes' Physiology of Domestic Animals*, 10th ed., ed. M. J. Swenson, pp. 320–339. Ithaca: Cornell University Press.

Eadie, J. M., and Mann, S. O. 1970. Development of the rumen microbial population: high starch diets and instability. In *Physiology of Digestion and Metabolism in the Ruminant*, ed. A. T. Phillipson, pp. 335–347. Newcastle upon Tyne, England: Oriel Press.

Eckerlin, R. H., and Stevens, C. E. 1973. Bicarbonate secretion by the glandular saccules of the llama stomach. *Cornell Vet.* 63: 436–445.

Eden, A. 1940. Coprohagy in the rabbit: origin of "night" faeces. *Nature (Lond.)* 145: 628–629.

Edmund, A. G. 1969. Dentition. In *Biology of the Reptilia*, Vol. 1, ed. C. Gans, A. d'A. Bellairs, and T. S. Parsons, pp. 117–200. New York: Academic Press.

Egan, A. R., Boda, K., and Varady, J. 1986. Regulation of nitrogen metabolism and recycling. In *Control of Digestion and Metabolism in Ruminants*, ed. L. P. Milligan, W. L. Grovum, and A. Dobson, pp. 386–402. Eaglewood Cliffs, New Jersey: Prentice-Hall.

Eggermont, E. 1969. The hydrolysis of the naturally occurring glucosides by the human intestinal mucosa. *Eur. J. Biochem.* 9: 483–487.

Ehle, F. R., Robertson, J. B., and Van Soest, P. J. 1982. Influence of dietary fibers on fermentation in the human large intestine. *J. Nutr.* 112: 158–166.

Ehrlein, H. J. 1970. Untersuchungen über die Motorik des Labmagens der Ziege unter besonderer Berücksichtigung des Pylorus. *Zentralbl. Veterinärmed. Reihe A.* 17: 481–497.

Ehrlein, H. J. 1980. Forestomach motility in ruminants. *Publ. Wiss. Film Sekt. Med. Ser.* 5, Nr. 9/C 1328, pp. 1–29.

Ehrlein, H. J., and Engelhardt, W. v. 1971. Untersuchungen überdie Magenmotorik beim Lama. *Zentralbl. Veterinärmed. Reihe A.* 18: 181–191.

Ehrlein, H. J., Reich, H., and Schwinger, M. 1983. Colonic motility and transit of digesta during hard and soft feces formation in rabbits. *J. Physiol. (Lond.)* 338: 75–86.

Einhorn, A. H. 1977. Rumination syndrome. In *Pediatrics*, 16th ed., ed. A. M. Rudolph, H. L. Barnett, and A. H. Einhorn, pp. 987–988. New York: Appleton-Century-Crofts.

Eisenberg, J. F. 1981. *The Mammalian Radiations.* Chicago: University of Chicago Press.

Ellis, R. 1981. *The Book of Whales.* New York: Alfred A. Knopf.

el-Salhy, M. 1984. Immunocytochemical investigation of the gastro-entero-pancreatic (GEP) neurohormonal peptides in the pancreas and gastrointestinal tract of the dogfish *Squalusacanthias. Histochemistry* 80: 193–205.

el-Salhy, M. E., Wilander, L., Grimelius, L., Terenius, L., Lundberg, J. M., and Tatemoto, K. 1982. The distribution of polypeptide YY (PYY)–immunoreactive and pancreatic polypeptide (PP)–immunoreactive cells in the domestic fowl. *Histochemistry* 75: 25–30.

Elsden, S. R., Hitchcock, M. W. S., Marshall, R. S., and Phillipson, A. T. 1946. Volatile acid in the digesta of ruminants and other animals. *J. Exp. Biol.* 22: 191–202.

Elliot, W. B. 1978. Chemistry and immunology of reptilian venoms. In *Biology of the Reptilia*, Vol. 8, ed. C. Gans, pp. 163–436. New York: Academic Press.

Elliott, T. R., and Barclay-Smith, E. 1904. Antiperistalsis and other muscular activities of the colon. *J. Physiol. (Lond.)* 31: 272–304.

Ellison, S. A. 1967. Proteins and glycoproteins of saliva. In *Handbook of Physiology*, Section 6, *Alimentary Canal*, Vol. 2, *Secretion*, ed. C. F. Code and W. Heidel, pp. 531–559. Washington, D. C.: American Physiological Society.

Eloff, A. K., and van Hoven, W. 1985. Volatile fatty acid production in the hindgut of *Procavia capensis. Comp. Biochem. Physiol.* 80A: 291–295.

el-Shazly, K. 1952a. Degradation of protein in the rumen of the sheep. 1. Some volatile fatty acids, including branched-chained isomers, found *in vivo. Biochem. J.* 51: 640–647.

el-Shazly, K. 1952b. Degradation of protein in the rumen of the sheep. 2. The action of rumen micro-organisms on amino acids. *Biochem. J.* 51: 647–653.

Else, P. L., and Hulbert, A. J. 1981. Comparison of the "mammal machine" and the "reptile machine": energy production. *Am. J. Physiol.* 204: R3–R9.

Elson, C. O., and Beagley, K. W. 1994. Cytokines and immune mediators. In *Physiology of the Gastrointestinal Tract.* 3rd ed., Vol. 1, ed. L. R. Johnson, D. H. Alpers, J. Christensen, E. D. Jacobson, and J. H. Walsh. pp. 243–266. New York: Raven Press.

Engelhardt, W. v. 1995. Absorption of short-chain fatty acids from the large intestine. In: *Physiological and Clinical Aspects of Short-chain Fatty Acids*, ed. J. H. Cummings, J. L. Rombeau and T. Sakata, pp. 149–169. Cambridge: Cambridge University Press.

Engelhardt, W. v., Ali, K. E., and Wipper, E. 1979. Absorption and secretion in the tubiform forestomach (compartment 3) of the llama. *J. Comp. Physiol.* 132: 337–341.

Engelhardt, W. v., Burmester, M., Hansen, K., Becker, G., and Rechkemmer, G. 1993. Effects of amiloride and ouabain on short-chain fatty acid transport in guinea pig large intestine. *J. Physiol. (Lond.)* 460: 455–466.

Engelhardt, W. v., and Hauffe, R. 1975. Role of the omasum in absorption and secretion of water and electrolytes in sheep and goats. In *Digestion and Metabolism in the Ruminant*, ed. I. W. McDonald and A. C. I. Warner, pp. 216–230. Armidale: University of New England Publishing Unit.

Engelhardt, W. v., Hinderer, S., Rechkemmer, G., and Becker, G. 1984. Urea secretion into the colon of sheep and goat. *Quart. J. Exp. Physiol.* 69: 469–475.

Engelhardt, W. v., Hinderer, S., and Wipper, E. 1978. Factors affecting the endogenous urea-N secretion and utilization in the gastrointestinal tract. In *Ruminant Digestion and Feed Evaluation*, ed. D. F. Osbourn, D. E. Beever, and D. J. Thomson, pp. 4.1–4.12. London: Agriculture Research Council.

Engelhardt, W. v., and Höller, H. 1982. Salivary and gastric physiology of camelids. *Verh. Dtsch. Zool. Ges.* 1982: 195–204.

Engelhardt, W. v., and Rechemmer, G. 1983. The physiological effects of short-chain fatty acids in the hindgut. In *Fibre in Human and Animal Nutrition*, ed. G. Wallace and L. Bell, Wellington: Bulletin 20, Roy. Soc. New Zealand, pp. 149–155.

Engelhardt, W. v., Wolter, S., Lawrenz, H., and Hemsley, J. A. 1978. Production of methane in two non-ruminant herbivores. *Comp. Biochem. Physiol.* 60A: 309–311.

Epstein, M. L., and Dahl, J. L. 1982. Development of enkephalinin in the rectum and ganglion of Remak of the chick. *Peptides* 3: 77–82.

Epstein, M. L., Lindberg, I., and Dahl, J. L. 1980. Development of enkephalin in chick brain, gut, adrenal gland and Remak's ganglion. *Soc. Neurosci. Abstr.* 6: 618.

Erdman, S., and Cundall, D. 1984. The feeding apparatus of the salamander *Amphiuma tridactylum*: morphology and behavior. *J. Morphol.* 181: 175–204.

Evans, H. E. 1986. Reptiles: introduction and anatomy. In *Zoo and Wild Animal Medicine*, ed. M. E. Fowler, pp. 107–132. Philadelphia: W. B. Saunders Co.

Faichney, G. J. 1969. Production of volatile fatty acids in the sheep caecum. *Aust. J. Agric. Res.* 20: 491–498.

Faichney, G. J. 1972. An assessment of chromic oxide as an indigestible marker for digestion studies in sheep. *J. Agric. Sci. (Camb.)* 79: 493–499.

Faichney, G. J. 1975. The use of markers to partition digestion within the gastrointestinal tract of ruminants. In *Digestion and Metabolism in the Ruminant*, ed. I. W. McDonald and A. C. I. Warner, pp. 277–291. Armidale: University of New England Publishing Unit.

Faichney, G. J., and Griffiths, D. A. 1978. Behavior of solute and particle markers in the stomach of sheep given a concentrate diet. *Br. J. Nutr.* 40: 71–82.

Faichney, G. J., and White, G. A. 1988. Rates of passage of solutes, microbes and particulate matter through the intestinal tract of ewes fed at a constant rate through digestion. *Aust. J. Agric. Res.* 39: 481–492.

Falkmer, S., Fahrenkrug, J., Alumets, J., Hakanson, R., and Sundler, F. 1980. Vasoactive intestinal polypeptide (VIP) in epithelial cells of the gut mucosa of an elasmobranchian cartilaginous fish, the ray. *Endocrinol. Japon.* 1: 31–35.

Falkmer, S., and Östberg, Y. 1977. Comparative morphology of pancreatic islets in animals. In *The Diabetic Pancreas*, ed. B. W. Volk and K. F. Wellmann, pp. 15–59. New York: Plenum.

Fänge, R., and Grove, D. J. 1979. Digestion. In *Fish Physiology*, Vol. VIII, ed. W. S. Hoar, D. J. Randall, and J. R. Bretts, pp. 161–260. New York: Academic Press.

Fänge, R., Lundblad, G., and Lind, J. 1976. Lysozyme and chitinase in blood and lymphomyeloid tissues of marine fish. *Marine Bio.* 36: 277–282.

Fänge, R., Lundblad, G., Lind, J., and Slettengren, K. 1979. Chitinolytic enzymes in the digestive system of marine fishes. *Mar. Biol.* 53: 317–321.

Fanning, J. C., Tyler, M. J., and Shearman, D. J. C. 1982. Converting a stomach to a uterus: the microscopic structure of the stomach of the gastric brooding frog *Rheobatrachussilus*. *Gastroenterology* 82: 62–70.

Farlow, J. O. 1987. Speculations about the diet and digestive physiology of herbivorous dinosaurs. *Paleobiology* 13: 60–72.

Farner, D. S. 1943. Biliary amylase in the domestic fowl. *Biol. Bull.* 84: 240–243.

Farner, D. S. 1960. Digestion and the digestive system. In *Biology and Comparative Physiology of Birds*, ed. A. J. Marshall, pp. 411–467. New York: Academic Press.

Feldman, M., and Fordtran, J. S. 1978. Rumination in adults: Rumination in infants. In *Gastrointestinal Disease*, 2nd ed., ed. M. H. Sleisenger, and J. S. Fordtran, pp. 207–208. Philadelphia: W. B. Saunders Company.

Fenna, L., and Boag, D. A. 1974. Filling and emptying of the galliform caecum. *Can. J. Zool.* 52: 537–540.

Ferraris, R. P., and Diamond, J. M. 1989. Specific regulation of intestinal nutrient transporters by their dietary substrates. *Annu. Rev. Physiol.* 51: 125–141.

Ferreira, H. G., Harrison, F. A., and Keynes, R. D. 1964. Studies with isolated rumen epithelium of the sheep. *J. Physiol. (Lond.)* 175: 28–29.

Finegold, S. M., Sutter, V. L., and Mathisen, G. E. 1983. Normal indigenous intestinal flora. In *Human Intestinal Microflora in Health and Disease*, ed. D. J. Hentges, pp. 3–32. New York: Academic Press.

Fioramonti, J., and Bueno, L. 1977. Electrical activity of the large intestine in normal and megacolon pigs. *Ann. Rech. Vétér.* 8: 275–283.

Fioramonti, J., and Ruckebusch, Y. 1976. La motricité caecalechez le lapin. III. Dualité de l'excrétion fécale. *Ann. Rech. Vétér.* 7: 281–295.

Fioramonti, J., and Ruckebusch, Y. 1979. Diet and caecal motility in sheep. *Ann. Rech. Vétér.* 10: 593–599.

Fiorenza, V., Yee, Y. S., and Zfass, A. M. 1987. Small intestinal motility: normal and abnormal function. *Am. J. Gastroenterology* 82: 1111–1115.

Fish, G. R. 1951. Digestion in *Tilapia esculenta*. *Nature (Lond.)* 167: 900–901.

Fishelson, L., Montgomery, W. L., and Myrberg, A. A. Jr. 1985. A unique symbiosis in the gut of tropical herbivorous surgeonfish (*Acanthuridae: Teleosei*) from the Red Sea. *Science* 229: 49–51.

Fitzgerald, R. J., Gustafsson, B. E., and McDaniel, E. G. 1964. Effects of coprophagy prevention on intestinal microflora in rats. *J. Nutr.* 84: 155–160.

Flannery, T. F. 1984. The kangaroos: 15 million years of bounders. In *Vertebrate Zoogeography and Evolution in Australasia*, ed. M. Archer and G. Clayton, pp. 817–836. Perth: Hesperian Press.

Fleisher, D. R. 1979. Infant rumination syndrome. *Am. J. Dis. Child.* 133: 266–269.

Flemström, G. 1977. Active alkalinization by amphibian gastric fundic mucosa in vitro. *Am. J. Physiol.* 233: E1–E12.

Flemström, G., and Frenning, B. 1968. Migration of acetic acid and sodium acetate and their effects on the gastric transmucosal ion exchange. *Acta Physiol. Scand.* 74: 521–532.

Flemström, G., and Garner, A. 1980. Stimulation of HCO_3 transport by gastric inhibitory peptide (GIP) in proximal duodenum of the bullfrog. *Acta Physiol. Scand.* 109: 231–232.

Flemström, G., Heylings, J. R., and Garner, A. 1982. Gastric and duodenal HCO_3^- transport in vitro: effects of hormones and local transmitters. *Am. J. Physiol.* 242: G100–G110.

Flower, W. H. 1872. Lectures on the comparative anatomy of the organs of digestion of the mammalia. *Medical Times and Gazette*, Feb. 24–Dec. 14.

Floyd, J. C. Jr., Fajans, S. S., Pek, S., and Chance, R. E. 1977. A newly recognized pancreatic polypeptide; plasma levels in health and disease. *Recent Prog. Horm. Res.* 33: 519–570.

Foley, W. J., Boulskila, A., Shkolnik, A., and Choshniak, I. 1992. Microbial digestion in the herbivorous lizard *Uromastyx aegyptius* (Agamidae). *J. Zool. (Lond.)* 226: 387–398.

Foley, W. J., and Hume, I. D. 1987a. Digestion and metabolism of high-tannin *Eucalyptus* foliage by the brushtail possum (*Trichosurus vulpecula*) (Marsupialia: Phalangeridae). *J. Comp. Physiol.* B157: 67–76.

Foley, W. J., and Hume, I. D. 1987b. Nitrogen requirements and urea metabolism on two arboreal marsupials, the greater glider (*Petauroides volans*) and the brushtail possium (*Trichosurus vulpecula*), fed eucalyptus foliage. *Physiol. Zool.* 60: 241–250.

Foley, W. J., and Hume, I. D. 1987c. Passage of digesta markers in two species of arboreal folivorous marsupials–the greater glider (*Petauroides volans*) and the brushtail possum (*Trichosurus vulpecula*). *Physiol. Zool.* 60: 103–113.

Foley, W. J., Hume, I. D., and Cork, S. J. 1989. Fermentation in the hindgut of the greater glider (*Petauroides volans*) and brushtail possum (*Trichosurus vulpecula*)–two arboreal folivores. *Physiol. Zool.* 62: 1126–1143.

Foltmann, B. 1981. Gastric proteinases: structure, function, evolution and mechanism of action. In *Essays in Biochemistry*, Vol. 17, ed. P. N. Campbell and R. D. Marshall, pp. 52–84. New York: Academic Press.

Foltmann, B., Jensen, A. L., Lonblad, P., Smidt, E., and Axelsen, N. H. 1981. A developmental analysis of the production of chymosin and pepsin in pigs. *Comp. Biochem. Physiol.* 68B: 9–13.

Fontaine-Perus, J., Chanconie, M., Polak, J. M., and Le Douarin, N. M. 1981. Origin and development of VIP and Substance P containing neurons in the embryonic avian gut. *Histochemistry* 71: 313–323.

Fonty, G. 1991. The rumen anaerobic fungi. In *Rumen Microbial Metabolism and Microbial Digestion*, ed. J. P. Jouany, pp. 53–70. Paris: Institut National de la Recherche Agronomique.

Fordtran, J. S. 1973. Diarrhea. In *Gastrointestinal Disease*, ed. M. H. Sleisenger and J. S. Fordtran, pp. 291–319. Philadelphia: W. B. Saunders.

Forman, G. L. 1972. Comparative morphological and histochemical studies of the stomachs of selected American bats. *Univ. of Kansas Sci. Bull.* 49: 594–729.

Forte, G. M., Limlomwongse, L., and Forte, J. G. 1969. The development of intracellular membranes concomitant with the appearance of HCl secretion in oxyntic cells of the metamorphosing bullfrog tadpole. *J. Cell Sci.* 4: 709–727.

Foster, E. S., Dudeja, P. K., and Brasitus, T. A. 1990. Contribution of Cl^-/OH^- exchange to electroneutral NaCl absorption in rat distal colon. *Am. J. Physiol.* 258: G261–267.

Fouchereau-Peron, M., Laburthe, M., Besson, J., Rosselin, G., and Le Gal, Y. 1980. Characterization of the vasoactive intestinal polypeptide (VIP) in the gut of fishes. *Comp. Biochem. Physiol.* 65A: 489–492.

Francis-Smith, K., and Wood-Gush, D. G. 1977. Coprophagia as seen in thoroughbred foals. *Equine Vet. J.* 9: 155–157.

Frankel, W. L., Zhang, W., Singh, A., Klurfeld, D. M., Don, S., Sakata, T., Modlin, I., and Rombeau, J. L. 1994. Mediation of the trophic effects of short-chain fatty acids on the rat jejunum and colon. *Gastroenterology* 106: 375–380.

Frankignoul, M., and Jeuniaux, C. 1965. Distribution des chitinases chez les mammifères rongeurs. *Ann. Soc. R. Zool. Belg.* 95: 1–8.

Frechkop, J. 1955. Sous-ordre des suiformes. In *Traite de Zoologie*, Vol. 12, ed. P. P. Grasse, pp. 507–567. Paris: Masson.

Freeland, W. J., and Winter, J. W. 1975. Evolutionary consequences of eating: *Trichosurus vulpecula* and the genus *Eucalyptus. J. Chem. Ecol.* 1: 439–455.

Freudenberger, D. O., and Hume, I. D. 1992. Ingestive and digestive responses to dietary fibre and nitrogen by two macropodid marsupials (*Macropus robustus erubescens* and *M. r. robustus*) and a ruminant (*Capra hircus*). *Aust. J. Zool.* 40: 191–194.

Freudenberger, D. O., and Hume, I. D. 1993. Effects of water restriction on digestive function in two macropodid marsupials from divergent habitats and the feral goat. *J. Comp. Physiol.* B163: 247–257.

Freudenberger, D. O., Wallis, I. R., and Hume, I. D. 1989. Digestive adaptations of kangaroos, wallabies and rat-kangaroos. In *Kangaroos, Wallabies and Rat-Kangaroos*, ed. G. Grigg, P. Jarman, and I. Hume, pp. 179–187. Sydney: Surrey Beatty.

Frizzell, R. A., Halm, D., Musch, M., Stewart, C., and Field, M. 1984. Potassium transport by flounder intestinal mucosa. *Am. J. Physiol.* 246: F946–F951.

Frizzell, R. A., Markscheid-Kaspi, L., and Schultz, S. 1974. Oxidative metabolism of rabbit ileal mucosa. *Am. J. Physiol.* 226: 1142–1148.

Frizzell, R. A., Smith, P. L., Vosburgh, E., and Field, M. 1979. Coupled sodium-chloride influx across brush border of flounder intestine. *J. Membrane Biol.* 46: 27–39.

Froment, G. E., and Bischoff, K. B. 1979. *Chemical Reactor Analysis and Design.* New York: John Wiley & Sons, Inc.

Fruton, J. S., and Simmonds, S. 1958. *General Biochemistry*, 2nd ed. New York: John Wiley & Sons, Inc.

Fujita, T., Yui, R., Swanaga, T., Nishiitsutsuji-Uwo, J., Endo, Y., and Yanaihara, N. 1981. Evolutionary aspects of "brain-gut peptides": an immunohistochemical study. *Peptides* 2, Suppl. 2: 123–131.

Gäbel, G., and Martens, H. 1991. Transport of Na^+ and Cl^- across the forestomach epithelium: mechanisms and interactions with short-chain fatty acids. In *Physiological Aspects of Digestion and Metabolism in Ruminants*, ed. T. Tsuda, Y. Sasaki, and R. Kawashima, pp. 129–151. San Diego: Academic Press.

Galef, B. G., Jr. 1979. Investigation of the functions of coprophagy in juvenile rats. *J. Comp. Physiol. Psychol.* 93: 295–305.

Gamble, J. I. 1954. *Chemical Anatomy, Physiology and Pathology of Extracellular Fluid.* 6th ed. Cambridge: Harvard University Press.

Ganapathy, V., and Leibach, F. H. 1982. Peptide transport in intestinal and renal brush border membrane vesicles. *Life Sci.* 30: 2137–2146.

Ganapathy, V., and Leibach, F. H. 1985. Is intestinal peptide transport energized by a proton gradient? *Am. J. Physiol.* 249: GI53–GI60.

Garland, T., Jr., and Huey, R. B. 1987. Testing symmorphosis: Does structure match functional requirements? *Evol.* 41: 1404–1409.

Garner, A., and Flemström, G. 1978. Gastric HCO_3^- secretion in the guinea pig. *Am. J. Physiol.* 234: E535–E541.

Garza, M. D. T., and Hernandez, R. M. 1986. Two new species of *Nyctherus* (Heterotrichidae, Protozoa) from the cecum of the iguana *Ctenosaura pectinata* from Islas Marias, Nayarit, Mexico. *Rev. Biol. Tropica* 34: 225–229.

Gas, N., and Noaillac-Depeyre, J. 1974. Renouvellement de l'épithélium intestinal de la carpe (*Cyprinus carpio* L.). Influence de la saison. *C. R. Séances Acad. Sci.* [D] (Paris) 279: 1085–1088.

Gasaway, W. C. 1976a. Seasonal variation in diet, volatile fatty acid production and size of the cecum of rock ptarmigan. *Comp. Biochem. Physiol.* 53A: 109–114.

Gasaway, W. C. 1976b. Volatile fatty acids and metabolizable energy derived from cecal fermentation in the willow ptarmigan. *Comp. Biochem. Physiol.* 53A: 115–121.

Gasaway, W. C., and Coady, J. W. 1974. Review of energy requirements and rumen fermentation in moose and other ruminants. *Naturaliste Can.* 101: 227–262.

Gasaway, W. C., Holleman, D. F., and White, R. G. 1975. Flow of digesta in the intestine and cecum of the rock ptarmigan. *Condor* 77: 467–474.

Gay, C. V., Schraer, H., and Shanabrook, V. M. 1981. Sites of carbonic anhydrase in avian gastric mucosa identified by electron microscope autoradiography. *Am. J. Physiol.* 241: G382–G388.

Geis, A. D. 1957. Coprophagy in the cottontail rabbit. *J. Mammal.* 38: 136.

Geisterdoerfer, P. 1973. Sur quelques particularities histologiques de l'intestin de *Chalinura mediterranea* (Macourideae Gadiformes). *C. R. Séances Acad. Sci.* [D] 276: 331–333.

Gerencser, G. A. 1988. Sodium and chloride transport across the molluskan gut. *Comp. Biochem. Physiol.* 90A: 621–626.

Gershon, M. D., and Erde, S. M. 1980. The nervous system of the gut. *Gastroenterology* 80: 1571–1594.

Gershon, M. D., Kirchgessner, A. L., and Wade, P. R. 1994. In *Physiology of the Gastrointestinal Tract.* 3rd. ed., Vol. 2, ed. L. R. Johnson, D. H. Alpers, J. Christensen, E. D. Jacobson, and J. H. Walsh, pp. 381–422. New York: Raven Press.

Giannella, R. A., Broitman, S. A., and Zamcheck, N. 1971. Vitamin B uptake by intestinal microorganisms: mechanism and relevance to syndromes of intestinal bacterial overgrowth. *J. Clin. Invest.* 50: 1100–1107.

Gibson, R., and Barker, P. L. 1979. The decapod hepatopancreas. *Oceanogr. Mar. Biol. Annu. Rev.* 17: 285–346.

Gingerich, P. D., Smith, B. H., and Simons, E. L. 1990. Hind limbs of eocene *Basilsaurus*: evidence of feet in whales. *Science* 249: 154–157.

Glinsky, M. J., Smith, R. M., Spires, H. R., and Davis, C. L. 1976. Measurement of volatile fatty acid production rates in the cecum of the pony. *J. Anim. Sci.* 42: 1469–1470.

Goda, T., and Koldovsky, O. 1988. Dietary regulation of small intestinal disaccharidases. In *World Review of Nutrition and Dietetics*, ed. G. H. Bourne, Vol 57, pp. 275–339, Basel: Karger Medical Pub, Co.

Goodall, E. D., and Kay, R. N. B. 1965. Digestion and absorption in the large intestine of the sheep. *J. Physiol. (Lond.)* 176: 12–23.

Goodrich, T. D., and Morita, R. Y. 1977. Bacterial chitinase in the stomach of marine fishes from Yaquina Bay, Oregon, U.S.A. *Mar. Biol.* 41: 355–360.

Gordon, J. G. 1968. Rumination and its significance. *Wld. Rev. Nutr. Diet.* 9: 251–273.

Gordon, H. A., and Bruckner, G. 1984. Anomalous lower bowel function and related phenomena in germ-free animals. In *The Germ-Free Animal in Biomedical Research*, ed. M. E. Coates and B. E. Gustafsson, pp. 193–213. London: Laboratory Animals Ltd.

Gordon, I. J., and Illius, A. W. 1994. The functional significance of the browser-grazer dichotomy in African ruminants. *Oecologia* 98: 167–175.

Gossling, J., Loesche, W. J., and Nace, G. W. 1982. Large intestine bacterial flora of nonhibernating and hibernating leopard frogs (*Rana pipiens*). *Appl. Environ. Microbiol.* 44: 59–66.

Goyal, R. K., and Paterson, W. G. 1989. Esophageal motility. In *Handbook of Physiology*, Section 6: *Alimentary Canal*, Vol. 1: *Motility and Circulation*, ed. S. G. Schultz, J. D. Wood, and B. B. Rauner, pp. 939–973. Washington, D. C.: American Physiological Society.

Graham, D. Y. 1989. *Campylobacter pylori* and peptic ulcer disease. *Gastroenterology* 96: 615–625.

Grajal, A., and Parra, O. 1995. Passage rates of digesta markers in the gut of the hoatzin *Opisthocamus hoazin*, a folivorous bird with foregut fermentation. *Condor* (In Press).

Grajal, A., Strahl, S. D., Parra, R., Dominguez, M. G., and Neher, A. 1989. Foregut fermentation in the hoatzin, a neotropical bird. *Science* 245: 1236–1238.

Grassé, P.-P. 1955. *Traité de Zoologie*, Vol. 17. Paris: Massonet Cie.

Graur, D., and Higgins, D. G. 1994. Molecular evidence for the inclusion of cetaceans within the order Artiodactyla. *Mol. Biol. and Evol.* 11: 357–364.

Gray, F. V. 1947. The digestion of cellulose by sheep at successive levels of alimentary tract. *J. Exp. Biol.* 24: 15–19.

Gregory, P. C. 1982. Forestomach motility in the chronically vagotomized sheep. *J. Physiol. (Lond.)* 328: 431–447.

Griffiths, M. 1978. *The Biology of the Monotremes*. New York: Academic Press.

Griffiths, M., and Davies, D. 1963. The role of the soft pellets in the production of lactic acid in the rabbit stomach. *J. Nutr.* 80: 171–180.

Griffiths, R. W., Pang, P. K., Srivastvaa, A. K., and Pickford, G. E. 1973. Serum composition of freshwater stingrays (*Potamotrygonidae*) adapted to fresh and dilute sea water. *Biol. Bull.* 144: 304–320.

Gross, J. E., Wang, Z., and Wunder, B. A. 1985. Effects of food quality and energy needs: changes in gut morphology and capacity of *Microtus ochrogaster*. *J. Mamm.* 66: 661–667.

Grovum, W. L., and Chapman, H. W. 1982. Pentagastrin in the circulation acts directly on the brain to depress motility of the stomach in sheep. *Regul. Pept.* 5: 35–42.

Grovum, W. L., and Williams, V. J. 1973. Rate of passage of digesta in sheep. 3. Differential rates of passage of water and dry matter from the reticulo-rumen, abomasum and caecum and proximal colon. *Br. J. Nutr.* 30: 231–240.

Guard, C. L. 1980. The reptilian digestive system: general characteristics. In *Comparative Physiology: Primitive Mammals*, ed. K. Schmidt-Nielsen, L. Bolis, and C. R. Taylor, pp. 43–51. New York: Cambridge University Press.

Gustafsson, B. 1948. Germ-free rearing of rats: general technique. *Acta Pathol. Microbiol. Scand.* Suppl. 73: 1–130.

Guppy, M. 1986. The hibernating bear: Why is it so hot, and why does it cycle urea through the gut. *Trends Biochem. Sci.* 11: 274–276.

Hackstein, J. H. P., and Stumm, C. K. 1994. Methane production in terrestrial arthropods. *Proc. Natl. Acad. Sci. USA.* 191: 5441–5445.

Haenlein, E. F. W., Holdren, R. D., and Yoon, Y. M. 1966. Comparative responses of horses and sheep to different physical forms of alfalfa hay. *J. Anim. Sci.* 25: 740–743.

Haga, R. 1960. Observations on the ecology of the Japanese pika. *J. Mammal.* 41: 200–212.

Hagen, P., and Robinson, K. W. 1953. The production and absorption of volatile fatty acids in the intestine of the guinea-pig. *Aust. J. Exp. Biol. Med Sci.* 31: 99–104.

Hagey, L. R. 1992. *Bile Biodiversity in Vertebrates: Chemistry and Evolutionary Implications*. Ph.D. Thesis, University of California, San Diego, California.

Hagemeister, H., and Dirksen, G. 1980. Adaptations to changes in dietary composition, level and frequency of feeding. In *Digestive Physiology and Metabolism in Ruminants*, ed. Y. Ruckebusch and P. Thivend, p. 590. Lancaster: MTP Press Limited.

Hajra, A., Tripathi, S. D., Nath, D., Chaterjee, J. G., and Karmaker, H. C. 1987. Comparative digestibility of dietary plant fiber in grass carp *Ctenophayngodon idella* (Val), *Proc. Nat. Acad. Sci. India*, 57(B): 231–236.

Haltenorth, T. 1963. Klassifikation der Säugetiere: Artiodactyla. *Handb. d. Zoologie* 8: 1–167.

Hamid, A., Sakata, T., and Kakimoto, D. 1976. Microflora in the alimentary tract of the gray mullet. I. Isolation and identification of bacteria. *Mem. Fac. Fish. Kagoshima Univ.*, Vol. 25, pp. 59–65.

Hamilton, J., and Coe, M. 1982. Feeding, digestion and assimilation of a population of giant tortoises (*Geochelone gigantea*, Schweigger) on Aldabra Atoll. *J. Arid Environ.* 5: 127–144.

Hamilton, W. J. 1955. Coprophagy in the swamp rabbit. *J. Mammal.* 36: 303–304.

Hammond, K., and Diamond, J. 1994. Limits to dietary nutrient uptake and intestinal nutrient uptake in lactating mice. *Physiol. Zool.* 67: 282–303.

Hammond, K. A., and Wunder, B. A. 1991. The role of diet quality and energy need in the nutritional ecology of a small herbivore, *Microtus ochrogaster. Physiol. Zool.* 64: 541–567.

Hammond, P. B., Dziuk, H. E., Usenik, E. A., and Stevens, C. E. 1964. Experimental intestinal obstruction in calves. *J. Comp. Pathol. Ther.* 74: 210–222.

Hamosh, M. 1979. A review. Fat digestion in the newborn: Role of lingual lipase and preduodenal digestion. *Pediat. Res.* 13: 615–622.

Harder, W. 1950. Zur Morphologie and Physiologie des Blinddarmsder Nagetiere. *Verh. Dtsch. Zool. Ges.* 2: 95–109.

Harder, W. 1975a. *Anatomy of Fishes*, Part I, Stuttgart: E. Schweizerbart'sche Verlagsbuchhandlung.

Harder, W. 1975b. *Anatomy of Fishes*, Part II, Stuttgart: E. Schweizerbart'sche Verlagsbuchhandlung.

Harding, R. S. O., and Strum, S. C. 1976. The predatory baboons of Kekopey. *Nat. Hist.* 85: 45–53.

Harig, J. M., Knaup, S. M., Shoshara, J., Dudeja, P. K. Ramaswamy, K., and Brasitus, T. A. 1990. Transport of N-butyrate into human colonic luminal membrane vesicles. *Gastroenterology* 98: A543.

Harris, D. R. 1980. Tropical savanna environments: definition, distribution, diversity and development. In *Human Ecology in Savanna Environments*, ed. D. R. Harris, pp. 3–27. London: Academic Press.

Harrison, F. A. 1962. Bile secretion in sheep. *J. Physiol. (Lond.)* 162: 212–224.

Harrison, F. A., Keynes, F. D., and Zurich, L. 1968. The active transport of chloride across the rumen epithelium of the sheep. *J. Physiol. (Lond.)* 194: 48–49.

Harrop, C. J. F., and Hume, I. D. 1980. Digestive tract and digestive function in monotremes and nonmacropod marsupials. In *Comparative Physiology: Primitive Mammals*, ed. K. Schmidt-Nielsen, L. Bolis, and C. R. Taylor, pp. 63–77. New York: Cambridge University Press.

Hartley, B. S., and Shotton, D. M. 1971. Pancreatic elastase. In *The Enzymes*, 3rd ed, Vol. 3, ed. P. Boyer, pp. 323–373. New York: Academic Press.

Haslewood, G. A. D. 1967. Bile salt evolution. *J. Lipid Res.* 8: 535–550.

Haslewood, G. A. D. 1978. *The Biological Utility of Bile Salts*. Amsterdam: North Holland Publishing Co.

Hatch, M. 1987. Short-chain fatty acid transport and its effect on ion transport by rabbit cecum. *Am. J. Physiol.* 253: G171–G178.

Hawker, P. C., Mashiter, K. E., and Turnberg, L. A. 1978. Mechanisms of transport of Na, Cl, and K in the human colon. *Gastroenterology* 74: 1241–1247.

Haysson, V., and Lacey, R. C. 1985. Basal metabolic rates in mammals: taxonomic differences in the allometry of BMR and body mass. *Comp. Biochem. Physiol.* 81A: 741–754.

Heard and Annison. 1986. Gastrointestinal absorption of vitamin B-6 in the chicken *(Gallus domesticus). J. Nutr.* 116: 107–120.

Hediger, M. A., and Rhoads, D. B. 1994. Molecular physiology of sodium-glucose cotransporters. *Physiol. Rev.* 74: 993–1026.

Heisinger, J. F. 1965. Analysis of the reingestion rhythm of confined cottontails. *Ecology* 46: 197–201.

Heller, R., Cerceasov, V., and Engelhardt, W. v. 1986. Retention of fluid and particles in the digestive tract of the llama *(Lama guanacoe* f. *glama). Comp. Biochem. Physiol.* 83A: 687–691.

Heller, R., Gregory, P. C., and Engelhardt, W. v. 1984. Pattern of motility and flow of digesta in the forestomach of the Llama *(Lama guanacoe* f. *glama). J. Comp. Physiol. B. Metab. Transp. Funct.* 154: 529–533.

Hemmingsen, A. M. 1960. Energy metabolism as related to body size and respiratory surfaces, and its evolution. *Rep. Steno Mem. Hosp.* 9: 1–110.

Hendrix, T. R. 1987. Alimentary tract motility: stomach, small intestine, colon and biliary tract. In *Undergraduate Teaching Program*, Unit 10 B.: American Gastroenterology Association.

Henning, S. J. 1985. Ontogeny of enzymes in the small intestine. *Annu. Rev. Physiol.* 47: 231–245.

Henning, S. J. 1987. Functional development of the gastrointestinal tract. In *Physiology of the Gastrointestinal Tract*, 2nd ed., ed. L. R. Johnson, pp. 285–300. New York: Raven Press.

Henning, S. J., and Hird, F. J. R. 1972a. Diurnal variations in the concentrations of volatile fatty acids in the alimentary tracts of wild rabbits. *Br. J. Nutr.* 27: 57–64.

Henning, S. J., and Hird, F. J. R. 1972b. Ketogenesis from butyrate and acetate by the caecum and colon of rabbits. *Biochem. J.* 130: 785–790.

Henning, S. J., Rubin, D. C., and Shulman, R. J. 1994. Ontogeny of the intestinal mucosa. In *Physiology of the Gastrointestinal Tract*, 3rd ed., ed. L. R. Johnson, pp. 571–610. New York: Raven Press.

Herd, R. M., and Dawson, T. J. 1984. Fiber digestion in the emu, *Dromaius novaehollandiae*, a large bird with a simple gut and high rates of digesta passage. *Physiol. Zool.*, 57: 70–84.

Herdt, T. 1992. Digestion: the fermentation processes. In *Textbook of Veterinary Physiology*, ed. J. Cunningham, pp. 316–344. Philadelphia: W. B. Saunders.

Hernandez, A., and Martinez del Rio, C. 1992. Intestinal disaccharidases in five species of phylostomoid bats. *Comp. Biochem. Physiol.* 103B: 105–111.

Herpol, C. 1964. Activité protéolytique de l'appareil gastriqued'oiseaux granivores et carnivores. *Ann. Biol. Anim. Biochim. Biophys.* 4: 239–244.

Herrara, E. A. 1985. Coprophagy in the capybara, *Hydrochoerus hydrochaeris. J. Zool. Lond.* 207A: 616–619.

Herschel, D. A., Argenzio, R. A., Southworth, M., and Stevens, C. E. 1981. Absorption of volatile fatty acid, Na, and H_2O by the colon of the dog. *Am. J. Vet. Res.* 42: 1118–1124.

Herwig, R. P., Staley, J. T., Nerini, M. K., and Braham, H. W. 1984. Baleen whales: preliminary evidence for forestomach microbial fermentation. *Appl. Environ. Microbiol.* 47: 421–423.

Hespell, R. B. 1981. Ruminal microorganisms: their significance and nutritional value. *Dev. Industr. Microbiol.* 22: 266.

Hewitt, J. E. A., and Schelkopf, R. L. 1955. pH values and enzymatic activity of the digestive tract of the chicken. *Am. J. Vet. Res.* 16: 576–579.

Heywood, L. H., and Wood, A. K. W. 1985. Thoracic oesophageal motor activity during eructation in sheep. *Quart. J. Exp. Physiol.* 70: 603–613.

Heywood, L. H., and Wood, A. K. W. 1988. Retrograde oesophageal contractions in the dog. *Quart. J. Exp. Physiol.* 73: 87–94.

Hickling, C. F. 1966. On the feeding process in the white amur, *Ctenopharyngodon idella. J. Zool.* 148: 408–419.

Hill, K. J. 1965. Abomasal secretory function in the sheep. In *Physiology of Digestion in the Ruminant*, ed. R. W. Dougherty, pp. 221–230. Washington: Butterworths.

Hill, M. J. 1983. Lipids, intestinal flora, and large bowel cancer. In *Dietary Fats and Health.* ed., E. G. Perkins and W. J. Visek, pp. 868–880. Champaign IL: American Oil Chemists Society.

Hill, W. C. O. 1952. The external and visceral anatomy of the olive colobus monkey *Procolobus verus. Proc. Zool. Soc. Lond.* 122: 127–186.

Hill, W. C. O. 1966a. *Primates: Comparative Anatomy and Taxonomy*, vol. 6, Cercopithecoidea. New York: John Wiley & Sons.

Hill, W. C. O. 1966b. *Primates: Comparative Anatomy and Taxonomy*, Vol. 7, Cynopithecinae. New York: John Wiley & Sons.

Hill, W. C. O., Porter, A., Bloom, R. T., Seago, J., and Southwide, M. D. 1957. Field and laboratory studies on the naked mole rat, *Heterocephalus glaber. Proc. Zool. Soc. Lond.* 128: 455–514.

Hill, W. C. O., and Rewell, R. E. 1948. The cecum of primates. Its appendages, mesenteries and blood supply. *Trans Zool. Soc. Lond.* 26: 199–257.

Hinderer, S. 1978. Kinetik des Harnstoff-Stoffwechsels beim Llama bei proteinarmen Diäten. Ph.D. thesis, Universität Hohenheim, Stuttgart.

Hintz, H. F., Hogue, D. E., Walker, E. F., Lowe, J. E., and Schryver, H. F. 1971. Apparent digestion in various segments of the digestive tract of ponies fed diets with varying roughage-grain ratios. *J. Anim. Sci.* 32: 245–248.

Hirano, T., and Mayer-Gostan, N. 1976. Eel esophagus as an osmoregulatory organ. *Proc. Natl. Acad. Sci. USA* 73: 1348–1350.

Hirji, K. N. 1982. Fine structure of the oesophageal and gastric glands of the red-legged pan frog *Kassina maculata* Dumeril. *S. Afr. J. Zool.* 17: 28–31.

Hladik, C. M., Charles-Dominique, P., Valdebouze, P., Delort-Laval, J., and Flanzy, J. 1971. La caecotrophie chezun Primate phyllophage du genre *Lepilemur* et les correlationsavec les particularités de son appareil digestif. *C. R. Séances Acad. Sci.* [D] (Paris), 272: 3191–3194.

Hoar, W. S. 1983. Nutrition and digestion. In *General Comparative Physiology*, 3rd ed., pp. 407–453. Englewood Cliffs, New Jersey: Prentice-Hall.

Hobson, P. N., and Wallace, R. J. 1982. Microbial ecology and activities in the rumen. *CRC Critical Rev. Microbiol.* 9: 165–225.

Hodgkiss, J. P. 1984. Evidence that enteric cholinergic neurones project orally in the intestinal nerve of the chicken. *Quart. J. Exp. Physiol.* 69: 797–807.

Hofer, R., and Schiemer, F. 1981. Proteolytic activity in the digestive tract of several species of fish with different feeding habits. *Oecologia* 48: 342–345.

Hofmann, A. F. 1994. Intestinal absorption of bile acids and biliary constituents: the intestinal component of the enterohepatic circulation and integrated system. In *Physiology of the Gastrointestinal Tract.* 3rd ed., Vol. 2, ed. L. R. Johnson, D. H. Alpers, J. Christensen, E. D. Jacobson, and J. H. Walsh. pp. 1845–1866. New York: Raven Press.

Hofmann, A. F., Palmer, K. R., Yoon, Y. B., Hagey, L. R., Gurantz, D., Huijghebaert, S. Converse, J. L., Cecchette, S., and Michelotti, E. 1985. The biological utility of bile acid conjugation with glycine and taurine. In *Advances in Glucuronide Conjugation with Glycine*, ed. S. Matern, K. W. Bock, W. Gerok, pp. 245–264. Lancaster U.K.: MTP Press.

Hofmann, A. F., and Mysels, K. J. 1988. Bile salts as biological surfactants. *Colloids and Surfaces* 30: 145–173.

Hofmann, R. R. 1968. Comparisons of the rumen and omasum structure in East African game ruminants in relation to their feeding habits. *Symp. Zool. Soc. Lond.* 21: 179–194.

Hofmann, R. R. 1969. Zur Topographie und Morphologie des Wiederkäuermagens im hinblick auf seine funktion. *Zentralbl. f. Veterinärmed.*, Beiheft 10.

Hofmann, R. R. 1973. *The Ruminant Stomach: Stomach Structure and Feeding Habits of East African Game Ruminants*. Nairobi, Kenya: East African Literature Bureau.

Hofmann, R. R. 1983. Adaptive changes of gastric and intestinal morphology in response to different fibre content in ruminent diets. *Roy. Soc. N. Z. Bull.* 20: 51–58.

Hofmann, R. R. 1991. The comparative morphology and functional adaptive differentiation of the large intestine of domestic mammals. In *Digestive Physiology of the Hindgut*, ed. K. D. Gunther and M. Kirchgessner, pp. 7–17. Hamburg: Verlag Paul Parey.

Hogan, J. P., and Phillipson, A. T. 1960. The rate of flow of digesta and their removal along the digestive tract of the sheep. *Br. J. Nutr.* 14: 147–155.

Hogben, C. A. M., Kent, T. H., Woodward, P. A., and Sill, A. J. 1974. Quantitative histology of the gastric mucosa: man, dog, cat, guinea pig, and frog. *Gastroenterology* 67: 1143–1154.

Hogben, C. A. M., Tocco, D. J., Brodie, B. B., and Shanker, L. S. 1959. On the mechanism of intestinal absorption of drugs. *J. Exp. Pharmacol. Therap.* 125: 275–282.

Höller, H. 1970a. Untersuchungen über Sekret und Sekretion der Cardiadrüsenzone im Magen des Schweines. I. Sekretions volumina und rhythmik, Eigenschaften der Sekrete. *Zentralbl. Veterinärmed. Reihe A* 17: 685–711.

Höller, H. 1970b. Untersuchungen über Sekret und Sekretion der Cardiadrüsenzone im Magen des Schweines. II. Versuche zur Beeinflussung der Spontan sekretion der isolierten Cardiadrüsenzone, Flüssigkeits- und Elektrolyt sekretion in denmit Vërschiedenen Flüssigkeiten gefüllten isolierten kleinenMagen. *Zentralbl. Veterinärmed. Reihe A* 17: 857–873.

Höller, H., Breve, G., Lechner-Doll, M., and Schultze, E. 1989. Concentrations of volatile fatty acids and acetate production in the forestomach of grazing camels. *Comp. Biochem. Physiol.* 93B: 413–416.

Holmes, J. H. G., Bayley, H. S., Leadbeater, R. A., and Horney, F. D. 1974. Digestion of protein in small and large intestine of the pig. *Br. J. Nutr.* 32: 479–489.

Holmgren, S., and Jonsson, A. C. 1988. Occurrence and effects on motility of bombesin related peptides in the gastrointestinal tract of the Atlantic cod *Gadus morhua*. *Comp. Biochem Physiol.* 89C: 249–256.

Holmgren, S., and Nilsson, S. 1982. Neuropharmacology of adrenergic neurons in teleost fish. *Comp. Biochem. Physiol.* 72C: 289–302.

Holmgren, S., and Nilsson, S. 1983. Bombesin-, gastrin/CCK-, 5-hydroxytryptamine-, neurotensin-, somatostatin-, and VIP-like immunoreactivity and catecholamine fluorescence in the gut of the elasmobranch, *Squalus acanthias*. *Cell Tissue Res.* 234: 595–618.

Holmgren, S., Vaillant, C., and Dimaline, R. 1982. VIP-, substance P-, gastrin/CCK-, bombesin-, somatostatin- and glucagon-like immunoreactivities in the gut of the rainbow trout, *Salmo gairdneri*. *Cell Tissue Res.* 223: 141–153.

Holmquist, A. L., Dockray, C. J., Rosenquist, G. L., and Walsh, J. H. 1979. Immunochemical characterization of cholecystokinin-like peptides in lamprey gut and brain. *Gen. Comp. Endocr.* 37: 474–481.

Holstein, B., and Humphrey, C. S. 1980. Stimulation of gastric acid secretion and suppression of VIP-like immunoreactivity by bombesin in the Atlantic codfish, *Gadus morhua*. *Acta Physiol. Scand.* 109: 217–223.

Holtenius, K., and Björnhag, G. 1985. The colonic separation mechanism in the guinea-pig (*Cavia porcellus*) and the chinchilla (*Chinchilla laniger*). *Comp. Biochem. Physiol.* 82A: 537–542.

Honde, C., and Bueno, L. 1984. Evidence for central neuro-peptidergic control of rumination in sheep. *Peptides* 5: 81–83.

Honigmann, H. 1936. Studies on nutrition in mammals, Part I. *Proc. Zool. Soc. Lond.,* 106: 517–530.

Hoogkamp-Korstanje, J. A. A., Lindner, J. G. E. M., Marcelis, J. H., den Daas-Slagt, H., and deVos, N. M. 1979. Composition and ecology of the human intestinal flora. *Antonie Van Leeuwenhoek J. Microbiol. Serol.* 45: 33–40.

Hoover, W. H., and Clarke, S. D. 1972. Fiber digestion in the beaver. *J. Nutr.* 102: 9–16.

Hoover, W. H., and Heitmann, R. N. 1972. Effects of dietary fiber levels on weight gain, cecal volume and volatile fatty acid production in rabbits. *J. Nutr.* 102: 375–379.

Hoover, W. H., and Heitmann, R. N. 1975. Cecal nitrogen metabolism and amino acid absorption in the rabbit. *J. Nutr.* 105: 245–252.

Hopfer, U. 1987. Membrane transport mechanisms for hexoses and amino acids in the small intestine. In *Physiology of the Gastrointestinal Tract*, ed. L. R. Johnson, J. Christensen, M. J. Jackson, E. D. Jacobson, and J. H. Walsh, pp. 1499–1526. New York: Raven Press.

Hoppe, P. P. Rumen fermentation and body weight in African ruminants. In *Proc. 13th Int. Congr. Game Biologists,* ed. T. J. Peterle, pp. 141–150. Washington, D.C.: Wildlife Society.

Horn, M. H. 1989. Biology of marine herbivorous fishes. *Oceanogr. Mar. Biol. Annu. Rev.* 27: 167–272.

Horn, M. H. 1992. Herbivorous fishes: feeding and digestive mechanisms. In *Plant-Animal Interactions in Marine Benthos*, ed. D. M. John, S. J. Hawkins, and J. H. Price, Vol. 46, pp. 339–362. Oxford: Clarendon Press.

Horn, M. H., and Messer, K. S. 1992. Fish guts as chemical reactors: a model of the alimentary canals of marine herbivorous fishes. *Mar. Biol.* 113: 527–535.

Hörnicke, H. 1981. Utilization of caecal digesta by caecotrophy (soft faeces ingestion) in the rabbit. *Livest. Prod. Sci.* 8: 361–366.

Hörnicke, H., and Björnhag, G. 1980. Coprophagy and related strategies for digesta utilization. In *Digestive Physiology and Metabolism in Ruminants*, ed. Y. Ruckebusch and P. Thivend, pp. 707–730. Westport, Connecticut: AVI Publishing Co.

Horsley, R. W. 1977. A review of the bacterial flora of teleosts and elasmobranches, including methods for its analysis. *J. Fish Biol.* 10: 529–553.

Hotton, N. 1955. A survey of adaptive relationships of dentition to diet in the North American Iguanidae. *Am. Midl. Nat.* 53: 88–114.

Houpt, T. R., and Houpt, K. A. 1968. Transfer of urea nitrogen across the rumen wall. *Am. J. Physiol.* 214: 1296–1303.

Hourdry, J., Chabot, J.-G., Menard, D., and Hugon, J. S. 1979. Intestinal brush border enzyme activities in developing amphibian *Rana catesbeiana. Comp. Biochem. Physiol.* 63A: 121–125.

Hsi-Chiang, L., and Visek, W. J. 1991a. Large intestinal pH and ammonia in rats: dietary fat and protein interactions. *J. Nutr.* 121: 832–843.

Hsi-Chiang, L., and Visek, W. J. 1991b. Colon mucosal cell damage by ammonia in rats. *J. Nutr.* 121: 887–893.

Hukuhara, T., Naitoh, T., and Kameyama, H. 1975. Observations on the gastrointestinal movements of the tortoise (*Geoclemys reevsii*) by means of the abdominal window technique. *Jpn. J. Smooth Musc. Res.* 11: 39–46.

Hukuhara, T., and Neya, T. 1968. The movements of the colon of rats and guinea pigs. *Jpn. J. Physiol.* 18: 551–562.

Hulbert, A. J., and Dawson, T. J. 1974. Water metabolism in perameloid marsupials from different environments. *Comp. Biochem. Physiol.* 47A: 617–633.

Hulbert, A. J., and Else, P. L. 1981. Comparison of the "mammal machine" and the "reptile machine": energy use and thyroid activity. *Am. J. Physiol.* 241: R350–R356.

Humbert, W., Kirsch, R., and Meister, M. F. 1984. Scanning electron microscopic study of the oesophageal mucous layer in the eel, *Anguilla anguilla* L. *J. Fish Biol.* 25: 117–122.

Hume, I. D. 1974. Nitrogen and sulphur retention and fibre digestion in euros, red kangaroos and sheep. *Aust. J. Zool.* 22: 13–23.

Hume, I. D. 1977. Production of volatile fatty acids in two species of wallaby and in sheep. *Comp. Biochem. Physiol.* 56A: 299–304.

Hume, I. D. 1978. Evolution of the Macropodidae digestive system. *Aust. Mammalogy* 2: 37–42.

Hume, I. D. 1982. *Digestive Physiology and Nutrition of Marsupials.* New York: Cambridge University Press.

Hume, I. D. 1989. Optimal digestive strategies in mammalian herbivores. *Physiol. Zool.* 62: 1145–1163.

Hume, I. D., Carlisle, C. H., Reynolds, K., and Pass, M. A. 1988. Effects of fasting and sedation on gastrointestinal function in two potoroine marsupials. *Aust. J. Zool.* 36: 411–420.

Hume, I. D., Karasov, W. H., and Darken, B. W. 1993. Acetate, butyrate, and proline uptake in the caecum and colon of prairie voles (*Microtus ochrogaster*). *J. Exp. Biol.* 176: 285–297.

Hume, I. D., Morgan, K. R., and Kenagy, G. J. 1993. Digesta retention and digestive performance in sciurud and microtine rodents: effects of hindgut morphology and body size. *Physiol. Zool.* 66: 396–411.

Hume, I. D., Rübsamen, K., and Engelhardt, W. v. 1980. Nitrogen metabolism and urea kinetics in the rock hyrax (*Procavia habessinica*). *J. Comp. Physiol. B* 138: 307–314.

Hume, I. D., and Sakaguchi, E. 1991. Patterns of digesta flow and digestion in foregut and hindgut fermenters. In *Physiological Aspects of Digestion and Metabolism in Ruminants*, ed. T. Tsuda, Y. Sasaki, and R. Kawashima, pp. 427–451. San Diego: Academic Press.

Hume, I. D., and Warner, A. C. I. 1980. Evolution of microbial digestion in mammals. In *Digestive Physiology and Metabolism in Ruminants*, ed. Y. Ruckebusch and P. Thivend, pp. 684–685. Westport, Connecticut: AVI Publishing Co.

Hungate, R. E. 1966. *The Rumen and Its Microbes.* New York: Academic Press.

Hungate, R. E. 1968. Ruminal fermentation. In *Handbook of Physiology*, Section 6, *Alimentary Canal*, Vol. 5, *Bile; Digestion; Ruminal Physiology*, Section 6, ed. C. F.

Code and W. Heidel, pp. 2725–2745. Washington, D. C.: American Physiological Society.

Hungate, R. E., Phillips, G. D., McGregor, A., Hungate, D. P., and Buechner, H. K. 1959. Microbial fermentation in certain mammals. *Science* 130: 1192–1194.

Husar, S. L. 1975. *A Review of the Literature of the Dugong (Dugong Dugong)*. Wildlife Research Report No. 4., Washington D. C.: Fish and Wildlife Service.

Hydén, S. 1961. The use of reference substances and the measurement of flow in the alimentary tract. In *Digestive Physiology and Nutrition of Ruminants*, ed. D. Lewis, pp. 35–47. London: Butterworths.

Hyodo-Taguchi, Y. 1970. Effect of X-irradiation on DNA synthesis and cell proliferation in the intestinal epithelial cells of goldfish at different temperatures with special reference to recovery process. *Radiation Res.* 41: 568–578.

Illius, A. W., and Gordon, I. J. 1991. Prediction of intake and digestion in ruminants by a model of rumen kinetics integrating animal size and plant characteristics. *J. Agric. Sci. (Camb.)* 116: 145–157.

Illius, A. W., and Gordon, I. J. 1992. Modelling the nutritional ecology of ungulate herbivores: evolution of body size and competitive interactions. *Oecologia* 89: 428–434.

Imon, M., and White, J. 1984. Association between HCO_3^- absorption and K^+ uptake by *Amphiuma* jejunum: relations among HCO_3^- absorption, luminal K^+, and intracellular K^+ activity. *Am. J. Physiol.* 246: G732–G744.

Imoto, S., and Namioka, S. 1978. VFA metabolism in the pig. *J. Anim. Sci.* 47: 479–487.

Ingles, L. G. 1961. Reingestion in the mountain beaver. *J. Mammal.* 42: 411–412.

Ito, S., and Winchester, R. J. 1960. Electron microscopic observations of the bat gastric mucosa. *Anat. Rec.* 136: 338–339.

Ito, S., and Winchester, R. J. 1963. The fine structure of the gastric mucosa in the bat. *J. Cell Biol.* 16: 541–577.

Iverson, J. B. 1980. Colic modifications in iguanine lizards. *J. Morph.* 163: 79–93.

Izraely, H., Choshniak, I., Stevens, C. E., Demment, M. W., and Shkolnik, A. 1989. Factors determining the digestive efficiency of the domesticated donkey (*Equus asinus asinus*). *Quart. J. Exp. Physiol.* 74: 1–6.

Jackson, D. C. 1971. Mechanical basis for lung volume variability in the turtle *Pseudemys scripta elegans. Am. J. Physiol.* 220: 754–758.

Jackson, M. J., Shiau, Y. F., Bane, S., and Fox, M. 1974. Intestinal transport of weak electrolytes: evidence in favor of a three-compartment system. *J. Gen. Physiol.* 63: 187–213.

Jackson, S. 1992. Do seabird gut sizes and mean retention times reflect adaptations to diet and foraging method? *Physiol. Zool.* 65: 674–697.

Jackson, S., and Place, A. R. 1990. Gastrointestinal transit and lipid assimilation efficiencies in three species of sub-Antarctic seabird. *J. Exp. Zool.* 255: 141–154.

Jacobs, L. R. 1988. Fiber and colon cancer. *Gastroenterol. Clin. North Am.* 17: 747–760.

Jacobs, L. R., and Lupton, J. R. 1986. Relationship between colonic lumen pH, cell proliferation, and colon carcinogenesis. *Cancer Res.* 46: 1727–1734.

Jacobshagen, E. 1937. Mittel- und Endarm (Rumpfdarm). In *Handbuch der vergleichenden Anatomie der Wirbeltiere*, Vol. 3, ed. L. Bolk, E. Göppert, E. Kallius, and W. Lubosch, pp. 563–724. Berlin: Schwarzenburg.

Jain, D. K. 1976. Pancreatic and intestinal amylase level in frugivorous, carnivorous, and omnivorous birds. 1976. *Acta. Biol. Acad. Sci. Hung.* 27: 317–319.

James, P. S., and Smith, M. W. 1976. Methionine transport by pig colonic mucosa measured during early post-natal development. *J. Physiol. (Lond.)* 262: 151–168.

James, R. G. 1934. Studies on the amphibian digestive system. I. Histological changes in the alimentary tract of anuran larvae during involution. *J. Exp. Zool.* 67: 73–91.

Janis, C. 1976. The evolutionary strategy of the equidae and the origins of rumen and cecal digestion. *Evolution* 30: 757–774.

Janis, C. 1982. Evolution of horns in ungulates: ecology and paleoecology. *Biol. Rev.* 57: 261–318.

Janis, C., and Fortelius, M. 1988. On the means whereby mammals achieve increased functional durability of their dentitions, with special reference to the limiting factors. *Biol. Rev.* 63: 197–230.

Janis, C. M., Gordon, I. J., and Illius, A. W. 1994. Modelling equid/ruminant competition in the fossil record. *Hist. Biol.* 8: 15–29.

Janssens, P. A., and Messer, M. 1988. Changes in nutritional metabolism during weaning. In *The Developing Marsupial: Models for Biomedical Research*, ed. C. H. Tyndale-Biscoe and P. A. Janssens, pp. 162–175. Berlin: Springer.

Jarvis, L. G., Morgan, G., Smith, M. W., and Wooding, F. B. P. 1977. Cell replacement and changing transport function in the neonatal pig colon. *J. Physiol.* 273: 717–729.

Jeuniaux, C. 1962. Digestion de la chitine chez les oiseaux et les mammifères. *Ann. Soc. R. Zool. Belg.* 92: 27–45.

Jeuniaux, C. 1961. Chitinase: an addition to the list of hydrolases in the digestive tract of vertebrates. *Nature (Lond.)* 192: 135–136.

Jeuniaux, C. 1962a. Digestion de la chitine chez les oiseaux et les mammiferes. *Ann. Soc. R. Zool. Belg.* 92: 27–45.

Jeuniaux, C. 1962b. Recherche de polysaccharidases dans l'estomac de'un paresseux. *Chloepus hoffmanni. Arch. Int. Physiol. Biochem.* 70: 407–408.

Jeuniaux, C. 1963. *Chitine et Chitinolyse*. Paris: Masson.

Jeuniaux, C. 1971a. Chitinous structures. In *Comprehensive Biochemistry*. Amsterdam: Elsevier.

Jeuniaux, C. 1971b. On some biochemical aspects of regressive evolution in animals. In *Biochemical Evolution and the Origin of Life*, ed. R. A. Muzzerelli and E. R. Pariser, pp. 542–549. Cambridge, Massachusetts: M. I. T. Press.

Jeuniaux, C., and Cornelius, C. 1978. Distribution and activity of chitinolytic enzymes in the digestive tract of birds and mammals. In *Proceedings of the First International Conference on Chitin/Chitosan*, ed. R. A. A. Muzzarelli and E. R. Pariser, pp. 542–549. Cambridge, Massachusetts: M. I. T. Press.

Jilge, B. 1979. Zur circadianen Caecotrophie des Kaninchens. *Z. Versuchstierkd.* 21: 302–312.

Johnson, J. L., and McBee, R. H. 1967. The porcupine cecal fermentation. *J. Nutr.* 91: 540–546.

Jones, W. T., and Mangan, J. L. 1977. Complexes of the condensed tannins of sainfoin (*Onobrychis vicifolia* Scop.) with fraction 1 leaf protein and with submaxillary mucoprotein and their reversal by polyethylene glycol and pH. *J. Sci. Fd. Agric.* 28: 126–136.

Jonsson, A. C., Holmgren, S., and Holstein, B. 1987. Gastrin/CCK-like immunoreactivity in endocrine cells and nerves in the gastrointestinal tract of the cod, *Gadus morhua. General and Comp. Endocrin.* 66: 190–201.

Josefsson, L., and Lindberg, T. 1965. Intestinal dipeptidases. I. Spectrophotometric determination and characterization of dipeptidase activity in pig intestinal mucosa. *Biochim. Biophys. Acta* 105: 149–161.

Junqueira, L. C. U., and de Moraes, F. F. 1965. Comparative aspects of the vertebrate major salivary glands biology. In *Funktionelle und morphologische Organisation der Zelle.* Sekretion and Exkretion, ed. W. Bothermann, pp. 36–48. New York: Springer-Verlag.

Junqueira, L. C. U., Malnic, G., and Monge, C. 1966. Reabsorptive function of the ophidian cloaca and large intestine. *Physiol. Zool.* 39: 151–159.

Justice, K. E., and Smith, F. A. 1992. A model of dietary fiber utilization by small mammalian herbivores, with empirical results for *Neotoma*. *Am. Nat.* 139: 398–416.

Kandatsu, M., Yoshihara, I., and Yoshida, T. 1959. Studies on cecal digestion. II. Excretion of hard and soft feces and fecal composition in rabbits. *Jpn. J. Zootechn. Sci.* 29: 366–371.

Kanou, T. 1984. Morphological studies of the mucous membrane of the small intestine of vertebrates with an emphasis on comparative anatomy. *Kawaski Med. J.* 10: 49–61.

Kapoor, B. G., and Khawna, B. 1993. The potential spectrum of the gut in teleost fishes. *Adv. Fish Biol.* 1: 221–226.

Kapoor, B. G., Smit, H., and Verighina, I. A. 1975. The alimentary canal and digestion in teleosts. *Adv. Mar. Biol.* 13: 109–239.

Karasawa, Y., and Maeda, M. 1994. Role of ceca in the nitrogen nutrition of the chicken on a moderate protein diet or low protein diet plus urea. *Brit. Poultry Sci.* 35: 383–391.

Karasov, W. H. 1988. Nutrient transport across vertebrate intestine. In *Advances in Environmental Physiology*, ed. R. Gilles, pp. 131–172. Berlin: Springer-Verlag.

Karasov, W. H. 1990. Digestion in birds: Chemical and physiological determinants and ecological implications. *Studies in Avian Biology.* No. 13, pp. 391–415.

Karasov, W. H. 1992a. Daily energy expenditure and cost of activity in mammals. *Am. Zool.* 32: 238–248.

Karasov, W. H. 1992b. Tests of the adaptive modulation hypothesis for dietary control of intestinal nutrient transport. *Am. J. Physiol.* 263: R496–R502.

Karasov, W. H., and Diamond, J. M. 1985. Digestive adaptations for fueling the cost of endothermy. *Science* 228: 202–204.

Karasov, W. H., and Diamond, J. M. 1988. Interplay between physiology and ecology in digestion. *Bioscience* 38: 602–611.

Karasov, W. H., and Hume, I. D. Vertebrate gastrointestinal system. In *Handbook of Comparative Physiology* ed. W. H. Dantzler. Oxford: Oxford University Press. (In press).

Karasov, W. H., and Levey, D. J. 1990. Digestive system trade-offs and adaptations of frugivorous passerine birds. *Physiol. Zool.* 63: 1248–1270.

Karasov, W. H., Phan, D., Diamond, J. M., and Carpenter, F. L. 1986. Food passage and intestinal absorption in hummingbirds. *Auk* 103: 453–464.

Karasov, W. H., Solberg, D. H., and Diamond, J. M. 1985. What transport adaptations enable mammals to absorb sugars and amino acids faster than reptiles? *Am. J. Physiol.* 249: G271–285.

Karpov, L. V. 1919. O perevarivanii nekrotorykh rastitelnykh izhivotnykh belkou gusiiym zheludochwom sokum. *Fiziol. Zh. SSSR* 2: 185–196.

Karr, M. R., Little, C. O., and Mitchell, G. E. Jr. 1966. Starch disappearance from different segments of the digestive tract of steers. *J. Anim. Sci.* 25: 652–654.

Kaske, M., Hatipoglu, S., and Engelhardt, W. v. 1992. The influence of density and size of particles on rumination and passage from the reticulo-rumen of sheep. *Br. J. Nutr.* 67: 235–244.

Kasper, H. 1962. Methode für Prüfung der Resorption in Caecum und Colon. *Z. Versuchstierkd.* 1: 104–106.

Kauffman, G. L., Reeve, J. J., and Grossman, M. I. 1980. Gastric bicarbonate secretion: effect of topical and intravenous 16-16 dimethyl prostaglandin E_2. *Am. J. Physiol.* 239: G44–G48.

Kaufmann, W., Hagemeister, H., and Dirksen, G. 1980. Adaptations to changes in dietary composition, level and frequency of feeding. In *Digestive Physiology and Metabolism in Ruminants*. ed. Y. Ruckebusch and P. Thivend, p. 590. Lancaster, U.K.: M T P Press Ltd.

Kaushik, S. J. 1986. Environmental effects on feed utilization. In *Fish Physiology and Biochemistry*, Vol. 2, Nos. 1–4, pp. 131–140. Amsterdam: Kugler Publications.

Kawai, S., and Ikeda, S. 1971. Studies on digestive enzymes of fishes: I. Carbohydrases in digestive organs of several fishes. *Bull. Jpn. Soc. Sci. Fish.* 37: 333–337.

Kawai, S., and Ikeda, S. 1973a. Studies on digestive enzymes of fishes. III. Development of the digestive system of rainbow trout after hatching and the effect of dietary change on the activities of digestive enzymes in the juvenile stage. *Bull. Jap. Soc. Sci. Fish.* 39: 819–823.

Kawai, S., and Ikeda, S. 1973b. Studies on digestive enzymes of fishes. IV. Development of the digestive enzymes of carp and black sea bream after hatching. *Bull. Jap. Soc. Sci. Fish.* 39: 877–881.

Kay, R. N. B. 1960. I. The rate of flow and composition of various salivary secretions in sheep and calves. II. The development of parotid salivary secretion in young goats. *J. Physiol. (Lond.)* 150: 515–545.

Kay, R. N. B., and Pfeffer, E. 1970. Movements of water and electrolytes into and from intestine of sheep. In *Physiology of Digestion and Metabolism in the Ruminant*, ed. A. T. Phillipson, pp. 390–402. Newcastle upon Tyne, England: Oriel Press.

Kear, J. 1972. Feeding habits of birds. In *Biology of Nutrition*, Vol. 18, ed. R. N. T-W-Fiennes, pp. 471–503. New York: Pergamon Press.

Kelly, K. A. 1981. Motility of the stomach and gastroduodenal junction. In *Physiology of the Gastrointestinal Tract*, Vol. 1, ed. L. R. Johnson, J. Christensen, M. I. Grossman, E. D. Jacobson, and S. G. Schultz, pp. 393–410. New York: Raven Press.

Kempton, T. J., Murray, R. M., and Leng, R. A. 1976. Methane production and digestibility measurements in the grey kangaroo and sheep. *Aust. J. Biol. Sci.* 29: 209–214.

Kenagy, G. J., and Hoyt, D. F. 1980. Reingestion of faeces in rodents and its daily rhythmicity. *Oecologia* 44: 403–409.

Kenagy, G. J., Masman, D., Sharbaugh, S. M., and Nagy, J. A. 1990. Energy expenditures during lactation in relation to litter size in free-living, golden-mantled ground squirrels. *J. Anim. Ecol.* 59: 73–88.

Kenchington, R. A. 1972. Observations on the digestive system of the dugong, *Dugong dugon* Erxleben. *J. Mammal.* 53: 884–887.

Kennedy, P. M., and Hume, I. D. 1978. Recycling of urea nitrogen to the gut of the tammar wallaby (*Macropus eugenii*). *Comp. Biochem. Physiol.* 61A: 117–121.

Kennelly, J. J., Ahearne, F. X., and Sauer, W. C. 1981. Volatile fatty acid production in the hindgut of swine. *Can. J. Anim. Sci.* 61: 349–361.

Kerry, K. R. 1969. Intestinal disaccharidase activity in a monotreme and eight species of marsupials (with an added note on the disaccharidases of five species of sea birds). *Comp. Biochem. Physiol.* 29: 1015–1022.

Kerry, K. R., and Messer, M. 1968. Intestinal glycosidases of three species of seals. *Comp. Biochem. Physiol.* 25: 437–446.

Keys, J. E. Jr., and DeBarthe, J. V. 1974. Site and extent of carbohydrate, dry matter, energy and protein digestion and the rate of passage of grain diets in swine. *J. Anim. Sci.* 39: 57–62.

Kidder, D. E., and Manners, M. J. 1978. *Digestion in the Pig.* Bristol, England: Scientechnica.

Kim, Y. S., and Erickson, R. H. 1985. Role of peptidases of the human small intestine in protein digestion. *Gastroenterology* 88: 1071–1073.

Kimmel, J. R., Hayden, L. J., and Pollock, H. G. 1975. Isolation and characterization of a new pancreatic polypeptide hormone. *J. Biol. Chem.* 24: 9369–9376.

Kimmel, J. R., Pollock, H. G., and Hayden, L. J. 1978. Biological activity of the avian PP in the chicken. In *Gut Hormones*, ed. S. R. Bloom, pp. 234–241. London: Churchill Livingstone.

Kimmich, G. A. 1981. Intestinal absorption of sugar. In *Physiology of the Gastrointestinal Tract*, ed. L. R. Johnson, J. Christensen, M. J. Jackson, E. D. Jacobson, and J. H. Walsh, pp. 1035–1061. New York: Raven Press.

King, P. A., Goldstein, S. R., Goldstein, J. M., and Goldstein, L. 1986. Taurine transport by the flounder (*Pseudopleuronectes americanus*) intestine. *J. Exp. Zool.* 238: 11–16.

King, I. S., Sepulveda, F. V., and Smith, M. W. 1981. Cellular distribution of neutral and basic amino acid transport in rabbit ileal mucosa. *J. Physiol. (Lond.)* 319: 355–368.

King, K. W., and Moore, W. E. C. 1957. Density and size as factors affecting passage rate of ingesta in the bovine and human digestive tracts. *J. Dairy Sci.* 40: 528–536.

Kinnear, J. E., and Main, A. R. 1975. The recycling of urea nitrogen by the wild tammar wallaby (*Macropus eugenii*) a "ruminant-like" marsupial. *Comp. Biochem. Physiol.* 51A: 593–610.

Kintner, W. B., and Wilson, T. H. 1965. Autoradiographic study of sugar and amino acid absorption by everted sacs of hamster intestine. *J. Cell Biol.* 25: 19–39.

Kirsch, R. 1978. Role of the esophagus in osmoregulation in teleost fishes. In *Osmotic and Volume Regulation*, ed. C. B. Jorgensen and E. Skadhauge, pp. 138–154. New York: Academic Press.

Kitamikado, M., and Tachino, S. 1960. Studies on the digestive enzymes of rainbow trout—I. Carbohydrases. *Bull. Jpn. Soc. Sci. Fish.* 26: 679–684.

Kleiber, M. 1961. *The Fire of Life.* New York: John Wiley & Sons.

Knickelbein, R., Aronson, P., Schron, C., Seifter, J., and Dobbins, J. 1985. Sodium and chloride transport across rabbit ileal brush border. II. Evidence for Cl-HCO$_3$ exchange and mechanism of coupling. *Am. J. Physiol.* 249: G236–G245.

Knight, I. T., Grimes, J. J., and Colwell, R. R. 1988. Bacterial hydrolysis of urea in tissues of carcharinid sharks. *Can. J. Fish Aquat. Sci.* 45: 357–360.

Koch, M. O., and Hall, M. C. 1992. Mechanism of ammonium transport by intestinal segments following urinary division: evidence for ionized NH_4^+ transport via K^+ pathways. *J. Urol.* 148: 453–457.

Kochva, E. 1978. Oral glands of the reptilia. In *Biology of the Reptilia*, Vol. 8, ed. C. Gans and K. A. Gans, pp. 43–162. New York: Academic Press.

Koefed-Johnson, V., and Ussing, H. H. 1958. On the nature of the frog skin potential. *Acta Physiol. Scand.* 42: 298–308.

Koike, T. I., and McFarland, L. Z. 1966. Urography in the unanesthetized hydropenic chicken. *Am. J. Vet Res.* 27: 1130–1133.

Koldovsky, O. 1970. Digestion and absorption during development. In *Physiology of the Perinatal Period*, ed. U. Stave, pp. 379–415. New York: Appleton-Century-Crofts.

Koldovsky, O. 1981. Developmental, dietary and hormonal control of intestinal disaccharidases in mammals (including man). In *Carbohydrate Metabolism and its Disorders*, ed. J. P. Randle, D. F. Steiner, W. J. Shelan, and W. P. Whelan, pp. 418–522. London: Academic Press.

Komori, S., and Ohashi, H. 1982. Some characteristics of transmission from non-adrenergic, non-cholinergic excitatory nerves to the smooth muscle of the chicken. *J. Auton. Nerv. Syst.* 6: 199–210.

Kormaniik, G. A. 1992. Ion and osmoregulation in prenatal elasmobranches: evolutionary implications. *Amer. Zool.* 32: 294–302.

Kostelecka-Myrcha, A., and Myrcha, A. 1965. Effect of the kind of indicator on the results of investigations of the rate of passage of foodstuffs through the alimentary tract. *Acta Theriol.* 10: 229–242.

Kostuch, T. E., and Duke, G. E. 1975. Gastric motility in great horned owls (*Bubo virginianus*). *Comp. Biochem. Physiol.* 51A: 201–205.

Koteja, P. 1991. On the relation between basal and field metabolic rates in birds and mammals. *Functional Ecol.* 5: 56–64.

Krause, W. J. 1970. Brunner's glands of the echidna. *Anat. Rec.* 167: 473–487.

Krawielitzki, K., Schadereit, R., Wünsche, J., Völker, T., and Bock, H.-D. 1983. Untersuchungen über Resorption und Verwertung von ins Zäkum wachsender Schweine infundierten Aminosäuren. *Arch. Tierernähr.* Berlin 33: 731–742.

Krawielitzki, K., Schadereit, R., Zebrowska, T., Wünsche, J., and Bock, H.-D. 1984. Untersuchungen über Resorption und Verwertung von ins Zäkum wachsender Schweine infundierten Aminosäuren. *Arch. Tierernähr.* Berlin 34: 1–18.

Kretchmer, N. 1981. Food: a selective agent in evolution. In *Nutrition and Evolution*, ed. D. N. Walcher and N. Kretchmer, pp. 37–48. New York: Niasson Publ.

Krieger, D. T. 1983. Brain peptides: what, where, and why? *Science* 222: 975–985.

Krockenberger, A. K. 1993. Energetics and nutrition during lactation in the koala. Ph.D. Thesis, University of Sydney, Australia.

Kronfeld, D. S. 1973. Diet and the performance of racing sled dogs. *J. Am. Vet. Med. Assoc.* 162: 470–473.

Kronfeld, D. S., and Van Soest, P. J. 1976. Carbohydrate nutrition. In *Comparative Animal Nutrition*, ed. M. Rechcigl, Jr., pp. 23–73. Basel: S. Karger.

Kuhn, H.-J. 1964. Zur Kenntnis von Bau and Funktion des Magens der Schlankaffen (*Colobinae*). *Folia Primatol.* 2: 193–221.

Landry, S. O. 1970. The rodentia as omnivores. *Quart. Rev. Biol.* 45: 351–372.

Lange, R., and Staaland, H. 1970. Adaptations of the caecum-colon structure of rodents. *Comp. Biochem. Physiol.* 35: 905–919.

Langer, M., Van Noorden, S., Polak, J., and Pearse, A. G. E. 1979. Peptide hormone-like immunoreactivity in the gastrointestinal tract and endocrine pancreas of eleven teleost species. *Cell Tissue Res.* 199: 493–508.

Langer, P. 1974. Stomach evolution in the artiodactyla. *Mammalia* 38: 295–314.

Langer, P. 1988. *The Mammalian Herbivore Stomach*. Stuttgart: Fischer.

Langer, P., and Chivers, D. J. 1994. Classification of foods for comparative analysis of the gastro-intestinal tract. In *The Digestive System in Mammals: Food, Form and Function*. ed. D. J. Chivers and P. Langer, pp. 74–86. New York: Cambridge University Press.

Langer, P., Dellow, D. W., and Hume, I. D. 1980. Stomach structure and function in three species of macropodine marsupials. *Aust. J. Zool.* 28: 1–18.

Langslow, D. R., Kimmel, J. R., and Pollock, H. G. 1973. Studies of the distribution of a new avian pancreatic polypeptide and insulin among birds, reptiles, amphibians and mammals. *Endocrinology* 93: 558–565.

Lanyon, J. M., and Marsh, H. 1995. Digesta passage time in the dugong *Dugong dugong*. Aust. J. Zool. 43: 119–127.

Larsen, L. O. 1984. Feeding in adult toads: physiology, behavior, ecology. *Vidensk. Meddr. Dansk. Naturh. Foren.* 145: 97–116.

Larsson, L. I., and Rehfeld, J. F. 1977. Evidence for a common evolutionary origin of gastrin and cholecystokinin. *Nature (Lond.)* 269: 335–338.

Larsson, L. I., and Rehfeld, J. F. 1978. Evolution of CCK-like hormones. In *Gut Hormones*, ed. S. R. Bloom, pp. 68–73. London: Churchill Livingstone.

Larsson, L.-I., Sundler, F., Hakanson, R., Rehfeld, J. F., and Stadil, F. 1974. Distribution and properties of gastrin cells in the gastrointestinal tract of the chicken. *Cell Tissue Res.* 154: 409–422.

Lauff, M., and Hofer, R. 1984. Proteolytic enzymes in fish development and the importance of dietary enzymes. *Aquaculture* 37: 335–346.

Law, A. T., Jamalullail, S. R. 1981. Digestibility of carpet grass (*Axonopus compressus*) in grass carp. *Ctenopharyngodon idella* (Cuvier and Valenciennes). *Pertinka* 4(1): 91–93.

Laws, B. M., and Moore, J. H. 1963. The lipase and esterase activities of the pancreas and small intestine of the chick. *Biochem. J.* 87: 632–638.

Lechner-Doll, M., Kaske, M., and Engelhardt W. v. 1991. Factors affecting the mean retention time of particles in the forestomach of ruminants and camelids. In: *Physiological Aspects of Digestion and Metabolism in Ruminants*, ed. T. Tsuda, Y. Sasaki, and R. Kawashima, pp. 455–482. San Diego: Academic Press.

Lee, K. Y., Park, H. J., Chang, T.-M., and Chey, W. Y. 1983. Cholinergic role on release and action of motilin. *Peptides* 4: 375–380.

Lee, R. F., Hirota, J., Nevenzel, J. C., Sauerheber, R., Lewis, A., and Benson, A. A. 1972. Lipids in the marine environment. *Calif. Mar. Res. Comm., Calif. COFI Rept.* 16: 95–102.

Lee, P. P. N., Hong, G. X., and Pang, S. F. 1991. Melatonin in the gastrointestinal tract. In *Role of Melatonin and Pineal Peptides in Neuroimmunodulation*, ed. F. Fraschini and R. J. Reiter, pp. 127–136. New York: Plenum Press.

Lee, P. P. N., and Pang, S. F. 1993. Melatonin and its receptors in the gastrointestinal tract. *Biol. Signals* 2: 181–193.

Lee, W. B., and Houston, D. C. 1993. The effect of the diet quality on gut anatomy in British voles (Microtinae). *J. Comp. Physiol.* B163: 337–339.

Leeson, C. R. 1967. Structure of salivary glands. In *Handbook of Physiology*, Section 6, *Alimentary Canal*, Vol. 2, *Secretion*, ed. C. F. Code and W. Heidel, pp. 463–495. Washington, D. C.: American Physiological Society.

Léger, C. 1979. La lipase pancréatique. In *Nutrition des Poissons*, ed. M. Fontaine, pp. 69–77. Paris: Actes duColloque C. N. E. R. N. A.

Lehinger, A. L. 1975. *Biochemistry*. New York: Worth Publishers, Inc.

Leng, R. A., Corbett, J. L., and Brett, D. J. 1968. Rates of production of volatile fatty acids in the rumen of grazing sheep and their relation to ruminal concentrations. *Br. J. Nutr.* 22: 57–68.

Leng, R. A., and Leonard, G. J. 1965. Measurement of the rates of production of acetic, propionic and butyric acids in the rumen of sheep. *Br. J. Nutr.* 19: 469–484.

Leon, B. 1980. Fermentation and the production of volatile fatty acids in the alimentary tract of the rock hyrax, *Procavia capensis*. *Comp. Biochem. Physiol.* 65A: 411–420.

Leon, M. A. 1974. Maternal pheromone. *Physiol. Behav.* 13: 441–453.

Leopold, A. S. 1953. Intestinal morphology of gallinaceous birds in relation to food habits. *J. Wildlife Mgt.* 17: 197–203.

Lerner, J. 1984. Cell membrane amino acid transport processes in the domestic fowl (*Gallus domesticus*). *Comp. Biochem. Physiol.* 78A: 205–215.

Le Roith, D., Shiloach, J., and Roth, J. 1982. Is there an earlier phylogenetic precursor that is common to both the nervous and endocrine systems? *Peptides* 3: 211–215.

Lesel, R., Fromageot, C., and Lesel, M. 1986. Cellulose digestibility in grass carp, *Ctenopharyngodon idella*, and goldfish, *Carassius auratus. Aquaculture* 54: 11–17.

Lester, R., and Grim, E. 1975. Substrate utilization and oxygen consumption by canine jejunal mucosa *in vitro. Am. J. Physiol.* 229: 139–143.

Levenson, S. M., Crowley, L. V., Horowitz, R. E., and Malm, O. J. 1959. The metabolism of carbon-labeled urea in the germ-free rat. *J. Biol. Chem.* 234: 2061–2062.

Levenspiel, O. 1972. *Chemical Reaction Engineering*, 2nd ed. New York: John Wiley & Sons.

Levitt, M. D., and Bond, J. H. 1970. Volume, composition, and source of intestinal gas. *Gastroenterology* 59: 921–929.

Lichtenberger, L. M., Dial, E. J., Romero, J. J., Lechago, J., Jarboe, L. A., and Wolfe, M. M. 1995. Role of luminal ammonia in development of gastropathy and hypergastrinemia in the rat. *Gastroenterology* 108: 320–329.

Liddle, R. A. 1993. Cholecystokinin. In *Gut peptides: Biochemistry and Physiology*. ed. J. H. Walsh, and J. H. Dockray, pp. 175–215. New York: Raven Press.

Liem, K. F., and Greenwood, P. H. 1981. A functional approach to the phylogeny of the pharyngognath teleosts. *Am. Zool.* 21: 83–101.

Lillegraven, J. A., Kraus, M. J., and Brown, T. M. 1979. Paleogeography of the world of the Mesozoic. In *Mesozoic Mammals*, ed. J. A. Lillegraven, Z. Kielan-Jaworoska, and W. A. Clemens, pp. 277–308. Berkley: University of California Press.

Lind, J., Munck, B. G., and Olsen, O. 1980a. Effects of dietary intake of sodium chloride on sugar and amino acid transport across isolated hen colon. *Am. J. Physiol. (Lond.)*. 305: 327–336.

Lind, J., Munck, B. G., Olsen, O., and Skadhauge, E. 1980b. Effects of sugars, amino acids and inhibitors on electrolyte transport across hen colon. *J. Physiol. (Lond.)* 305: 315–325.

Lindsay, G. J. G. 1984. Distribution and function of digestive tract chitinolytic enzymes in fish. *J. Fish Biol.* 24: 529–536.

Lintern-Moore, S. 1973. Utilization of dietary urea by the Kangaroo Island wallaby–*Protemnodon eugenii* (Desmarest). *Comp. Biochem. Physiol.* 46A: 345–351.

Livingston, H. G., Payne, W. J.A., and Friend, M. T. 1962. Urea excretion in ruminants. *Nature (Lond.)* 194: 1057–1058.

Lobel, P. S. 1981. Trophic biology of herbivorous reef fishes: alimentary pH and digestive capabilities. *J. Fish Biol.* 19: 365–397.

Lochmiller, R. L., Hellgren, E. C., Gallagher, J. F., Varner, L. W., and Grant, W. E. 1989. Volatile fatty acids in the gastrointestinal tract of the collared peccary (*Tayassu tajacu*). *J. Mammal.* 70: 189–191.

Lomolino, M. V., and Ewel, K. C. 1984. Digestive efficiencies of the West Indian manatee (*Trichechus manatus*). *Florida Scientist* 47: 176–179.

Lönnberg, E., 1902. On some points of relation between the morphological structure of the intestine and the diet of reptiles. *Bih. Suensk. Vet. Ak. Handl.* 28: 1–51.

Lönnberg, E. 1902. On some remarkable digestive adaptations in diprotodont marsupials. *Proc. Zool. Soc. Lond.* 73: 12–31.

Lucas, M. L. 1976. The association between acidification and electrogenic events in the rat proximal jejunum. *J. Physiol. (Lond.)* 257: 645–662.

Lucas, M. L. 1984. Weak-electrolyte absorption and the acid microclimate. In *Intestinal Absorption and Secretion*, ed. E. Skadhauge and K. Heintze, pp. 39–54. Lancaster, U.K.: M T P Press.

Lucas, M. L., Schneider, W., Haberich, F. J., and Blair, J. A. 1975. Direct measurement by pH microelectrode of the pH-microclimate in rat proximal jejunum. *Proc. R. Soc. (Lond.)* Ser. B, 192: 39–43.

Luciano, L., Konitz, H., and Reale, E. 1989. Localization of cholesterol in the colonic epithelium of the guinea pig: regional differences and functional implications. *Cell and Tissue Res.* 258: 339–347.

Luckey, T. D., Pleasants, J. R., Wagner, M., Gordon, H. A., and Reyniers, J. A. 1955. Some observations on vitamin metabolism in germ-free rats. *J. Nutr.* 57: 169–182.

Luick, B. R., and Penner, M. H. 1991. Nominal response of passage rates to fiber particle size in rats. *J. Nutr.* 121: 1940–1947.

Lundgren, O. 1984. Microcirculation of the gastrointestinal tract and pancreas. In *Handbook of Physiology*, Section 2, *The Cardiovascular System*, Vol. 4, *Microcirculation, Part 2*, ed. E. M. Renkin, C. C. Michel, and S. R. Geiger, pp. 799–863. Bethesda, Maryland: American Physiological Society.

Luppa, H. 1977. Histology of the digestive tract. In *Biology of the Reptilia*, Vol. 6., ed. C. E. Gans and T. S. Parsons, pp. 225–314. New York: Academic Press.

Macfarlane, W. V., Howard, B., Haines, H., Kennedy, P. J., and Sharpe, C. M. 1971. Hierarchy of water and energy turnover of desert mammals. *Nature (Lond.)* 234: 483–484.

Mackie, R. I. 1987. Microbial digestion of forages in herbivores. In *The Nutrition of Herbivores*, ed. J. B. Hacker and J. H. Ternouth, pp. 233–265. Sydney: Academic Press.

Mackie, R. I., and Wilkins, C. A. 1988. Enumeration of anaerobic bacterial microflora of the equine gastrointestinal tract. *Appl. Environ. Microb.* 54: 2155–2160.

Madsen, H. 1939. Does the rabbit chew its cud? *Nature (Lond.)* 143: 981.

Maffia, M., Ahearn, G. A., Vilella, S., Zonno, V., and Storelli, C. 1993. Ascorbic acid transport by intestinal brush border membrane vesicles in the teleost *Augilla aguilla*. *Am. J. Physiol.* 264: R1248–1253.

Magee, D. F. 1961. An investigation into the external secretion of the pancreas in sheep. *J. Physiol. (Lond.)* 158: 132–143.

Malagelada, J-R., and Azpiroz, F. 1989. Determinants of gastric emptying and transit in the small intestine. In *Handbook of Physiology*, Section 6: *Alimentary Canal*, Vol. 1: *Motility and Circulation*, ed. S. G. Schultz, J. D. Wood, and B. B. Rauner, pp. 909–937. Washington, D. C.: American Physiological Society.

Marshall, B. J. 1986. *Campylobacter pyloridis* and gastritis. *J. Infect. Dis.* 153: 650–657.

Marshall, B. J., McGechie, D. B., Rogers, P. A., and Glancy, P. A. 1985. Pyloric *Campylobacter* infection and gastroduodenal disease. *Med. J. Aust.* 142: 439–444.

Marshall, J. A., and Dixon, K. E. 1978. Cell proliferation in the intestinal epithelium of *Xenopus laevis* tadpoles. *J. Exp. Zool.* 203: 31–40.

Martens, H., and Gabel, G. 1988. Transport of Na and Cl across the epithelium of ruminant forestomachs: rumen and omasum. *Comp. Biochem. Physiol.* 90A: 569–575.

Mårtensson, P. E., Nordøy, E. S., and Blix, A. S. 1994. Digestibility of krill (*Euphausia superba* and *Thysanoessa sp.*) in minke whales (*Balaenoptera acutorostrata*) and crabeater seals (*Lobodon carcinophagus*). *Br. J. Nutr.* 72: 713–716.

Martin, M. M. 1991. The evolution of cellulose digestion in insects. *Phil. Trans. R. Soc. Lond.* B333: 281–288.

Martin, R. 1971. Étude autoradiographique du renouvellement de l'épithélium intestinal de l'axolotl (amphibien urodele). *C. R. Acad. Sci. [D] (Paris)* 272: 2816–2819.

Martinez, V., Jimenez, M., Gonalons, E., and Vergara, P. 1993. Effects of cholecystokinin and gastrin on gastroduodenal motility and coordination in chickens. *Life Sci.* 52: 191–198.

Martinez del Rio, C. 1990. Dietary, phylogenetic, and ecological correlates of intestinal sucrase and maltase activity in birds. *Physiol. Zool.* 63: 987–1011.

Martinez del Rio, C., Karasov, W. H., and Levey, D. J. 1989. Physiological basis and ecological consequences of sugar preferences in cedar waxwings. *Auk* 106: 64–71.

Martinez del Rio, C., and Stevens, B. R. 1988. Intestinal brush border membrane-bound disaccharidases of the American alligator, *Alligator mississippiensis. Comp. Biochem. Physiol.* 91B: 751–754.

Marty, J., and Vernay, M. 1984. Absorption and metabolism of the volatile fatty acids in the hind-gut of the rabbit. *Br. J. Nutr.* 51: 265–277.

Mascola, N., Rajendran, V. M., and Binder, H. J. 1991. Mechanism of short-chain fatty acid uptake by apical vesicles of rat distal colon. *Gastroenterology* 101: 331–338.

Materazzi, G., and Menghi, G. 1975. Il canale alimentare in embrioni ed avannotti di *Salmo irideus. Ann. Ist. e Mus. Zool. Napoli* 21: 21–37.

Mathias, J. R., and Sninsky, C. A. 1985. Motility of the small intestine: a look ahead. *Am. J. Physiol.* 248: G495–G500.

Mathiesen, S. D., Aagnes, T., and Sørmo, W. 1990. Microbial symbiotic digestion in minke whales (*Balaenoptera acutorostrata*). Paper SC/42/NHMi9 presented to the International Whaling Scientific Committee.

Mathison, G. W., Okine, E. K., Vaage, A. S., Kaske, M., and Milligan, L. P. 1994. Current understanding of the contribution of the propulsive activities in the forestomach to the flow of digesta. Proc. VIII Symp. Ruminant Physiol., Willingen, Germany (In press).

Mattson, F. H., and Volpenheim, R. A. 1968. Hydrolysis of primary and secondary esters of glycerol by pancreatic juice. *J. Lipid Res.* 9: 79–84.

Mayer, E. A. 1994. The physiology of gastric storage and emptying. In *Physiology of the Gastrointestinal Tract.* 3rd ed., ed. L. R. Johnson, D. H. Alpers, J. Christensen, E. D. Jacobson, and J. H. Walsh, pp. 929–976. New York: Raven Press.

McAvoy, J. W., and Dixon, K. E. 1977. Cell proliferation and renewal in the small intestinal epithelium of metamorphosing and adult *Xenopus laevis. J. Exp. Zool.* 202: 128–138.

McBee, R. H. 1977. Fermentation in the hindgut. In *Microbial Ecology of the Gut*, ed. R. T. J. Clarke and T. Bauchop, pp. 185–222. New York: Academic Press.

McBee, R. H., and McBee, V. H. 1982. The hindgut formation in the green iguana, *Iguana iguana.* In *Iguanas of the World: Their Behavior, Ecology and Conservation*, ed. G. M. Burghardt, and A. S. Rand, pp. 77–83. Park Ridge, New Jersey: Noyes.

McBee, R. H., and West, G. C. 1969. Cecal fermentation in the willow ptarmigan. *Condor* 71: 54–58.

McBride, B. W., and Kelly, J. M. 1990. Energy cost of absorption and metabolism in the ruminant gastrointestinal tract and liver: a review. *J. Anim. Sci.* 68: 2997–3010.

McBride, B. W., Milligan, L. P., and Turner, B. V. 1984. Endoscopic observation of digesta transfer from the reticulo-rumen to the omasum of cattle. *Can. J. Anim. Sci.* 64 (Supp.): 84–85.

McCuistion, W. R. 1966. Coprophagy, a quest for digestive enzymes. *Vet. Med. Small Anim. Clin.* 61: 445–447.

McDonald, T. J., Jörnvall, H., Nilsson, G., Vagne, M., Ghatei, M., Bloom, S. R., and Mutt, V. 1979. Characterization of a gastrin releasing peptide from porcine nonantral gastric tissue. *Biochem. Biophys. Res. Commun.* 90: 227–233.

McGeachin, R. L., and Bryan, J. A. 1964. Amylase in the cotton-mouth water moc-casin, *Agkistrodon piscivorus. Comp. Biochem. Physiol.* 13: 473–475.

McLeay, L. M., and Titchen, D. A. 1970. Effects of pentagastrin in gastric secretion and motility in the sheep. *Proc. Aust. Physiol. Pharmacol. Soc.* 1: 33–34.

McLeay, L. M., and Titchen, D. A. 1975. Gastric, antral and fundic pouch secretion in sheep. *J. Physiol. (Lond.)* 248: 595–612.

McMahon, T. A., and Bonner, J. T. 1983. *On Size and Life.* New York: Scientific American Books, Inc.

McMillan, G. L., and Churchill, E. P. 1947. The gross and histological structure of the digestive system of the little brown bat. *Proc. So. Dak. Acad. Sci.* 26: 103–109.

McNab, B. K. 1978. The comparative energetics of neotropical marsupials. *J. Comp. Physiol.* 125: 115–128.

McNeil, N. I. 1984. The contribution of the large intestine to energy supplies in man. *Am. J. Clin. Nutr.* 39: 338–342.

McNeil, N. I., Cummings, J. H., and James, W. P. T. 1978. Short-chain fatty acid absorption by the human large intestine. *Gut* 19: 819–822.

McNeil, N. I., Ling, K. L. E., and Wager, J. 1987. Mucosal surface pH of the large intestine of the rat and of the normal and inflamed large intestine in man. *Gut* 28: 707–713.

Mead, G. C. 1974. Anaerobic utilization of uric acid by some group D streptococci. *J. Gen. Micro.* 82: 421–423.

Mead, G. C., and Adams, B. W. 1975. Some observations on the caecal microflora of the chick during the first two weeks of life. *Br. Poult. Sci.* 16: 169–176.

Meddings, J. B. 1989. Lipid permeability of the intestinal microvillus membrane may be modulated by membrane fluidity in the rat. *Biochim. Biophys. Acta.* 984: 158–166.

Meldrum, L. A., and Burnstock, G. 1985. Investigations into the identity of the non-adrenergic, non-cholinergic excitatory transmitter in the smooth muscle of chicken rectum. *Comp. Biochem. Physiol.* 81C: 307–309.

Menking, M., Wagnitz, J. G., Burton, J. J., Coddington, R. D., and Sotos, J. F. 1969. Rumination: a near fatal psychiatric disease of infancy. *N. Engl. J. Med.* 280: 802–804.

Merrett, T. G., Bar-Eli, E., and Van Vunakis, H. 1969. Pepsinogens A, C, and D from the smooth dogfish. *Biochemistry* 8: 3696–3702.

Mertens, D. R., and Ely, L. O. 1979. A dynamic model of fiber digestion and passage in the ruminant for evaluating forage quality. *J. Anim. Sci.* 49: 1085–1095.

Messer, M., Crisp, E. A., and Czolij, R. 1989. Lactose digestion in suckling macropo-dids. In *Kangaroos, Wallabies and Rat-kangaroos,* ed. G. Grigg, P. Jarman, and I. Hume, pp. 217–221. Sydney: Surrey Beatty.

Mgbenka, B. O., and Lovell, R. T. 1986. Intensive feeding of grass carp in ponds. *Progressive Fish Culturist* 48: 238–241.

Micha, J. C., Dandrifosse, G., and Jeuniaux, C. 1973a. Activitésdes chitinases gastriques de reptiles en fonction du pH. *Arch. Int. Physiol. Biochim.* 81: 629–637.

Micha, J. C., Dandrifosse, G., and Jeuniaux, C. 1973b. Distribution et localisation tissulaire de la synthèse des chitinases chez les vertébrés inférieurs. *Arch. Int. Physiol. Biochim.* 81: 439–451.

Mikel'Saar, M. E., Tjuri, M. E., Väljaots, M. E., and Lencner, A. A. 1984. Anaerobe Inhalts- und Wandmikroflora des Magen-Darm-Kanals. *Nahrung* 28: 727–733.

Mikhail, M., Brugere, H., Le Bars, H., and Colvin, H. W. 1988. Stimulated esophageal groove closure in adult goats. *Am. J. Vet. Res.* 49: 1713–1715.

Milinkovitch, M. C., Guillermo, O., and Meyer, A. 1993. Revised phylogeny of whales suggested by mitochondrial ribosomal DNA sequences. *Nature (Lond.)* 361: 346–348.

Milligan, L. P., and McBride, B. W. 1985. Shifts in animal energy requirements across physiological and alimentation status. *J. Nutr.* 115: 1374–1382.

Milton, K., and McBee, R. H. 1983. Rates of fermentative digestion in the howler monkey, *Alouatta palliata* (Primates: Ceboidea). *Comp. Biochem. Physiol.* 74A: 29–31.

Minchin, K. 1973. Notes on the weaning of a young koala (*Phascolarctos cinereus*). *Rec. South Aust. Mus.* 6: 1–3.

Minhas, B. S., and Field, M. 1994. Localization of bicarbonate transport along the crypt-villus axis in the rabbit ileum. *Gastroenterology* 106: 1562–1567.

Minhas, B. S., Sullivan, S. K., and Field, M. 1990. Ileal HCO_3^- secretion *in vitro*. Effects of Na^+ and Cl^-. *Gastroenterology.* 98: A548.

Minnich, J. E. 1970. Water and electrolyte balance of the desert iguana, *Dipsosaurus dorsalis*, in its natural habitat. *Comp. Biochem. Physiol.* 35: 921–933.

Mircheff, A. K., Van Os, G. H., and Wright, E. M. 1980. Pathways for alanine transport in intestinal basolateral membrane vesicles. *J. Membr. Biol.* 52: 83–92.

Mishra, S. K., and Raviprakash, V. 1981. Non-adrenergic inhibitory and non-cholinergic excitatory neural involvement in DMPP and nicotine action in fowl rectum. *Arch. Int. Pharmacodyn.* 253: 210–219.

Mitchell, H. K., and Isbel, E. R. 1942. Intestinal bacterial synthesis as a source of B vitamin for the rat. *Univ. Tex. Publ. 4237, Part 2*, pp. 125–134.

Mitchell, P. C. 1901. On the intestinal tract of birds; with remarks on the valuation and nomenclature of zoological characters. *Trans. Linn. Soc. Lond.*, Second Series, Zoology 8: 173–275.

Mitchell, P. C. 1905. On the intestinal tract of mammals. *Trans. Zool. Soc. Lond.* 17: 437–536.

Miyamoto, Y., Tiruppathi, C., Ganapathi, B., and Leibach, F. H. 1989. Active transport of taurine in rabbit jejunal brush-border membrane vesicles. *Am. J. Physiol.* 257: G65–G72.

Mohsin, M. S. 1962. Comparative morphology and histology of the alimentary canals in certain groups of teleosts. *Acta Zool.* 43: 79–133.

Moin-Un-Nisa, Q., Zain, B. K., and Zain-Ul-Abedin, M. 1970. Intestinal transport of amino acids in a lizard during hibernation and activity. *Comp. Biochem. Physiol.* 36: 569–577.

Moir, R. J. 1965. The comparative physiology of ruminant-like animals. In *Physiology of Digestion in the Ruminant*, ed. R. W. Dougherty, pp. 1–14. Washington, D. C.: Butterworths.

Moir, R. J. 1968. Ruminant digestion and evolution. In *Handbook of Physiology*, Section 6, *Alimentary Canal*, Vol. 5, *Bile; Digestion; Ruminal Physiology*, ed. C. F. Code and W. Heidel, pp. 2673–2694. Washington, D. C.: American Physiological Society.

Mommsen, T. P., and Walsh, P. J. 1992. Biochemical and environmental perspectives on nitrogen metabolism in fish. *Experientia* 46: 583–593.

Montgomery, G. G., and Sunquist, M. E. 1978. Habitat selection and use by two-toed and three-toed sloths. In *Ecology of Arboreal Folivores*, ed. G. G. Montgomery, pp. 329–359. Washington, D. C.: Smithsonian Institution Press.

Montgomery, W. L., and Gerking, S. D. 1980. Marine macroalgae as foods for fishes: an evaluation of potential food quality. *Env. Biol. Fish* 5: 143–153.

Montgomery, W. L., and Pollak, P. E. 1988. Gut anatomy and pH in a Red Sea surgeonfish, *Acanthuris nigrofuscus*. *Mar. Ecol. Progr. Series* 44: 7–13.

Moreto, M., and Planas, J. 1989. Sugar and amino acid transport properties of the chicken ceca. *J. Exp. Zool.* Supp. 3: 111–116.

Morii, H. 1972. Bacteria in the stomach of marine little toothed whales. *Bull. Jpn. Soc. Sci. Fish.* 38: 1177–1183.

Morii, H. 1979. The viable counts of microorganisms, pH values, amino acid contents, ammonia contents and volatile fatty acid contents in the stomach fluid of marine little toothed whales. *Bull. Fac. Fish. Nagasaki Univ.* 47: 55–60.

Morii, H., and Kanazu, R. 1972. The free volatile fatty acids in the blood and stomach fluid from porpoise, *Neomeris phocaenoides. Bull. Jpn. Soc. Sci. Fish.* 38: 1035–1039.

Morot, M. Ch. 1882. Des pelotes stomachales des léporidés. *Mém. Soc. Centr. Méd. Vét. (Paris).* 12: 139–239.

Morris, A., and Nicholson, G. 1987. Ingestion of *Campylobacter pyloridis* causes gastritis with raised fasting gastric pH. *Am. J. Gastroenterology* 82: 192–199.

Mortensen, A., and Tindall, A. 1981. Caecal decomposition of uric acid in captive and free-ranging willow ptarmigan (*Lagopus lagopus*). *Acta Physiol. Scand.* 111: 129–133.

Mousa, H. M., Ali, K. E., and Hume, I. D. 1983. Effects of water deprivation on urea metabolism in camels, desert sheep and desert goats fed dry desert grass. *Comp. Biochem. Physiol.* 74A: 715–720.

Moyle, D. I., Hume, I. D., and Hill, D. M. 1995. Digestive performance and selective digesta retention in the long-nosed bandicoot, *Perameles nasuta*, a small omnivorous marsupial. *J. Comp. Physiol.* 164: 552–560.

Munck, B. G. 1983. Comparative aspects of amino acid transport in guinea pig, rabbit and rat small intestine. In *Intestinal Transport*, ed. M. Gilles-Baillie and R. Gillies, pp. 260–283. Berlin: Springer-Verlag.

Munck, B. G. 1989. Amino acid transport across the hen colon: interactions between leucine and lysine. *Am. J. Physiol.* 256: G532–G539.

Munck, L. K. 1993. Cotransport of 2-methyl-aminoisobutyric acid and chloride in rabbit small intestine. *Am. J. Physiol.* 265: G979–G986.

Munck, L. K., and Munck, B. G. 1990. Chloride-dependence of amino acid transport in rabbit ileum. *Biochim. Biophys. Acta* 1027: 17–20.

Murer, H., Hopfer, U., and Kinne, R. 1976. Sodium/proton antiport in brush-border-membrane vesicles isolated from rat small intestine and kidney. *Biochem. J.* 154: 597–604.

Murray, R. M., Bryant, A. M., and Leng, R. A. 1976. Rates of production of methane in the rumen and large intestine of sheep. *Br. J. Nutr.* 36: 1–14.

Murray, R. M., Marsh, H., Heinsohn, G. E., and Spain, A. V. 1977. The role of the midgut caecum and large intestine in the digestion of sea grasses by the dugong (Mammalia: Sirenia). *Comp. Biochem. Physiol.* 56A: 7–10.

Nagy, K. A. 1987. Field metabolic rate and food requirement scaling in mammals and birds. *Ecol. Monogr.* 57: 111–128.

Nagy, K. A., and Medica, P. A. 1986. Physiological ecology of desert tortoises in southern Nevada. *Herpetologica* 42: 73–92.

Nagy, K. A., and Peterson, C. C. 1988. Scaling of water flux in animals. *Univ. Calif. Publ. Zool.* 120: 1–172.

Nagy, K. A., Shoemaker, V. H., and Costa, W. R. 1976. Water, electrolyte and nitrogen budgets of jack-rabbits (*Lepus californicus*) in the Mojave Desert. *Physiol. Zool.* 49: 351–363.

Nakaya, M., Takahashi, I., Suzuki, T., Takeuchi, S., Arai, H., Wakabayashi, K., and Ito, Z. 1983. Regulation of interdigestive contractions in the denervated stomach. *Gastroenterol. Jpn.* 18: 417–427.

Nellans, H., Frizzell, R., and Schultz, S. 1973. Coupled sodium-chloride influx across the brush border of rabbit ileum. *Am. J. Physiol.* 225: 467–475.

Nelson, J. S. 1984. *Fishes of the World,* 2nd ed. New York: John Wiley & Sons.

Nicholson, T. 1982. The mode of action of intravenous pentagastrin injections on forestomach motility of sheep. *Quart. J. Exp. Physiol.* 67: 537–542.

Nicol, S. C. 1978. Rates of water turnover in marsupials and eutherians: A comparative review, with new data on the Tasmanian devil. *Aust. J. Zool.* 26: 465–473.

Nilsson, A. 1970. Gastrointestinal hormones in the holocephalian fish *Chimaera monstrosa* (L.). *Comp. Biochem. Physiol.* 32: 387–390.

Nilsson, A. 1973. Secretin-like and cholecystokinin-like activity in *Myxine glutinosa* L. *Acta Regiae Soc. Sci. Litt. Gothub.* 8: 30–32.

Nilsson, A., Carlquist, M., Jörnvall, H., and Mutt, V. 1980. Isolation and characterization of chicken secretin. *Eur. J. Biochem.* 112: 383–388.

Nilsson, A., and Fänge, R. 1969. Digestive proteases in the holocephalian fish *Chimaera monstrosa* (L.). *Comp. Biochem. Physiol.* 31: 147–165.

Nilsson, A., and Fänge, R. 1970. Digestive proteases in the cyclostome *Myxine glutinosa* (L.). *Comp. Biochem. Physiol.* 32: 237–250.

Noaillac-Depeyre, J., and Hollande, E. 1981. Evidence for somatostatin, gastrin and pancreatic polypeptide-like substances in the mucosa cells of the gut in fishes with and without stomach. *Cell Tissue Res.* 216: 192–203.

Norman, D. B., and Weishampel, D. B. 1985. Ornithopod feeding mechanisms: their bearing on the evolution of herbivory. *Am. Nat.* 126: 151–164.

Novacek, M. K. 1992. Mammalian phylogeny: shaking the tree. *Nature (Lond.)* 356: 121–125.

Nowak, R. M., and Paradiso, J. I. 1983. *Walker's Mammals of the World,* Vols. 1 and 2, 4th ed. Baltimore: John's Hopkins University Press.

Obara, Y., Dellow, D. W., and Nolan, J. V. 1991. The influence of energy-rich supplements on nitrogen kinetics in ruminants. In *Physiological Aspects of Digestion and Metabolism in Ruminants,* ed. T. Tsuda, Y. Sasaki, and R. Kawashima, pp. 515–539. San Diego: Academic Press.

Obst, B. S., and Diamond, J. M. 1989. Interspecific variation in sugar and amino acid transport by the avian cecum. *J. Exp. Zool. Suppl.* 3: 117–126.

Oftedal, O. T. 1984. Milk composition, milk yield and energy output at peak lactation: a comparative review. *Symp. Zool. Soc. Lond.* 51: 33–85.

Ogimoto, K., and Imai, S. 1981. *Atlas of Rumen Microbiology.* Tokyo: Japan Scientific Societies Press.

Ohmart, R. D., McFarland, L. Z., and Morgan, J. P. 1970. Urographic evidence that urine enters the rectum and ceca of the roadrunner (*Geococcyx californianus*) Aves. *Comp. Biochem. Physiol.* 35: 487–489.

Ohwaki, K., Hungate, R. E., Lotter, L., Hofmann, R. R., and Maloiy, G. 1974. Stomach fermentation in East African colobus monkeys in their natural state. *Appl. Micro-Biol.* 27: 713–723.

Okutani, K. 1966. Studies of chitinolytic systems in the digestive tracts of *Lateolabrax japonicus. Bull. Misaki Mar. Biol. Inst. Kyoto Univ.* 10: 1–47.

Olsen, H. M., and Madsen, H. 1943. Investigations on pseudo-rumination rabbits. *Vidensk. Meddr. Dans. Naturhist. Foren.* 107: 37–58.

Olsen, M. A., Nordøy, E. S., Blix, A. S., and Mathiesen, S. D. 1994. Functional anatomy of the gastrointestinal system of Northeastern Atlantic minke whales (*Balaenoptera acutorostrata*). *J. Zool. (Lond.)* 234: 55–74.

Oltmer, S. 1993. Die Resorption von kurzkettigen Fettsauren und Electrolyten aus dem

Caecum und Colon des Meerschweinchens. *PhD Thesis*, School of Veterinary Medicine, Hannover.

Oppel, A. 1897. *Lehrbuch der vergleichenden mikroskopischen Anatomie der Wirbeltiere*. Zweiter Teil. Schlund und Darm. Jena: Gustav Fischer.

Orskov, E. R., Fraser, C., and McDonald, I. 1971. Digestion of concentrates in sheep. Effects of rumen fermentation on barley and maize diets on protein digestion. *Br. J. Nutr.* 26: 477–486.

Orton, R. K., Hume, I. D., and Leng, R. A. 1985. Effects of exercise and level of dietary protein on digestive function in horses. *Equine Vet. J.* 17: 386–390.

Osawa, R., Bird, P. S., Harbrow, D. J., Ogimoto, K., and Seymour, G. J. 1993. Microbiological studies of the intestinal microflora of the koala, *Phacolarctos cinereus* I. Colonization of the caecal wall by tannin-protein-complex-degrading enterobacteria. *Aust. J. Zool.* 41: 599–609.

Osawa, R., Blanchard, W. H., and O'Callaghan, P. G. 1993. Microbiological studies of the intestinal microflora of the koala, *Phascolarctos cinereus. Aust. J. Zool.* 41: 611–620.

Osborne, T. B., and Mendel, L. B. 1911. Feeding experiments with isolated food substances. Part II. Carnegie Institute of Washington, Pub. No. 156.

Östberg, Y., Van Noorden, S., Pearse, A. G. E., and Thomas, N. W. 1976. Cytochemical, immunofluorescence, and ultrastructural investigations on polypeptide hormone containing cells in the intestinal mucosa of a cyclostome, *Myxine glutinosa. Gen. Comp. Endocr.* 28: 213–227.

Ostrom, J. H. 1963. Further comments on herbivorous lizards. *Evolution* 17: 368–369.

Ottaviani, G., and Tazzi, A. 1977. The lymphatic system. In *Biology of the Reptilia*, Vol. 6, ed. C. G. Gans and T. S. Parsons, pp. 315–462. New York: Academic Press.

Owaga, M. L. 1975. The feeding ecology of wildebeest and zebra in Athi-Kaputei plains. *E. Afr. Wildl. J.* 13: 375–383.

Owen, R. 1835. On the sacculated form of stomach as it exists in the genus *Semnopithecus*, F. Cuv. *Trans. Zool. Soc. Lond.* 1: 65–70.

Pairet, M., Bouyssou, T., and Ruckebusch, Y. 1986. Colonic formation of soft feces in rabbits: a role for endogenous prostaglandins. *Am. J. Physiol.* 250: G302–G308.

Pajor, A. M., Hirayama, B. A., and Wright, E. M. 1992. Molecular biology approaches to comparative study of Na-glucose transport. *Am. J. Physiol.* 263: R489–R495.

Pappenheimer, J. R. 1990. Paracellular intestinal absorption of glucose, creatinine and mannitol in normal animals: relation to body size and applications to man. *Am. J. Physiol.* 259: G290–G299.

Pappenheimer, J. R. 1993. On the coupling of membrane digestion with absorption of sugars and amino acids. *Am. J. Physiol.* 265: G409–G417.

Paris, H., Muat, J. C., and Castilla, C. 1977. Etude des acides gras volatils dans l'intestin trois especes de Poissons Teleosteens. *C. R. Seances Soc. Biol. Fil.* 171: 1297–1301.

Parker, D. S. 1976. The measurement of production rates of volatile fatty acids in the caecum of the conscious rabbit. *Br. J. Nutr.* 36: 61–70.

Parker, J. W., and Clawson, A. J. 1967. Influence of level of total feed intake on digestibility, rate of passage and energetic efficiency of reproduction in swine. *J. Anim. Sci.* 26: 485–489.

Parker, T. J. 1885. On the intestinal spiral valve in the genus *Raia. Trans. Zool. Soc. Lond.* 11: 49–61.

Parmelee, J. T., and Renfro, J. L. 1983. Esophageal desalination of seawater in flounder: role of active sodium transport. *Am. J. Physiol.* 245: R888–R893.

Parmenter, R. R. 1981. Digestive turnover rates in freshwater turtles: the influence of temperature and body size. *Comp. Biochem. Physiol.* 70A: 235–238.

Parra, R. 1978. Comparison of foregut and hindgut fermentation in herbivores. In *The Ecology of Arboreal Folivores*, ed. G. G. Montgomery, pp. 205–229. Washington, D. C.: Smithsonian Institution Press.

Parsons, T. S., and Cameron, J. E. 1977. Internal relief of the digestive tract. In *Biology of the Reptilia*, Vol. 6, ed. C. G. Gans and T. S. Parsons, pp. 159–224. New York: Academic Press.

Patton, J. S. 1975. High levels of pancreatic nonspecific lipase in rattlesnake and leopard shark. *Lipids* 10: 562–564.

Patton, J. S., and Benson, A. A. 1975. A comparative study of wax ester digestion in fish. *Comp. Biochem. Physiol.* 52B: 111–116.

Patton, J. S., Nevenzel, J. C., and Benson, A. A. 1975. Specificity of digestive lipases in hydrolysis of wax esters and triglycerides studied in anchovies and other selected fish. *Lipids* 10: 575–583.

Pearse, A. G. E. 1969. The cytochemistry and ultrastructure of polypeptide hormone producing cells of the APUD series and the embryological, physiological and pathological implications of the concept. *J. Histochem. Cytochem.* 17: 303–313.

Pehrson, A. 1983. Caecotrophy in caged mountain hares (*Lepustimidus*). *J. Zool.* 199: 563–574.

Pel, R. 1989. Microbial interactions in anaerobic chitin-degrading mixed cultures. Ph.D. Thesis, Rijksuniversiteit, Groningen.

Peng, Y., Zhang, Y., Ye, Z., and Liu, S. 1983. Study of the stomachs of the snub-nosed monkeys (*Rhinopithecus*). *Zool. Res.* 4: 168–174.

Pennington, R. J., 1952. The metabolism of short-chain fatty acids in the sheep. 1. Fatty acid utilization and ketone body production by rumen epithelium and other tissues. *Biochem. J.* 51: 251–258.

Pennington, R. J., and Sutherland, T. M. 1956. The metabolism of short-chain fatty acids in the sheep. 4. The pathway of propionate metabolism in rumen epithelial tissue. *Biochem. J.* 63: 618–628.

Penry, D. L., and Jumars, P. A. 1987. Modeling animal guts as chemical reactors. *Am. Nat.* 129: 69–96.

Pernkopf, E. 1930. Beiträge zur vergleichenden Anatomie des vertebratenmagens. *Z. Anat.* 91: 329–390.

Pernkopf, E. 1937. B. Die Vergleichung der vercheidenen Formtypen des Vorderdarmes der Kranioten. In *Handbuch der Vergleichenden den Anatomie der Wirbeltiere*, Vol. 3, ed. L. Bolk, E. Goppert, E. Kallius, and W. Lubosh, pp. 477–562. Berlin: Urban und Schwartzenberg.

Pernkopf, E., and Lehner, J. 1937. A Vergleichende Beschreibung des Vorderdarmes bei den einzelnen Klassen der Kranioten. In *Handbuch der Vergleichenden den Anatomie der Wirbelttiere*, Vol. 3, ed. L. Bolk, E. Goppert, E. Kallius, and W. Lubosch, pp. 349–476. Berlin: Urban and Scwarzenberg.

Perrin, M. R., and Kokkin, M. J. 1986. Comparative gastric anatomy of *Cricetomys gambianus* and *Saccostomus campostris* (Cricetomyinae) in relation to *Mystromys albicaudatus* (Cricetinae). *S. Afr. J. Zool.* 15: 22–33.

Peterson, C. C., Nagy, K. A., and Diamond, J. 1990. Sustained metabolic scope. *Proc. Natl. Acad. Sci.* 87: 2324–2328.

Phaneuf, L. P., and Ruckebusch, Y. 1983. Physiological, pharmacological and therapeutic aspects of some gastrointestinal disorders in the horse. In: *Veterinary Pharmacology and Toxicology*, ed. Y. Ruckebusch, P. L. Toutain, and G. D. Koritz, pp. 371–380. Lancaster, U.K.: MTP Press.

Phillips, G. D. 1961a. Physiological comparisons of European and Zebu steers. I. Digestibility and retention times of food and rate of fermentation of rumen contents. *Res. Vet. Sci.* 2: 202–208.

Phillips, G. D. 1961b. Physiological comparisons of European and Zebu steers. II. Effects of restricted water intake. *Res. Vet. Sci.* 2: 209–216.

Phillips, J. E., Audsley, N., Lechleitner, R., Thomson, J., Meredith, J., and Chamberlin, M. 1988. Some major transport mechanisms of insect absorptive epithelia. *Comp. Biochem. Physiol.* 90A: 643–650.

Phillips, J. E., Hanrahan, J., Chamberlain, M., and Thomson, B. 1986. Mechanisms and control of reabsorption in insect hindgut. *Adv. Insect Physiol.* 19: 329–342.

Phillips, J. E., Thomson, R. B., Audsley, N., Peach, J. L., and Stagg, A. P. 1994. Mechanisms of acid-base transport in the locust excretory system. *Physiol. Zool.* 67: 95–119.

Phillips, S. F. 1988. The ileocolonic junction: physiology and clinical implications. *Viewpoints Digest. Dis.* 20 (5): 17–20.

Phillipson, A. T. 1947. The production of fatty acids in the alimentary tract of the dog. *J. Exp. Biol.* 23: 346–349.

Phillipson, A. T. 1970. Ruminant digestion. In *Dukes' Physiology of Domestic Animals*, 8th ed., ed. M. J. Swenson, pp. 424–483. Ithaca, New York: Cornell University Press.

Phillipson, A. T. 1977. Ruminant digestion. In *Dukes' Physiology of Domestic Animals*, 9th ed., ed. M. J. Swenson, pp. 250–286. Ithaca, New York: Cornell University Press.

Phillipson, A. T., and Ash, R. W. 1965. Physiological mechanisms affecting the flow of digesta in ruminants. In *Physiology of Digestion in the Ruminant*, ed. R. W. Dougherty, pp. 97–107. Washington, D. C.: Butterworths.

Physiology of the Gastrointestinal Tract, 1994. 3rd ed., Vol. 1, ed. L. R. Johnson, D. H. Alpers, J. Christensen, E. D. Jacobson, and J. H. Walsh, New York: Raven Press.

Pickard, D. W., and Stevens, C. E. 1972. Digesta flow through the rabbit large intestine. *Am. J. Physiol.* 222: 1161–1166.

Place, A. R. 1990. Chitin digestion in nestling Leach's storm petrels *Oceanodroma leucorhoa. Bull. Mt. Desert Island. Biol. Lab.* 20: 139–142.

Place, A. R. 1992a. Comparative aspects of lipid digestion and absorption: physiological correlates of wax ester digestion. *Am. J. Physiol.* 32: R464–R471.

Place, A. R. 1992b. Bile is essential for lipid assimilation in Leach's storm petrel *Oceanodroma leucorhoa. Am. J. Physiol.* 263: R389–R399.

Place, A. R., and Roby, D. D. 1986. Assimilation and deposition of dietary fatty alcohols in Leach's storm petrel, *Oceanodroma leucorhoa. J. Exp. Zool.* 240: 149–161.

Planas, J. M., Ferrer, R., and Moreto, M. 1987. Relation between α-methyl-D-glucoside influx and brush border surface area in enterocytes from chicken cecum and jejunum. *Pflugers Arch.* 408: 515–518.

Planas, J. M., Villa, C., Ferrer, R., and Moreto, M. 1986. Hexose transport by chicken cecum during development. *Plfügers Arch.* 407: 216–220.

Pochart, P., Lemann, F., Flourie, B., Pellier, P., Goderel, I., and Rambaud, J. C. 1993. Pyxigraphic sampling to enumerate methanogens and anaerobes in the right colon of healthy humans. *Gastroenterology.* 105: 1281–1285.

Pollak, P. E., and Montgomery, W. L. 1994. Giant bacterium (*Epulopiscium fishelsoni*) influences digestive enzyme activity of an herbivorous surgeonfish (*Acanyhurus nigrofuscus*). *Comp. Biochem. Physiol.* 108: 657–662.

Potter, G. D. 1989. Development of colonic function. In *Human Gastrointestinal Development* ed. E. Lebanthal, pp. 548–558. New York: Raven Press.

Pough, F. H. 1973. Lizard energetics and diet. *Ecology* 54: 837–844.

Powell, D. W. 1986. Ion and water transport in the intestine. In *Physiology of Membrane Disorders*, 2nd ed., ed. T. E. Andreoli, J. F. Hoffman, D. D. Fanestil, and S. G. Schultz, pp. 559–596. New York: Plenum Publishing Corporation.

Powell, D. W., and Fan, C.-C. 1984. Coupled NaCl transport: cotransport or parallel ion exchange? In *Mechanisms of Intestinal Electrolyte Transport and Regulation by Calciuam*, ed. M. Donowitz and E. W. G. Sharp, pp. 13–26. New York: Alan R. Liss, Inc.

Powell, D. W., Johnson, P., Bryson, J., Orlando, R., and Fan, C.-C. 1982. Effect of phenolphthalein on monkey intestinal water and electrolyte transport. *Am. J. Physiol.* 243: G268–G275.

Powell, D. W., and Tapper, E. J. 1979. Autonomic control of intestinal electrolyte transport. In *Frontiers of Knowledge in the Diarrheal Diseases*, ed. H. J. Janowitz and D. B. Sachar, pp. 37–52. Upper Montclair, New Jersey: Projects in Health, Inc.

Prahl, J. W., and Neurath, H. 1966. Pancreatic enzymes of the spiny pacific dogfish. I. Cationic chymotrypsinogen and chymotrypsin. *Biochemistry* 5: 2131–2146.

Prejs, A., and Plaszcyzyk, M. 1977. Relationships between food and cellulase activity in freshwater fishes. *J. Fish Biol.* 11: 447–452.

Prins, R. A. 1991. Rumen ciliates and their function. In *Rumen Microbial Metabolism and Microbial Digestion*, ed. J. P. Jouany, pp. 39–52. Paris: Institut National de la Recherche Agronomique.

Prior, R. L., Hintz, H. F., Lowe, J. E., and Visek, W. J. 1974. Urea recycling and metabolism of ponies. *J. Anim. Sci.* 38: 565–571.

Rahm, S. 1980. Makroskopische und mikroskopische Untersuchungen des Magen-Darm Traktes beim Klipp-, Busch- und Baumschliefer. Thesis, Zurich Univ.

Raikhlin, N. T., and Kvetnoy, I. M. 1976. Melantonin and entreochromaffin cells. *Acta Histochem.* 55: 19–24.

Rao, M., Dubinsky, W., Vosburgh, E., Field, M., and Frizzell, R. 1981. Sodium proton antiport in intestinal brush border vesicles of the flounder, *Pseudopleuronectes americanus. Bull. Mt. Desert Island Biol. Lab.* 21: 99–103.

Rapley, S., Lewis, W. H. P., and Harris, H. 1971. Tissue distribution, substrate specificities, and molecular sizes of human peptidases determined by separate gene loci. *Ann. Hum. Genet.* 34: 307–320.

Rauws, E. A., Langenberg, J. W., and Houthoff, H. J. 1988. *Campylobacter pyloridis*-associated chronic active antral gastritis: a prospective study of its prevalence and the effects of anti-bacterial and anti-ulcer treatment. *Gastroenterology* 94: 33–40.

Raven, H. C. 1950. *The Anatomy of the Gorilla.* New York: Columbia University Press.

Rawdon, B. B. 1984. Gastrointestinal hormones in birds: morphological, chemical, and developmental aspects. *J. Exp. Zoology* 232: 659–670.

Rawdon, B. B., and Andrew, A. 1981. An immunocytochemical survey of endocrine cells in the gastrointestinal tract of chicks at hatching. *Cell Tissue Res.* 220: 279–292.

Rechkemmer, G. 1988. Mechanismen und Regulation des Elektrolyttransports in verschiedenen Dickdarmabschnitten des Meerschweinchens. *Habilitationsschrift*, pp. 1–183, Tierarztliche Hochschule, Hannover.

Rechkemmer, G. 1991. Transport of weak electrolytes. In *Handbook of Physiology*, Section 6: *The Gastrointestinal System*, Vol. 4: *Intestinal Absorption and Secretion*, ed. S. G. Schultz, M. Field, R. A. Frizzell, and B. B. Rauner, pp. 371–388. Washington, D. C.: American Physiological Society.

Rechkemmer, G., and Engelhardt, W. v. 1988. Concentration and pH dependence of the short-chain fatty acid absorption in the proximal and distal colon of guinea pig (*Cavia porcellus*). *Comp. Biochem. Physiol.* 91A: 659–663.

Rechkemmer, G., and Engelhardt, W. v. 1993. Absorption and secretion of electrolytes and short-chain fatty acids in the guinea pig large intestine. In *Advances in Comparative and Environmental Physiology*, Vol. 16, ed. W. Clauss, pp. 139–167. Berlin: Springer-Verlag.

Rechkemmer, G., Wahl, M., Kuschinsky, W., and Engelhardt, W. v. 1986. pH microclimate at the luminal surface of the intestinal mucosa of guinea pig and rat. *Pflüegers Arch.* 407: 33–44.

Reddy, B. S. 1983. Experimental research on dietary lipids and colon cancer. In: *Dietary Fats and Health*, ed. E. G. Perkins and W. J. Visek, pp. 741–760. Champaign, Illinois: American Oil Chemists Society.

Reeck, G. R., Winter, W. P., and Neurath, H. 1970. Pancreatic enzymes of the African lungfish *Protopterus aethiopicus*. *Biochemistry* 9: 1398–1403.

Reeder, W. G. 1964. The digestive system. In *Physiology of the Amphibia*, Vol. 1, ed. J. A. Moore, pp. 99–149. New York: Academic Press.

Reeds, P. J. 1988. Regulation of protein metabolism. In *Control and Regulation of Animal Growth*, EAAP Publication No. 36, ed. J. F. Quirke and H. Schmid, pp. 25–44. Wageningen: Center for Agricultural Publishing and Documentation.

Reeds, P. J., Fuller, M. F., and Nicholson, B. A. 1985. Metabolic basis of energy expenditure with particular reference to protein. In *Substrate and Energy Metabolism*, ed. J. S. Garrow and D. Halliday, pp. 46–57. London: John Libbey.

Reenstra, W. W., Bettencourt, J. D., and Forte, J. G. 1987. Mechanisms of active Cl⁻ secretion by frog gastric mucosa. *Am. J. Physiol.* 252: 543–547.

Regal, P. J. 1977. Ecology and evolution of flowering plant dominance. *Science* 196: 622–629.

Regoeczi, E., Irons, L., Koj, A., and McFarlane, A. S. 1965. Isotopic studies of urea metabolism in rabbits. *Biochem. J.* 95: 521–535.

Reid, C. S. W. 1963. Diet and the motility of the forestomachs of sheep. *Proc. N. Z. Soc. Anim. Prod.* 23: 169–188.

Reid, C. S. W., and Cornwall, J. B. 1959. The mechanical activity of the reticulo-rumen of cattle. *Proc. N. Z. Soc. Anim. Prod.* 19: 23–25.

Reinecke, M., Carraway, R. E., Falkmer, S., Feurle, G. E., and Forssmann, W. G. 1980. Occurrence of neurotensin-immunoreactive cells in the digestive tract of lower vertebrates and deuterostomian invertebrates. *Cell Tissue Res.* 212: 173–183.

Reinecke, M., Schlüter, P., Yanaihara, N., and Forssmann, W. G. 1981. VIP immunoreactivity in enteric nerves and endocrine cells of the vertebrate gut. *Peptides* 2 (Suppl. 2): 149–156.

Remond, D., Chaise, J. P., Delval, E., and Poncet, C. 1992. Net flux of metabolites across the ruminal wall of sheep fed twice a day with orchard grass. *J. Anim. Sci.* 71: 2529–2538.

Rérat, A. 1978. Digestion and absorption of carbohydrates and nitrogenous matters in the hindgut of the omnivorous nonruminant animal. *J. Anim. Sci.* 46: 1808–1837.

Reshkin, S. J., and Ahearn, G. A. 1987. Basolateral glucose transport by intestine of teleost *Oreochromis mossambicus*. *Am. J. Physiol.* 252: R573–586.

Reshkin, S. J., and Ahearn, G. A. 1991. Intestinal glycyl-L-phenylalanine and L-phenylalanine transport in a euryhaline teleost. *Am. J. Physiol.* 260: R563–R569.

Reshkin, S. J., Vilella, S., Ahearn, G. A., and Storelli, C. 1989. Basolateral inositol transport by intestines of carnivorous and herbivorous teleosts. *Am. J. Physiol.* 256: G509–G516.

Reshkin, S. J., Vilella, S., Cassano, G., Ahearn, G. A., and Storelli, C. 1988. Basolateral amino acid and glucose transport by the intestine of the teleost (*Anguilla anguilla*). *Comp. Biochem. Physiol.* 91A: 779–788.

Reuss, L. 1991. Salt and water transport by gallbladder epithelium. In *Handbook of Physiology*, Section 6: *The Gastrointestinal System*, Vol. 4: *Intestinal Absorption and Secretion*, ed. S. G. Schultz, M. Field, R. A. Frizzell, and B. B. Rauner, pp. 303–322. Washington, D. C.: American Physiological Society.

Reynolds, D. A., Rajendran, V. M., and Binder, H. J. 1993. Bicarbonate-stimulated [^{14}C]butyrate transport in basolateral membrane vesicles of rat distal colon. *Gastroenterology* 105: 725–732.

Rice, D. W., and Wolman, A. A. 1990. The stomach of *Kogia breviceps*. *J. Mamm.* 71: 237–242.

Richard, P.-B. 1959. La caecotrophie chez le castor du Rhône (castor-fiber). *C. R. Séances Acad. Sci. (Paris)* 248: 1424–1426.

Richardson, K. C., and Creed, K. 1981. Aspects of haustral motility of the stomach of macropods. *Proc. Aust. Physiol. Pharmacol. Soc.* 12: 24P.

Richardson, K. C., and Wooller, R. 1990. Adaptations of the alimentary tracts of some Australian lorikeets to a diet of pollen and nectra. *Aust. J. Zool.* 38: 581–586.

Richardson, K. C., and Wyburn, R. S. 1983. Electromyographic events in the stomach of a small kangaroo, the tammer wallaby (*Macropus eugenii*) *J. Physiol. (Lond.)* 342: 453–463.

Richardson, K. C., and Wyburn, R. S. 1988. Electromyography of the stomach and small intestine of the tammer wallaby, *Macropus eugennii*, and the quokka, *Setonix brachyurus*. *Aust. J. Zool.* 36: 363–371.

Ridgway, S. H. 1972. *Mammals of the Sea*. Springfield: Charles C. Thomas.

Rimmer, D. W., and Wiebe, W. J. 1987. Fermentative microbial digestion in herbivorous fishes. *J. Fish Biol.* 31: 229–236.

Roberts, M. C. 1975. The development and distribution of mucosal enzymes in the small intestine of the foetus and the young foal. *J. Reprod. Fert.* Suppl. 23: 717–723.

Robertshaw, D. 1984. *The Evolution of Homeothermy: from Dinosaurs to Man*. Special Report 329, Agricultural Experiment Station, University of Missouri-Columbia.

Robertson, D. R. 1982. Fish feces as fish food on a Pacific coral reef. *Mar. Ecol. Prog. Ser.* 7: 253–265.

Robinson, I. M., Allison, M. J., and Bucklin, J. A. 1981. Characterization of the cecal bacteria of normal pigs. *Appl. Environ. Microbiol.* 41: 950–955.

Robinson, G. B. 1960. The hydrolysis of dipeptides by rat intestinal extracts. *Biochim. Biophys. Acta* 44: 386–387.

Roby, D. D., Place, A. R., and Ricklefs, R. E. 1986. Assimilation and deposition of wax esters in planktivorous sea birds. *J. Exp. Zool.* 238: 29–41.

Roche, M., and Ruckebusch, Y. 1978. A basic relationship between gastric and duodenal motilities in chickens. *Am. J. Physiol.* 235: E670–E677.

Roediger, W. E. W. 1991. Cellular metabolism of short-chain fatty acids in colonic epithelial cells. In *Short-Chain Fatty Acids: Metabolism and Clinical Importance*, ed. A. F. Roche, pp. 67–71. Columbus, Ohio: Ross Laboratories.

Roediger, W. E. W., Duncan, A., Kapaniris, O., and Millard, S. 1993. Reducing sulfur compounds of the colon impair colonocyte nutrition: implications for ulcerative colitis. *Gastroenterology* 104: 802–809.

Roman, C., and Gonella, J. 1981. Extrinsic control of digestive tract motility. In *Physiology of the Gastrointestinal Tract*, Vol. 1, ed. L. R. Johnson, J. Christensen, M. I.

Grossman, E. D. Jacobson, and S. G. Schultz, pp. 289–334. New York: Raven Press.

Rombeau, J. L. 1991. Uses of short-chain fatty acids in experimental postoperative conditions. In *Short-Chain Fatty Acids: Metabolism and Clinical Importance*, ed. A. F. Roche, pp. 93–96. Columbus, Ohio: Ross Laboratories.

Rombout, J. H. W. M., and Reinecke, M. 1984. Immunohistochemical localization of neuropeptide hormones in endocrine cells and nerves of the gut of a stomachless teleost fish, *Barbusconchonius* (Cyprinidae). *Cell Tissue Res.* 237: 57–65.

Romer, A. S. 1966. *Vertebrate Paleontology*, 3rd ed. Chicago: University of Chicago Press.

Rose, C. J., Hume, I. D., and Farrell, D. J. 1987. Fibre digestion and volatile fatty acid production in domestic and feral pigs. In *Recent Advances in Animal Nutrition in Australia 1987*, ed. D. J. Farrell, pp. 347–360. Armidale, Australia: University of New England Press.

Rose, R. C. 1987. Intestinal absorption of water-soluble vitamins. In *Physiology of the Gastrointestinal Tract*, ed. L. R. Johnson, J. R. Christensen, M. J. Jackson, E. D. Jacobson, and J. H. Walsh, pp. 1581–1596. New York: Raven Press.

Rouk, C. S., and Glass, B. P. 1970. Comparative gastric histology of five North and Central American bats. *J. Mammal.* 51: 455–490.

Rübsamen, K., and Engelhardt, W. v. 1978. Bicarbonate secretion and solute absorption in forestomach of the llama. *Am. J. Physiol.* 234: E1–E6.

Rübsamen, K., and Engelhardt, W. v. 1981. Absorption of Na, H ions and short-chain fatty acids from the sheep colon. *Pflügers Arch. Eur. J. Physiol.* 391: 141–146.

Rübsamen, K., Hume, I. D., and Engelhardt, W. v. 1982. Physiology of the rock hyrax. *Comp. Biochem. Physiol.* 72A: 271–277.

Ruckebusch, Y. 1971. The effects of pentagastrin on the motility of the ruminant stomach. *Experientia* 27: 1185–1186.

Ruckebusch, Y. 1975. Motility of the ruminant stomach associated with states of sleep. In *Digestion and Metabolism in the Ruminant*, ed. I. W. McDonald and A. C. I. Warner, pp. 77–90. Armidale: The University of New England Publishing Unit.

Ruckebusch, Y. 1981. Motor functions of the intestine. In *Advances in Veterinary Science and Comparative Medicine*, Vol. 25, ed. C. E. Cornelius and C. F. Simpson, pp. 345–369. New York: Academic Press.

Ruckebusch, Y., Fargeas, J., and Dumas, J.-P. 1970. Recherchessur le comportement alimentaire des Ruminants. IX.-La mastication mérycique. *Rev. Méd. Vét.* 121: 345–357.

Ruckebusch, Y., and Fioramonti, J. 1976. The fusus coli of the rabbit as a pacemaker area. *Experientia* 32: 1023–1024.

Ruckebusch, Y., and Hörnicke, H. 1977. Motility of the rabbit colon and cecotrophy. *Physiol. Behav.* 18: 871–888.

Ruckebusch, Y., and Tomov, T. 1973. The sequential contractions of the rumen associated with eructation in sheep. *J. Physiol. (Lond.)* 235: 447–458.

Ruckebusch, Y., and Vigroux, P. 1974. Etude électromyographiquede la motricité du caecum chez le Cheval (*Equus caballus*). *C.R. Séances Soc. Biol. (Paris)* 168: 887–892.

Ruppin, H., Bur-Meir, S., Soergel, K., Wood, C., and Schmitt, M. G. 1980. Absorption of short-chain fatty acids by the colon. *Gastroenterology* 78: 1500–1507.

Ryan, C. A. 1965. Chicken chymotrypsin and turkey trypsin. Part 1: Purification. *Arch. Biochem. Biophys.* 110: 169–174.

Sack, W. O., and Ballantyne, J. H. 1965. Anatomical observations on a musk-ox calf

(*Ovibos moschatus*) with particular reference to thoracic and abdominal topography. *Can. J. Zool.* 43: 1033–1047.

Said, H. M., Mock, D. M., and Collins, J. C. 1989. Regulation of intestinal biotin transport in the rat: effect of biotin deficiency and supplementation. *Am. J. Physiol.* 261: R94–R97.

Said, H. M., and Mohammadkhani, R. 1993. Uptake of riboflavin across the brush border membrane of rat intestine: regulation of dietary vitamin levels. *Gastroenterology* 105: 1294–1298.

Saito, M., Murakami, E., Nishida, T., Fujisawa, Y., and Suda, M. 1975. Circadian rhythms of digestive enzymes in the small intestine of rats. 1. Patterns of rhythms in various regions of the small intestine, *J. Biochem.* 78: 275–480.

Sakaguchi, E., Becker, G., Rechkemmer, G., and Engelhardt, W. v. 1985. Volume, solute concentrations and production of short-chain fatty acids in the caecum and upper colon of the guinea pig. *Z. Tierphysiol. Tierernahrg. u. Futtermittelkde.* 54: 276–285.

Sakaguchi, E., Heller, R., Becker, G., and Engelhardt, W. v. 1986. Retention of digesta in the gastrointestinal tract of the guinea pig. *J. Anim. Physiol. Anim. Nutr.* 55: 434–50.

Sakaguchi, E., and Hume, I. D. 1990. Digesta retention and fibre digestion in brushtail possums, ringtail possums and rabbits. *Comp. Biochem. Physiol.* 96A: 351–354.

Sakaguchi, E., Itoh, H., Uchida, S., and Horigame, T. 1987. Comparison of fibre digestion and digesta retention time between rabbits, guinea-pigs, rats and hamsters. *Br. J. Nutr.* 58: 149–158.

Sakaguchi, E., Kaizu, K., and Nakamichi, M. 1992. Fibre digestion and fibre retention from different physical forms of the feed in the rabbit. *Comp. Biochem. Physiol.* 102A: 559–563.

Sakaguchi, E., and Matsumoto, T. 1985. Effect of monensin on feed utilization and gastrointestinal fermentation in the hamster (*Mesocricetus auratus*). *Br. J. Nutr.* 54: 147–155.

Sakaguchi, E., and Nabata, A. 1992. Comparison of fibre digestion and digesta retention time between nutrias (*Myocaster coypus*) and guinea-pigs (*Cavia porcellus*). *Comp. Biochem. Physiol.* 103A: 601–604.

Sakaguchi, E., Nippashi, K., and Endoh, G. 1992. Digesta retention and fibre digestion in maras (*Dolichotis patagonum*) and guinea-pigs. *Comp. Biochem. Physiol.* 101A: 867–870.

Sakaguchi, E., and Ohmura, S. 1992. Fibre digestion and digesta retention time in guinea-pigs (*Cavia porcellus*), degus (*Octodon degus*) and leaf-eared mice (*Phyllotis darwini*). *Comp. Biochem. Physiol.* 103A; 787–791.

Sakata, T. 1987. Stimulatory effect of short-chain fatty acids on epithelial cell proliferation in the rat intestine: a possible explanation for trophic effects of fermentable fibre, gut microbes and luminal trophic factors. *Br. J. Nutr.* 58: 95–103.

Sakata, T., and Yajima, T. 1984. Influence of short chain fatty acids on the epithelial cell division of digestive tract. *Quart. J. Exp. Physiol.* 69: 639–648.

Salanitro, J. P., Blake, I. G., Muirhead, P. A., Maglio, M., and Goodman, J. R. 1978. Bacteria isolated from the duoenum, ileum, and cecum of young chicks. *Appl. Environ. Microbiol.* 35: 782–790.

Salsbury, C. M., and Armatage, K. B. 1994. Resting and field metabolic rates of adult male yellow-bellied marmots, *Marmota flavicentris*. *Comp. Biochem. Physiol.* 108A: 579–588.

Salse, A., Crampes, F., and Raynaud, P. 1977. Measurement of N urea dietary value by intracaecal perfusion in rabbit. *Ann. Biol. Anim. Biochim. Biophys.* 17: 559–566.

Sanderson, I. R., and Walker, W. A. 1993. Uptake and transport of macromolecules by the intestine: possible role in clinical disorders (an update). *Gastroenterology* 104: 622–639.

Samulitus-Dos Santo, B. K., Goda, T., and Koldovsky, O. 1992. Dietary-induced increases in disaccharidase activity in the rat jejunum. *J. Nutr.* 67: 267–268.

Sanson, G. D. 1989. Morphological adaptations of teeth to diets and feeding in the Macropodoidea. In *Kangaroos, Wallabies and Rat-kangaroos*, ed. G. Grigg, P. Jarman, and I. Hume, pp. 151–168. Sydney: Surrey Beatty.

Sarna, S. K., Bardakjian, B. L., Waterfall, W. E., and Lind, J. F. 1980. Human colonic electrical control activity (ECA). *Gastroenterology* 78: 1526–1536.

Sauer, W. C., Stothers, S. C., and Parker, R. J. 1977. Apparent and true availabilities of amino acids in wheat and milling by-products for growing pigs. *Can. J. Anim. Sci.* 57: 775–784.

Savage, D. C. 1977. Interactions between the host and its microbes. In *Microbial Ecology of the Gut*, ed. R. T. J. Clarke and T. Bauchop, pp. 277–310. New York: Academic Press.

Savage, D. C. 1986. Gastrointestinal microflora in mammalian nutrition. *Annu. Rev. Nutr.* 6: 155–178.

Scapin, S., and Lambert-Gardini, S. 1979. Digestive enzymes in the exocrine pancreas of the frog *Rana esculenta*. *Comp. Biochem. Physiol.* 62A: 691–697.

Schalk, A. F., and Amadon, R. S. 1921. Gastric motility studies in the stomach of the goat and horse. *J. Am. Vet. Med. Assoc.* 59: 151–172.

Schalk, A. F., and Amadon, R. S. 1928. Physiology of the ruminant stomach. *Bull. 216, North Dakota Agric. Exp. Station*, Fargo.

Schmidt-Nielsen, B., Schmidt-Nielsen, K., Houpt, T. R., and Jarnum, S. A. 1957. Urea excretion in the camel. *Am. J. Physiol.* 188: 477–483.

Schmidt-Nielsen, K. 1984. *Scaling: Why is Animal Size so Important?* Cambridge: Cambridge University Press.

Schmidt-Nielsen, K. 1990. Food and energy. In *Animal Physiology: Adaptation and Environment,* 4th ed., pp. 138–159. New York: Cambridge University Press.

Schmidt-Nielsen, K., Crawford, E. C., Newsome, A. E., Rawson, K. S., and Hammal, H. T. 1967. Metabolic rate of camels: effects of body temperature and dehydration. *Am. J. Physiol.* 212: 341–346.

Schneyer, L. H., and Schneyer, C. A. 1967. Inorganic composition of saliva. In *Handbook of Physiology*, Section 6, *Alimentary Canal*, Vol. 2, *Secretion*, ed. C. F. Code and W. Heidel, pp. 497–530. Washington, D. C.: American Physiological Society.

Schurg, W. A., Frei, D. L., Cheeke, P. R., and Holtan, D. W. 1977. Utilization of whole corn plant pellets by horses and rabbits. *J. Anim. Sci.* 45: 1317–1321.

Seino, Y., Porte, D., and Smith, P. H. 1979. Immunohistochemical localization of somatostatin-containing cells in the intestinal tract: a comparative study. *Gen. Comp. Endocr.* 38: 229–233.

Seino, Y., Porte, D., Yanaihara, N., and Smith, P. H. 1979. Immunocytochemical localization of motilin-containing cells in the intestines of several vertebrate species and a comparison of antisera against natural and synthetic motilin. *Gen. Comp. Endocr.* 38: 234–237.

Seiki, H., and Taga, N. 1963. Microbiological studies on the decomposition of chitin in marine environments. V: Chitinoclastic bacteria as symbionts. *J. Oceanog. Soc. Japan*, 19: 38.

Self, R. F. L., Jumars, P. A., and Mayer, L. M. In vitro amino acid and glucose uptake rates across the wall of a surface deposit feeder. *J. Exp. Mar. Biol. Ecol.* (In Press).

Sellin, J. H. 1992. Short-chain fatty acid (SCFA) transport in rabbit proximal colon (PC): regulation by pH and Na. *Gastroenterology* 100: A702.

Sellin, J. H., and DeSoignie, R. 1984. Rabbit proximal colon: a distinct transport epithelium. *Am. J. Physiol.* 246: G603–G610.

Sellin, J. H., and DeSoignie, R. 1990. Short-chain fatty acid absorption in rabbit colon in vitro. *Gastroenterology* 99: 676–683.

Sellers, A. F., Lowe, J. E., and Brondum, J. 1979. Motor events in equine large colon. *Am. J. Physiol.* 237: E457–E464.

Sellers, A. F., Lowe, J. E., Drost, C. J., Rendano, V. T., Georgi, J. R., and Roberts, M. C. 1982. Retropulsion-propulsion in equine large colon. *Am. J. Vet. Res.* 43: 390–396.

Sellers, A. F., and Stevens, C. E. 1960. Pressure events in bovine esophagus and reticulorumen associated with eructation, deglutition and regurgitation. *Am. J. Physiol.* 199: 598–602.

Sellers, A. F., and Stevens, C. E. 1966. Motor functions of the ruminant forestomach. *Physiol. Rev.* 46: 634–661.

Sellers, A. F., and Titchen, D. A. 1959. Responses to localized distension of the oesophagus in decerebrate sheep. *Nature (Lond.)* 184: 645.

Shcherbina, M. A., and Kazlauskene, O. P. 1971. The reaction of the medium and the rate of absorption of nutrients in the intestines of the carp (*Cyprinus carpio* [L.]). *J. Ichthyol.* 11: 81–85.

Shcherbina, M. A., Mochul'skaya, V. F., and Erman, Y. Z. 1970. Study of the digestibility of the nutrients in artificial foods by pond fishes. Communication I. Digestibility of the nutrients of peanut oil meal, peas, barley and feed mixture by two-year-old carp. *J. Ichthyol.* 10: 662–667.

Shearman, D. J. C., Taylor, P., Tyler, M. J., O'Brien, P., Laidler, P., and Seamark, R. F. 1984. An update on the role of prostaglandins in the stomach and intestine of the gastric brooding frog *Rheobatrachus silus*. In: *Mechanisms of mucosal Protection in the Upper Gastrointestinal Tract*, ed. A. Allen et al., pp. 323–327. New York: Raven Press.

Shiau, Y., Boyle, J. T., Umstetter, C., and Koldovsky, O. 1980. Apical distribution of fatty acid esterification capacity along the villus-crypt unit of rat jejunum. *Gastroenterology* 79: 47–53.

Shirazi-Beechey, S. P., Hirayama, B. A., Wang, Y., Scott, D., Smith, M. W., and Wright, E. M. 1991. Ontogenic development of lamb intestinal sodium-glucose transporter is regulated by diet. *J. Physiol. (Lond.).* 437: 699–708.

Shirazi-Beechey, S. P., Smith, M. W., Wang, Y., and James, P. S. 1991. Postnatal development of lamb intestinal digestive enzymes is not regulated by diet. *J. Physiol. (Lond.)* 437: 691–698.

Shkolnik, A., Maltz, E., and Choshniak, I. 1980. The role of the ruminant digestive tract as a water reservoir. In *Digestive Physiology and Metabolism in Ruminants*, ed. Y. Ruckebusch and P. Thivend, pp. 731–742. Lancaster, U.K.: M T P Press.

Shparkovskii, I. A. 1988. Effect of cholecystokinin octapeptide on the motor function of the plaice [Russian]. *Fiziologicheskii Zhurnal SSSR Imeni I. M. Sechenova.* 74: 569–576.

Sibly, R. M. 1981. Strategies of digestion and defecation. *In Physiological Ecology: An Evolutionary Approach*, ed. C. R. Townsend and P. Callow, pp. 109–139. Sunderland, Massachusetts: Sinauer.

Siciliano-Jones, J., and Murphy, M. R. 1989. Production of volatile fatty acids in the rumen and cecum-colon of steers as affected by forage: concentrate and forage physical form. *J. Dairy Sci.* 72: 485–492.

Silanikove, N. 1990. Role of rumen and saliva in the homeostatic response to rehydration in cattle. *Am. J. Physiol.* 256: R816–R821.

Silk, D. B. A., Grimble, G. K., and Rees, R. G. 1985. Protein digestion and amino acid and peptide absorption. *Proc. Nut. Soc.* 44: 63–72.

Simnet, J., and Spray, G. 1961. The influence of diet on the vitamin B activity in the cecum, urine and faeces of rabbits. *Br. J. Nutr.* 15: 555–556.

Simon, G. L., and Gorbach, S. L. 1987. Intestinal flora and gastrointestinal function. In *Physiology of the Gastrointestinal Tract*, Vol. 2, 2nd ed., ed. L. R. Johnson, J. Christensen, M. J. Jackson, E. D. Jacobson, and J. H. Walsh, pp. 1729–1747. New York: Raven Press.

Simpson, G. G. 1945. The principals of classification and the classification of mammals. *Bull. Am. Mus. Nat. Hist.* 85: I-XLV, 1–350.

Sisson, S. 1975. Porcine digestive system. In *Sisson and Grossman's The Anatomy of the Domestic Animals*, 5th ed., ed. R. Getty, pp. 1268–1282. Philadelphia: W. B. Saunders Co.

Skadhauge, E. 1968. The cloacal storage of urine in the rooster. *Comp. Biochem. Physiol.* 24: 7–18.

Skadhauge, E. 1973. Renal and cloacal salt and water transport in the fowl (*Gallus domesticus*). *Dan. Med. Bull.* 20: 1–82.

Skadhauge, E. 1993. Basic characteristics and hormonal regulation of ion transport in avian hindguts. In: *Advances in Comparative and Environmental Physiology*, Vol. 16, ed. W. Clauss, pp. 67–93. Berlin: Springer-Verlag.

Skadhauge, E., Thomas, D. H., Chadwick, A., and Jallageas, M. 1983. Time course of adaptation to low and high NaCl diets in the domestic fowl. *Pflügers Arch.* 396: 301–307.

Skadhauge, E., Warui, C. N., Kamau, J. M. Z., and Maloiy, G. M. O. 1984. Function of the lower intestine and osmoregulation in the ostrich: preliminary anatomical and physiological observations. *J. Exp. Physiol.* 69: 809–818.

Skoczylas, R. 1978. Physiology of the digestive tract. In *Biology of the Reptilia*, Vol. 8, ed. C. G. Gans and K. A. Gans, pp. 589–717. New York: Academic Press.

Slade, L. M., Bishop, R., Morris, J. G., and Robinson, D. G. 1971. Digestion and absorption of ^{15}N-labeled microbial protein in the large intestine of the horse. *Br. Vet. J.* 127: XI–XII.

Slaytor, M. 1992. Cellulose digestion in termites and cockroaches: What role do symbionts play? *Comp. Biochem. Physiol.* 103 B: 775–784.

Slijper, E. J. 1962. *Whales*. New York: Basic Books, Inc.

Smith, C. R., and Richmond, M. E. 1972. Factors influencing pellet egestion and gastric pH in the barn owl. *Wilson Bull.* 84: 179–183.

Smith, H., Farinacci, N., and Breitweiser, A. 1930. The absorption and excretion of water and salts by marine teleosts. *Am. J. Physiol.* 93: 480–505.

Smith, H. W. 1943. The evolution of the kidney. In *Lectures on the Kidney*, Porter Lectures Series No. 9. Lawrence: University of Kansas Press.

Smith, H. W. 1965. The development of the flora of the alimentary tract in young animals. *J. Pathol. Bacteriol.* 90: 495–513.

Smith, P. L., Cascairo, M., and Sullivan, S. 1985. Sodium dependence of luminal alkalinization by rabbit ileal mucosa. *Am. J. Physiol.* 249: G358–G368.

Smith, P. L., and McCabe, R. D. 1984. Mechanism and regulation of transcellular potassium transport by the colon. *Am. J. Physiol.* 247: G445–G456.

Smith, R. H., and McAllan, A. B. 1971. Nucleic acid metabolism in the ruminant. *Br. J. Nutr.* 25: 181–190.

Smuts, D. B. 1935. The relation between the basal metabolism and the endogenous nitrogen metabolism, with particular reference to the estimation of the maintenance requirement of protein. *J. Nutr.* 9: 403–433.

Snipes, R. L. 1978. Anatomy of the rabbit cecum. *Anat. Embryol.* 155: 57–80.

Snipes, R. L. 1984. Anatomy of the cecum of the West Indian manatee, *Trichechus manatus* Mammalia, Sirenia. *Zoomorphology* 104: 67–78.

Snipes, R. L., Clauss, W., Weber, A., and Hörnicke, H. 1982. Structural and functional differences in various divisions of the rabbit colon. *Cell Tissue Res.* 225: 331–346.

Snipes, R. L., Hörnicke, H., Björnhag, G., and Stahl, W. 1988. Regional differences in hindgut structure in the nutria (*Myocaster coypus*). *Cell and Tissue Res.* 252: 435–447.

Soergel, K. H., and Hofmann, A. F. 1972. Absorption. In *Pathophysiology: Altered Regulatory Mechanisms in Disease*, ed. E. D. Frohlich, pp. 423–453. Philadelphia: J. B. Lippincott.

Sorrell, M. F., Frank, O., Thomson, A. D., Aquino, H., and Baker, H. 1971. Absorption of vitamins from the large intestine *in vivo*. *Nutr. Rep. Int.* 3: 143–148.

Southern, N. H. 1940. Coprophagy in the wild rabbit. *Nature (Lond.)* 145: 262.

Southern, N. H. 1942. Periodicity of refection in the wild rabbit. *Nature (Lond.)* 149: 553.

Spalinger, D. E., and Robbins, C. T. 1992. The dynamics of particle flow in the rumen of mule deer (*Odocoileus hemionus hemionus*) and elk (*Cervus elaphus nelsoni*). *Physiol. Zool.* 65: 379–402.

Sperber, I. 1968. Physiological mechanisms in herbivores for retention and utilization of nitrogenous compounds. In *Isotope Studies on the Nitrogen Chain–Symposium on the Use of Isotopes in the Studies of Nitrogen Metabolism in the Soil-plant-animal System*, pp. 209–219. Vienna: International Atomic Energy Agency.

Sperber, I., Björnhag, G., and Holtenius, K. 1992. A separation mechanism and fluid flow in the large intestine of the equine. *Pferdeheilkunde*, 1. Europaische Konferenz über die Ernahrung des Pferdes, Sanderausgabe, pp. 29–32.

Sperber, I., Björnhag, G., and Ridderstrale, Y. 1983. Function of proximal colon in lemming and rat. *Swed. J. Agric. Res.* 13: 243–256.

Steenbock, H., Seel, M. T., and Nelson, E. M. 1923. Vitamin B. *J. Biol. Chem.* 60: 399.

Stein, E. D., and Diamond, J. M. 1989. Do dietary levels of pantothenic acid regulate its intestinal uptake in mice? *J. Nutr.* 119: 1973–1983.

Steven, G. A. 1930. Bottom fauna and the food of fishes. *J. Mar. Biol. Assoc. U. K.*, New Series 16: 677–706.

Stevens, B. R., Kaunitz, J. D., and Wright, E. M. 1984. Intestinal transport of amino acids and sugars: advances using membrane vesicles. *Annu. Rev. Physiol.* 46: 417–433.

Stevens, C. E. 1964. Transport of sodium and chloride by the isolated rumen epithelium. *Am. J. Physiol.* 206: 1099–1105.

Stevens, C. E. 1970. Fatty acid transport through the rumen epithelium. In *Physiology of Digestion in the Ruminant*, ed. A. T. Phillipson, pp. 101–112. Newcastle: Oriel Press.

Stevens, C. E. 1973. Transport across rumen epithelium. In *Transport Mechanisms in Epithelia*, ed. H. H. Ussing and N. A. Thorn, pp. 404–426. New York: Academic Press.

Stevens, C. E. 1977. Comparative physiology of the digestive system. In *Duke's Physiology of Domestic Animals*, ed. M. J. Swenson, 9th ed., pp. 216–232. Ithaca: Cornell University Press.

Stevens, C. E. 1980. The gastrointestinal tract of mammals: major variations. In *Comparative Physiology: Primitive Mammals*, ed. K. Schmidt-Nielsen, L. Bolis, and C. R. Taylor, pp. 52–62. New York: Cambridge University Press.

Stevens, C. E. 1983. Comparative anatomy and physiology of the herbivore digestive tract. In *Proceedings of the Second Annual Dr. Scholl Conference on the Nutrition of Captive Wild Animals*, ed. T. P. Meehan, B. A. Thomas, and K. Bell, pp. 8–16. Chicago: Lincoln Park Zoological Society.

Stevens, C. E. 1990. Evolution of vertebrate herbivores. *Acta. Vet. Scand.* (Supplement) 86: 9–19.

Stevens, C. E., Argenzio, R. A., and Clemens, E. T. 1980. Microbial digestion: rumen versus large intestine. In *Digestive Physiology and Metabolism in Ruminants*, ed. Y. Ruckebusch and P. Thivend, pp. 685–706. Lancaster, U.K.: M T P Press Limited.

Stevens, C. E., Argenzio, R. A., and Roberts, M. C. 1986. Comparative physiology of the mammalian colon and suggestions or animal models of human disorders. *Clinics in Gastroenterology* 15: 763–786.

Stevens, C. E., Dobson, A., and Mammano, J. H. 1969. A transepithelial pump for weak electrolytes. *Am. J. Physiol.* 216: 983–987.

Stevens, C. E., and Sellers, A. F. 1968. Rumination. In *Handbook of Physiology*, Section 6, *Alimentary Canal*, Vol. 5, *Bile; Digestion; Ruminal Physiology*, ed. C. F. Code and W. Heidel, pp. 2699–2704. Washington, D. C.: American Physiological Society.

Stevens, C. E., Sellers, A. F., and Spurrell, F. A. 1960. Function of the bovine omasum in ingesta transfer. *Am. J. Physiol.* 198: 449–455.

Stevens, C. E., and Stettler, B. K. 1966a. Factors affecting the transport of volatile fatty acids across rumen epithelium. *Am. J. Physiol.* 210: 365–372.

Stevens, C. E., and Stettler, B. K. 1966b. Transport of fatty acid mixtures across rumen epithelium. *Am. J. Physiol.* 211: 264–271.

Stevens, C. E., and Stettler, B. K. 1967. Evidence for active transport of acetate across bovine rumen epithelium. *Am. J. Physiol.* 213: 1335–1339.

Stevenson, N. R., Ferrigni, F., Parnicky, K., Day, S., and Fierstein, J. S. 1975. Effect of changes in feeding schedule on the diurnal rhythms and daily activity levels of intestinal brush-border enzymes and transport systems. *Biochim. Biophys. Acta* 4060: 131–145.

Stickney, R. R., and Shumway, S. E. 1974. Occurrence of cellulose activity in the stomachs of fishes. *J. Fish Biol.* 6: 779–790.

Stingelin, Y., Wolffram, S., and Scharrer, E. 1986. Evidence for potassium/acetate cotransport in rat distal colon. In: *Ion Gradient-Coupled Transport*, ed. F. Alvarado and C. H. van Os, pp. 317–320. Amsterdam: Elsevier Science Publishers.

Storelli, C., Vilella, S., and Cassano, G. 1986. Na-dependent D-glucose and L-alanine transport in eel intestinal brush border membrane vesicles. *Am. J. Physiol.* 251: R463–R469.

Storelli, C., Vilella, S., Romano, M. P., Maffia, M., and Cassano, G. 1989. Brush border amino acid transport mechanisms in carnivorous eel intestine. *Am. J. Physiol.* 257: R506–R510.

Storer, R. W. 1971a. Classification of birds. In *Avian Biology*, Vol. 1, ed. D. S. Farner, J. R. King, and K. C. Parkes, pp. 1–18. New York: Academic Press.

Storer, R. W. 1971b. Adaptive radiation of birds. In *Avian Biology*, Vol. 1, ed. D. S. Farner, J. R. King, and K. C. Parkes, pp. 150–188. New York: Academic Press.

Straganov, N. S. 1963. The food selectivity of the amur fishes. In *Problems of Fisheries: Exploitation of Plant-eating Fishes*, pp. 181–191. Turkman U. S. S. R.: Ashhagad Academy of Science.

Strauss, E. W., and Ito, S. 1969. A fine structural study of lipid-uptake in the terminal intestine of the scup, *Stenotomus chrysops*. *Biol. Bull.* 137: 414–415.

Stroband, H. W. J., and Debets, F. M. H. 1978. The ultrastructure and renewal of the intestinal epithelium of the juvenile grass carp, *Ctenopharyngodon idella* (Val.). *Cell Tissue Res.* 187: 181–200.

Sturkie, P. D. 1970. Avian digestion. In *Dukes' Physiology of Domestic Animals*, 8th ed., ed. M. J. Swenson, pp. 526–537. Ithaca, New York: Cornell University Press.

Sullivan, S. K., and Field, M. 1991. Ion transport across mammalian small intestine. In *Handbook of Physiology*, Section 6: *The Gastrointestinal System*, Vol. 4: *Intestinal Absorption and Secretion*, ed. S. G. Schultz, M. Field, R. A. Frizzell, and B. B. Rauner, pp. 287–301. Washington, D. C.: American Physiological Society.

Sundler, F., Alumets, J., Hakanson, R., Carraway, R. E., and Lemman, S. E. 1977a. Ultrastructure of the gut neurotensin cell. *Histochemistry* 53: 25–34.

Sundler, F., Hakanson, R., Hammer, R. A., Alumets, J., Carraway, R. E., Leeman, S. E., and Zimmerman, E. A. 1977b. Immunohistochemical localization of neurotensin in endocrine cells of the gut. *Cell Tissue Res.* 178: 313–322.

Sunshine, P., and Kretchmer, N. 1964. Intestinal disaccharidases: absence in two species of sea lions. *Science* 144: 850–852.

Sutton, D. L. 1977. Digestion of duckweed by the grass carp (*Ctenopharyngoden idella*). *J. Fish Biol.* 11: 273–278.

Suzuki, K., Miyake, K., Hayama, S., and Ehara, A. 1991. Comparative study of the gastric mucosa in primates. In *Primatology Today*, ed. A. Ehara, T. Kimura, O. Takenaka, and M. Iwamoto, pp. 587–588. Amsterdam: Elsevier.

Svendsen, P. 1969. Etiology and pathogenesis of abomasal displacement in cattle. *Nord. Vet. Med.* 21, Suppl. I.

Symons, L. E. A., and Jones, W. O. 1966. The distribution of dipeptidase activity in the small intestine of the sheep (*Ovis aries*). *Comp. Biochem. Physiol.* 18: 71–82.

Takahashi, S., and Seifter, S. 1974. An enzyme with collagenolytic activity from dog pancreatic juice. *Israel J. Chem.* 12: 557–571.

Tamminga, S., and Doreau, M. 1991. Lipids and rumen digestion. In *Rumen Microbial Metabolism and Microbial Digestion*, ed. J. P. Jouany, pp. 151–164. Paris: Institut National de la Recherche Agronomique.

Tanaka, H. 1942. Chemie der Meerchildkroten II Uber die fermente der meerchildkrotenorgane. *J. Biochem. (Tokyo)* 36: 301–312.

Tanaka, M., Kawai, S., and Yamamoto, S. 1972. On the development of the digestive system and changes in activities of digestive enzymes during larval and juvenile stage. *Ayu. Bull. Jap. Soc. Sci. Fish.* 38: 1143–1152.

Tansy, M. F., Kendall, F. M., and Murphy, J. J. 1972. The reflex nature of the gastrocolic propulsive response in the dog. *Surg. Gynecol. Obstet.* 135: 404–410.

Tapper, E. J. 1983. Local modulation of intestinal ion transport by enteric neurons. *Am. J. Physiol.* 244: 457–468.

Tarpley, R. A., Sis, R. F., Albert, T. F., Dalton, L. M., and George, J. C. 1987. Observations on the anatomy of the stomach and duodenum of the bowhead whale, *Balaena musticetus*. *Am. J. Anat.* 180: 295–332.

Taverner, M. R., Hume, I. D., and Farrell, D. J. 1981a. Availability to pigs of amino acids in cereal grains. 1. Endogenous levels of amino acids in ileal digesta and faeces of pigs given cereal diets. *Br. J. Nutr.* 46: 149–158.

Taverner, M. R., Hume, I. D., and Farrell, D. J. 1981b. Availability to pigs of amino acids in cereal grains. 2. Apparent and true ileal availability. *Br. J. Nutr.* 46: 159–171.

Taylor, C. R., and Sale, J. B. 1969. Temperature regulation in the hyrax. *Comp. Biochem. Physiol.* 31: 903–907.

Taylor, C. R., and Weibel, E. R. 1981. Design of the mammalian respiratory system. I. Problem and strategy. *Respir. Physiol.* 44: 1–10.

Taylor, R. B. 1962. Pancreatic secretion in the sheep. *Res. Vet. Sci.* 3: 63–77.

Teather, R. M., Erfle, J. D., Boila, R. J., and Sauer, F. D. 1980. Effect of dietary nitrogen on the rumen microbial population in lactating dairy cattle. *J. Appl. Bacteriol.* 49: 231–238.

Tedman, R. A., and Hall, L. S. 1985. The morphology of the gastrointestinal tract and food transit time in the fruit bats *Pteropus alecto* and *P. poliocephalus* (Megachiroptera). *Aust. J. Zool.* 33: 625–640.

Tevis, L. J. 1953. Stomach contents of chipmunks and mantled squirrels in northeastern California. *J. Mammal.* 34: 316–324.

Thacker, E. J., and Brandt, C. S. 1955. Coprophagy in the rabbit. *J. Nutr.* 55: 375.

Thaysen, J. H., Thorn, N. A., and Schwartz, I. L. 1954. Excretion of sodium, potassium, chloride and carbon dioxide in human parotid saliva. *Am. J. Physiol.* 178: 155–159.

Thomas, B. A., and Spicer, R. A. 1987. *The Evolution and Palaeobiology of Land Plants*. Beckenham, Kent: Croom Helm.

Thomas, D. H., and Skadhauge, E. 1988. Transport function and control in bird ceca. *Comp. Biochem. Physiol.* 90A: 591–596.

Thomson, A. B. R., Hotke, C. A., and O'Brien, B. D. 1983. Intestinal uptake of fatty acids and cholesterol in four animal species and man: role of unstirred water layer and bile salt micelle. *Comp. Biochem. Physiol.* 75A: 221–232.

Thurston, J. P., Noirot-Timothée, C., and Arman, P. 1968. Fermentative digestion in the stomach of *Hippopotamus amphibius* (Artiodactyla: Suiformes), and associated ciliateprotozoa. *Nature (Lond.)* 218: 882–883.

Timson, C. M., Polak, J. M., Wharton, J., Ghatei, M. A., Bloom, S. R., Usellini, L., Capella, C., Solicia, E., Brown, M. R., and Pearse, A. G. E. 1979. Bombesin-like immunoreactivity in the avian gut and its localization to a distinct cell type. *Histochemistry* 61: 213–222.

Tindall, A. R. 1979. The innervation of the hind gut of the domestic fowl. *Br. Poult. Sci.* 20: 473–480.

Titchen, D. A. 1968. Nervous control of motility of the fore-stomach of ruminants. In *Handbook of Physiology*, Section 6, *Alimentary Canal*, Vol. 5, *Bile; Digestion; Ruminal Physiology*, ed. C. F. Code and W. Heidel, pp. 2705–2724. Washington, D. C.: American Physiological Society.

Titus, E., and Ahearn, G. A. 1988. Short-chain fatty acid transport in the intestine of a herbivorous teleost. *J. Exp. Biol.* 135: 77–94.

Titus, E., and Ahearn, G. A. 1991. Transintestinal acetate transport in a herbivorous teleost: anon exchange at the basolateral membrane. *J. Exp. Biol.* 156: 41–61.

Titus, E., and Ahearn, G. A. 1992. Vertebrate gastrointestinal fermentation: transport mechanisms for short-chain fatty acids. *Am. J. Physiol.* 262: R547–R553.

Titus, E., Karasov, W. H., and Ahearn, G. A. 1991. Dietary modulation of intestinal nutrient transport in the teleost fish, Tilapia. *Am. J. Physiol.* 261: R1568–R1574.

Tobey, N., Heizer, W., Yek, R., Huang, T., and Hoffner, C. 1985. Human intestinal brush border peptidases. *Gastroenterology* 88: 913–926.

Torrey, T. W. 1971. *Morphogenesis of the Vertebrates*, 3rd ed., pp. 44–45. New York: John Wiley & Sons.

Tracy, M. V. 1957. Chitin. *Rev. Pure Appl. Chem.* 7: 1–14.

Troyer, K. 1984a. Behavioral acquisition of the hindgut fermentation system by hatchling *Iguana iguana*. *Behav. Ecol. Sociobiol.* 14: 189–193.

Troyer, K. 1984b. Diet selection and digestion in *Iguana iguana*: the importance of age and nutrient requirements. *Oecologia* 61: 201–207.

Trust, T. J., Bull, L. M., Currie, B. R., and Buckley, J. T. 1979. Obligate anaerobic bacteria in the gastrointestinal microflora of the grass carp *Ctenopharyngodon idella*, goldfish (*Carassius auratus*), and rainbow trout (*Salmo gairdneri*). *J. Fish Res. Board Can.* 36: 1174–1179.

Trust, T. J., and Sparrow, R. A. H. 1974. The bacterial flora in the alimentary tract of salmonid fishes. *Can. J. Microbiol.* 20: 1219–1228.

Turcek, F. J. 1963. Beitrag zur Ökologie des Ziesels (*Citellus citellus* L.) II. *Biológia, Bratisl.* 18: 419–423.

Turnberg, L. A., Bieberdorf, F. A., Morawski, S. G., and Fordtran, J. S. 1970. Interrelationships of chloride, bicarbonate, sodium and hydrogen transport in the human ileum. *J. Clin. Invest.* 49: 557–567.

Udén, P., Rounsaville, T. R., Wiggans, G. R., and Van Soest, P. J. 1982. The measurement of liquid and solid digesta retention in ruminants, equines, and rabbits given timothy (*Phleum pratense*) hay. *Br. J. Nutr.* 48: 329–339.

Umesaki, Y., Yajima, T., Yokokura, T., and Mutai, M. 1979. Effect of organic acid absorption on bicarbonate transport in rat colon. *Pflügers Arch.* 379: 43–47.

Usellini, L., Buchan, A. M. J., Polak, J. M., Capella, C., Cornaggia, M., and Solcia, E. 1984. Ultrastructural localization of motilin in endocrine cells of human and dog intestine by the immunogold technique. *Histochemistry* 81: 363–368.

Ushiyama, H., Fujimori, T., Shibata, T., and Yoshimura, K. 1965. Studies on carbohydrases in the pyloric caeca of the salmon, *Oncorhynchus keta*. *Bull. Fac. Fish Hokkaido Univ.* 16: 183–188.

Utley, P. R., Bradley, N. W., and Boling, J. A. 1970. Effect of water restriction on nitrogen metabolism in bovine fed two levels of nitrogen. *J. Nutr.* 100: 551–556.

Vaillant, C., Dimaline, R., and Dockray, G. J. 1980. The distribution and cellular origin of vasoactive intestinal polypeptide in the avian gastrointestinal tract and pancreas. *Cell Tissue Res.* 211: 511–523.

Vaillant, C., Dockray, G. J., and Walsh, J. H. 1979. The avian proventriculus is an abundant source of endocrine cells with bombesin-like immunoreactivity. *Histochemistry* 64: 307–314.

Vallenas, A. P., Cummings, J. F., and Munnell, J. F. 1971. A gross study of the compartmentalized stomach of two New-World camelids, the llama and guanaco. *J. Morphol.* 134: 399–423.

Vallenas, A. P., and Stevens, C. E. 1971a. Volatile fatty acid concentrations and pH of llama and quanaco forestomach digesta. *Cornell Vet.* 61: 239–252.

Vallenas, A. P., and Stevens, C. E. 1971b. Motility of the llama and guanaco stomach. *Am. J. Physiol.* 220: 275–282.

Van Dyke, J. M., and Sutton, D. L. 1977. Digestion of duckweed (*Lemna spp.*) by the grass carp (*Ctenopharyngodon idella*). *J. Fish Biol.* 11: 273–278.

van Hoven, W., and Boomker, E. A. 1981. Feed utilization and digestion in the black wildebeest (*Connochaetes gnou*), Zimmerman, 1780 in the Golden Gate Highlands National Park. *S. Afr. J. Wildl. Res.* 11: 35–40.

van Hoven, W., Prins, R. A., and Lankhorst, A. 1981. Fermentative digestion in the African elephant. *S. Afr. J. Wildl. Res.* 11: 78–86.

van Lawick-Goodall, J. 1968. The behavior of free-living chimpanzees in the Gombe Stream Reserve. *Anim. Behav. Monogr.* 1: 161–311.

van Lennep, E. W., and Lanzing, W. J. R. 1967. The ultrastructure of glandular cells in the external dendritic organ of some marine catfish. *J. Ultrastruct. Res.* 18: 333–344.

van Lennep, E. W., and Young, J. A. 1979. Part II: Salt glands. In *Membrane Transport in Biology*, Vol. 4, *B. Transport Organs*, ed. G. Giebisch, D. C. Tosteson, H. H. Ussing, and M. T. Tosteson, pp. 675–692. New York: Springer-Verlag.

Van Noordan, S., and Polak, J. M. 1979. Hormones of the alimentary tract. In *Hormones and Evolution*, Vol. 2, ed. E. J. W. Barrington, pp. 791–828. New York: Academic Press.

Van Soest, P. J. 1982. *Nutritional Ecology of the Ruminant*. Corvallis: O & B Books, Inc.

Van Soest, P. J. 1994. *Nutritional Ecology of the Ruminant.* 2nd ed. Ithaca: Cornell University Press.

Van Soest, P. J., Sniffen, C. J., and Allen, M. S. 1988. Rumen dynamics. In *Aspects of Digestive Physiology in Ruminants*, ed. A. Dobson and M. J. Dobson, pp. 21–42. Ithaca: Comstock Publ. Assoc.

van Weel, P. B. 1974. "Hepatopancreas"? *Comp. Biochem. Physiol.* 47A: 1–9.

Vaughan, T. A. 1986. *Mammology.* Philadelphia: W. B. Saunders Co.

Vercellotti, J. R., Salyers, A. A., and Wilkins, T. D. 1978. Complex carbohydrate breakdown in the human colon. *Am. J. Clin. Nutr.* 31: S86–S89.

Vigna, S. R. 1979. Distinction between cholescystokinin-like and gastrin-like biological activities extracted from gastrointestinal tissues of some lower vertebrates. *Gen. Comp. Endocr.* 39: 512–520.

Vigna, S. R. 1983. Evolution of endocrine regulation of gastrointestinal function in lower vertebrates. *Amer. Zool.* 23: 729–738.

Vigna, S. R. 1984. Radioreceptor and biological characterization of cholecystokinin and gastrin in the chicken. *Am. J. Physiol.* 246: G296–G304.

Vigna, S. R. 1986. Functional evolution of gastrointestinal hormones. In *Evolutionary Biology of Primitive Fishes.* ed. R. E. Foreman, A. Gorbman, J. M. Dodd, and R. Olsson., pp. 401–412. New York: Plenum Publishing Corp.

Vigna, S. R., Fischer, B. L., Morgan, J. L. M., and Rosenquist, G. L. 1985. Distribution and molecular heterogeneity of cholecystokinin-like immunoreactive peptides in the brain and gut of the rainbow trout, *Salmo gairdneri. Comp. Biochem. Physiol.* 82C: 143–146.

Vigna, S. R., and Gorbman, A. 1977. Effects of cholecystokinin, gastrin, and related peptides on coho salmon gallbladder contraction in vitro. *Am. J. Physiol.* 232: E485–E491.

Vigna, S. R., and Gorbman, A. 1979. Stimulation of intestinal lipase secretion by porcine cholecystokinin in the hagfish, *Eptatretus stouti. Gen. Comp. Endocr.* 38: 356–359.

Vinardell, M. P., and Lopera, M. T. 1987. Jejunal and cecal 3-oxy-methy-D-glucose absorption in chicken using a perfusion system in vivo. *Comp. Biochem. Physiol.* 86A: 625–627.

Vispo, C., and Hume, I. D. 1995. The digestive tract and digestive function in the North American porcupine and beaver. *Can. J. Zool.* (In press).

von Euler, U. S., and Gaddum, J. H. 1931. An unidentified depressor substance in certain tissue extracts. *J. Physiol. (Lond.)* 72: 74–87.

Vonk, H. J. 1927. Die Verdauuing bei den Fischen. *Z. Vergl. Physiol.* 5: 445–546.

Vonk, H. J., and Western, J. R. H. 1984. *Comparative Biochemistry and Physiology of Enzymatic Digestion.* New York: Academic Press.

Vorontsov, N. N. 1962. The ways of food specialization and evolution of the alimentary tract in Muroidea. In *Symposium Theriologicum, 1960*, ed. J. Kratochvil and J. Pelikan, pp. 360–377. Praha: Czechoslavak Academy of Science.

Waldo, D. R., Smith, L. W., and Cox, E. L. 1972. Model of cellulose disappearance from the rumen. *J. Dairy Sci.* 55: 125–129.

Waldschmidt, S. R., Jones, S. M., and Porter, W. P. 1986. The effect of body temperature and feeding regime on activity, passage time, and digestive coefficient in the lizard *Uta stansburiana. Physiol. Zool.* 59: 376–383.

Walker, W. F. 1987. *Functional Anatomy of Vertebrates: An Evolutionary Perspective.* Philadelphia: W. B. Saunders.

Wallis, I. R. 1994. The rate of passage of digesta through the gastrointestinal tract of potoroine marsupials: more evidence about the role of the potoroine foregut. *Physiol. Zool.* 67: 771–795.

Walsh, J. H. 1981. Gastrointestinal hormones and peptides. In *Physiology of the Gastrointestinal Tract*, Vol. 1, ed. L. R. Johnson, J. Christensen, M. I. Grossman, E. D. Jacobson, and S. G. Schultz, pp. 59–144. New York: Raven Press.

Walsh, J. H. 1994. Gastrointestinal hormones. In *Physiology of the Gastrointestinal Tract*, 3rd ed., Vol. 2, ed. L. R. Johnson, D. H. Alpers, J. Christensen, E. D. Jacobson, and J. H. Walsh, pp. 1–128. New York: Raven Press.

Walsh, P. J., Danulat, E., and Mommsen, T. P. 1990. Variations in urea excretion in the Gulf toadfish (*Opsanas beta*). *Mar. Biol.* 106: 323–328.

Walter, W. G. 1939. Bedingte Magensaftsekretion bei der ente. *Acta Brev. Neerl. Physiol.* 9: 56–57.

Waring, H., Moir, R. J., and Tyndale-Biscoe, C. H. 1966. Comparative physiology of marsupials. *Adv. Comp. Physiol. Biochem.* 2: 237–376.

Warner, A. C. I. 1981. Rate of passage of digesta through the gut of mammals and birds. *Nutr. Abstr. and Rev. Ser. B* 51: 789–820.

Washburn, L. E., and Brody, S. 1937. Growth and Development with Special Reference to Domestic Animals. XLII. Methane, Hydrogen and Carbon Dioxide Production in the Digestive Tract of Ruminants in Relation to Respiratory Exchange. *Mo. Agric. Exp. Sta. Bull. No. 263*.

Wassersug, R. J., and Johnson, R. K. 1976. A remarkable pyloric cecum in the evermannellid genus *Coccorella* with notes on gut structure and function in alepisauroid fishes (Pisces, Myctophiformes). *J. Zool (Lond.)* 179: 273–289.

Watabe, J., and Bernstein, H. 1985. The mutagenicity of bile acids using a fluctuation test. *Mutat. Res.* 158: 45–51.

Watari, N. 1968. Fine structure of nervous elements in the pancreas of some vertebrates. *Z. Zellforsch. Mikroskop. Anat.* 85: 291–314.

Watson, J. S., and Taylor, R. H. 1955. Reingestion in the hare, *Lepus europaeus* Pal. *Science* 121: 314.

Webb, K. E. Jr., Matthews, J. C., and DiRienzo, D. B. 1992. Peptide absorption: a review of current concepts and future perspectives. *J. Anim. Sci.* 70: 3248–3257.

Weisbrodt, N. W. 1981. Motility of the small intestine. In *Physiology of the Gastrointestinal Tract*, Vol. 1, ed. L. R. Johnson, J. Christensen, M. I. Grossman, E. D. Jacobson, and S. G. Schultz, pp. 411–443. New York: Raven Press.

Wellard, G. A., and Hume, I. D. 1981. Digestion and digesta passage in the brushtail possum, *Trichosurus vulpecula* (Kerr). *Aust. J. Zool.* 29: 157–166.

Welsh, M., Smith, P., Fromm, M., and Frizzell, R. 1982. Crypts are the site of intestinal fluid and electrolyte secretion. *Science* 218: 1219–1221.

Wester, J. 1926. *Die Physiologie und Pathologie der Vormägenbeim Rinde*. Berlin: Richard Schoetz.

Westergaard, H. 1987. The passive permeability properties of in vivo perfused rat jejunum. *Biochim. Biophy. Acta.* 900: 129–138.

Wharton, C. H. 1950. Notes on the life history of the flying lemur. *J. Mammal.* 31: 161–273.

White, J. 1985. Omeprazole inhibits H^+ secretion by *Amphiuma* jejunum. *Am. J. Physiol.* 248: G256–G259.

White, N. A. 1990. Epidemiology and etiology of colic. In: *Equine Acute Abdomen* ed. N. A. White, pp. 49–64. Philadelphia: Lea and Febiger.

White, R. G., Hume, I. D., and Nolan, J. V. 1988. Energy expenditure and protein turnover in three species of wallabies (Marsupialia: Macropodidae). *J. Comp. Physiol.* B158: 237–246.

Whiteside, C. H., and Prescott, J. M. 1962. Activities of chicken pancreatic proteinases toward synthetic substrates. *Proc. Soc. Exp. Biol. Med.* 110: 741–744.

Wigglesworth, V. B. 1984. *Insect Physiology*, 8th ed. New York: Chapman and Hall.

Wikelski, M., Gall, B., and Trillmich, F. 1993. Ontogenetic changes in food intake and digestion rate of the herbivorous marine iguana (*Amblyrhynchus cristatus*, Bell). *Oecologia* 94: 373–379.

Williams, V. J. 1963. Rumen function in the camel. *Nature (Lond.)* 197: 1221.

Withers, P. C. 1992. *Comparative Animal Physiology*. Fort Worth: Saunders College Publishing.

Wolfe, J. A. 1978. A paleobotanical interpretation of tertiary climates in the Northern Hemisphere. *Am. Sci.* 66: 694–703.

Wolin, M. J. 1979. The rumen fermentation: a model for microbial interactions in anaerobic ecosystems. *Adv. Microbiol. Ecol.* 3: 49–77.

Wolin, M. J. 1981. Fermentation in the rumen and human large intestine. *Science* 213: 1463–1468.

Wood, J. D. 1994. Physiology of the enteric nervous system. In *Physiology of the Gastrointestinal Tract*, 3rd ed., Vol. 2, ed. L. R. Johnson, D. H. Alpers, J. Christensen, E. D. Jacobson, and J. H. Walsh, pp. 423–482. New York: Raven Press.

Woodall, P. F. 1987. Digestive tract dimensions and body mass of elephant shrews (Macroscelididae) and the effects of season and habitat. *Mammalia* 51: 537–545.

Woodall, P. F., and Currie, G. J. 1989. Food consumption, assimilation and rate of food passage in the cape rock elephant shrew, *Elephantulus edwardii* (Macroscelidea: Macroscelidinae). *Comp. Biochem. Physiol.* 92A: 75–79.

Wootton, J. F., and Argenzio, R. A. 1975. Nitrogen utilization within equine large intestine. *Am. J. Physiol.* 229: 1062–1067.

Wostmann, B. S., and Knight, P. L. 1961. Synthesis of thiamine in the digestive tract of the rat. *J. Nutr.* 74: 103–110.

Wostmann, B. S., Knight, P. L., and Reyniers, J. A. 1958. The influence of orally administered penicillin upon growth and liver thiamine of growing germ-free and normal stock rats fed a thiamine deficient diet. *J. Nutr.* 66: 577–586.

Wright, R. D., Florey, H. W., and Sanders, A. G. 1957. Observations on the gastric mucosa of Reptilia. *Qt. J. Exp. Physiol.* 42: 1–14.

Wrong, O. M. 1988. Bacterial metabolism of protein and endogenous nitrogen compounds. In *Role of the Gut Flora in Toxicity and Cancer*, ed. I. R. Rowland, pp. 227–262. New York: Academic Press.

Wrong, O. M., Edmonds, L. J., and Chadwick, V. S. 1981. *The Large Intestine*. Lancaster, U.K.: M T P Press.

Wrong, O. M., and Vince, A. 1984. Urea and ammonia metabolism in the human large intestine. *Proc. Nutr. Soc.* 43: 77–86.

Wrong, O. M., Vince, A. J., and Waterlow, J. C. 1985. The contribution of endogenous urea to faecal ammonia in man, determined by N labelling of plasma urea. *Clin. Sci.* 68: 193–199.

Wurth, M. A., and Musacchia, X. J. 1964. Renewal of intestinal epithelium in the freshwater turtle, *Chrysemys picta*. *Anat. Rec.* 148: 427–439.

Yamada, J., Campos, V. J. M., Kitamura, N., Pacheco, A. C., Yamashita, T., and Yanaihara, N. 1987. An immunohistochemical study of the endocrine cells in the gastrointestinal mucosa of the *Caiman latirostris*. *Arch. Histol. Jap.* 50: 229–241.

Yamamoto, M., and Hirano, T. 1978. Morphological changes in the esophageal epithelium of the eel, *Anguilla japonica*, during adaptation to seawater. *Cell Tiss. Res.* 192: 25–38.

Yang, M. G., Manoharan, K., and Mickelsen, O. 1970. Nutritional contribution of volatile fatty acids from the cecum of rats. *J. Nutr.* 100: 545–550.

Yoshida, T., Pleasants, J. R., Reddy, B. S., and Wostmann, B. S. 1968. Efficiency of digestion in germ-free and conventional rabbits. *Br. J. Nutr.* 22: 723–737.

Youson, J. H., and Horbert W. R. 1982. Transformation of the intestinal epithelium of the larval anadromous sea lamprey *Petromyzon marinus* L. during metamorphosis. *J. Morphol.* 171: 89–117.

Youson, J. H., and Langille, R. M. 1981. Proliferation and renewal of the epithelium in the intestine of young adult anadromous sea lampreys, *Petromyzon marinus* L. *Can. J. Zool.* 59: 2341–2349.

Yule, D. I., and Williams, J. A. 1994. Stimulus-secretion coupling in the pancreatic acinus. In: *Physiology of the Gastrointestinal Tract.* 3rd ed., Vol. 2, ed. L. R. Johnson, D. H. Alpers, J. Christensen, E. D. Jacobson, and J. H. Walsh, pp. 1353–1376. New York: Raven Press.

Zendzian, E. N., and Barnard, E. A. 1967a. Distributions of pancreatic ribonuclease, chymotrypsin, and trypsin in vertebrates. *Arch. Biochim. Biophys.* 122: 699–713.

Zendzian, E. N., and Barnard, E. A. 1967b. Reactivity evidence for homologies in pancreatic enzymes. *Arch. Biochim. Biophys.* 122: 714–720.

Zihler, F. 1982. Gross morphology and configuration of digestive tracts of Cichlidae (*Teleostei perciformes*): phylogenetic and functional significance. *Netherlands J. Zool.* 32(4): 544–571.

Ziswiler, V., and Farner, D. S. 1972. Digestion and the digestive system. In *Avian Biology*, Vol. 2, ed. D. S. Farner, J. R. King, and K. C. Parkes, pp. 343–430. New York: Academic Press.

Zoppi, G., and Schmerling, D. H. 1969. Intestinal disaccharidase activities in some birds, reptiles and mammals. *Comp. Biochem. Physiol.* 29: 289–294.

Index

digestion (endogenous enzymes) in, 159t, 160,
 163, 168, 171, 172, 173t, 174, 176t
microbial fermentation and metabolism in, 193t,
 211, 214
motility in, 111, 115
neuroendocrine control in, 280, 282, 284–286
nitrogen metabolism in, 214, 215
chimaeras (*Chimaera*), 12t, 25, 160, 172, 175
chimpanzees (*see also* apes), 75f, 76, 77f, 164t
chinchilla, 67, 68f, 70, 137, 139, 216, 217, 219
chipmunk, 124t, 315
chitin, 155, 156f, 157, 157t, 181, 204, 228, 304
chitobiose, 155
cholesterol, 166, 167, 182
Choloepus (*see* sloths)
Choysomys (*see* turtles)
Chrysemys (*see* terrapins)
chubs (*see also* sea chubs), 25, 28f, 31
Clemmys (*see* terrapins)
civets, 58
cloaca, 23, 46, 115, 214, 231, 232f, 299
Coccorella, 28
cod, 29, 171, 174, 282
codfish, 276
coelacanths (*see also* Crossopterygii), 229
Coendou (*see* porcupine)
colobus monkeys (*see also* *Procolobus*; monkeys),
 72, 73, 74f, 76, 77f, 190t, 204, 293, 312f
colugos (*see* flying lemurs)
Columbia (*see* pigeons)
conger eels, 12t, 31, 173t, 176t
Connochaetes (*see* wildebeest)
coprophagy, 129, 135, 215–220, 218t, 219f, 227, 310
Corvus (*see* crows)
Corytophanes (*see* chameleons)
Coturnix (*see* quail)
cows (*see* cattle)
cranes, 163
Cricetus (*see* hamsters)
crocodiles (*see also* Crocodilia), 12t, 36, 38, 40, 280,
 282, 286
Crocodilia (*see also* alligators, caimans, crocodiles,
 and gavials), 12t
 anatomy in, 36, 37f, 38, 40
 digesta transit in, 121, 126
 digestion (endogenous enzymes), 163, 176t, 296
 microbial fermentation and metabolism in, 193t,
  211
 neuroendocrine control in, 280, 281t, 282, 285,
  286
crop, 14, 16, 41, 228
Crossopterygii, 12t, 25, 29, 229, 289
Crotalus (*see* snakes)
crows, 159t
crustaceans, 4, 51, 126, 157, 167, 295
crypts of Lieberkühn, 18, 19f, 251
Crynodus (*see* sea carp)
cyclids, 31, 221
Cyclopes (*see* anteaters)
cyclostomes (*see also* hagfish; lampreys), 11, 289
 anatomy of, 12t, 13, 19, 20, 23–26, 27f
 body compartments, 229
 digestion (endogenous enzymes) in, 160, 168,
  171, 172, 174, 175

neuroendocrine control in, 276, 282–285, 301, 302
Cynocephalus (*see also* flying lemurs)
Cyprinus (*see* carp and goldfish)
cytokines, 286

Dasypus (*see* armadillos)
Dasyurus (*see* quoll)
Daubentonia (*see* aye-aye)
deer, 20, 83, 160, 197, 198t, 209
deglutition, 95, 96f, 103f, 106f
degu, 124t
Desmodus (*see* bats)
Diadophis (*see* snakes)
diarrhea, 185, 265
Diceros (*see* rhinoceros)
Didelphis (*see* opossum)
digestive strategies, 143, 144f
dik-diks, 86, 87f
dingo, 216
dinosaurs, 289, 305–308, 306f
Dopodomys (*see* kangaroo-rat)
dogfish (*see also* sharks), 96, 168, 171, 172, 173t,
 176t, 280
dogs (*see also* dingoes), 12t, 183
 absorption in, 186t
 anatomy of, 17f, 22f, 57f, 58, 96, 97
 coprophagy in, 216
 digesta transit in, 96, 127f, 129, 131, 135
 digestion (endogenous enzymes) in, 164t, 168,
  171, 173t, 174, 176t, 185
 electrolytes, secretion and absorption in, 233, 238,
  240–242, 244, 246, 250t, 257
 metabolism, 5f
 microbial fermentation and metabolism in, 191t,
  192t, 205, 206f, 207, 209, 210, 224, 235f
 motility in, 100, 112, 114f
 neuroendocrine control in, 272, 282–285
Dolichotis (*see* mara)
dolphins, 12t, 49f, 51, 52f, 53, 176t
donkeys (*see also* horses and ponies; zebra) 213, 219t
Dorosoma (*see* shad)
dove, 177
Dromaius (*see also* emu), 43f, 121t
ducks, 44, 256, 274, 286
dugong, 12t, 90, 91f, 92f, 129, 190t, 192t, 204
duikers, 143, 192, 198t

echidna, 12t, 49f, 50, 160, 162, 163, 166
eels, 12t, 26, 31, 32f, 173t, 176t, 182, 183, 229,
 254
Eidolon (*see* bats)
Elephe (*see* snakes)
elephants, 5t, 5f, 12t, 90, 91f, 147, 148, 165t, 166,
 176t, 192t, 205, 223
elephant shrews, 12t, 53, 54f, 124t
Elephantulus (*see* elephant shrews)
Elephas (*see* elephant)
elk, 83, 176t
Embiotocidae (*see* surf perch)
emesis, 97, 98, 264
emu
 anatomy in, 41, 43f, 309f
 motility in, 121t, 141, 145

motility in, 114f
goldfish, 172, 206
Gophurus (*see* tortoises)
gorillas (*see also* apes), 276
greater glider, 60, 62, 63f, 125t, 140, 193t, 199t, 213
grebe, 44
grouper, 172, 176t
grouse, 40, 42f, 44, 177, 200t
guanaco, 81, 82f
guinea fowl, 141
guinea pig
  absorption in, 183, 186t
  anatomy of, 67, 68f, 96, 97
  coprophagy in, 216, 217, 219
  digesta transit in, 124t, 137, 139
  digestion (endogenous enzymes) in, 160, 164, 176t, 185
  electrolyte secretion in, 234, 246, 257
  microbial fermentation and metabolism in, 192t, 199t, 211
  motility in, 110, 112, 114, 114f
  neuroendocrine control in, 272, 302
Gymnobelideus (*see* possum)
gulls, 41

hagfish (*see also* cyclostomes), 12t, 13, 23, 24, 160, 168, 172, 175, 229, 285, 302
hake, 171
hamsters, 67, 68f, 70f, 96, 97, 125t, 158t, 164, 176t, 191t, 193, 205, 211, 215
hares, 12t, 88, 137, 216, 217, 219t
hawks and buzzards (*see also* falcons), 41, 42f, 44, 98, 171
hedgehogs, 53, 54f, 96, 112, 127f, 129, 130, 157, 158t, 185, 186t
hemicellulose, 155, 156f, 197, 212, 225
hemirampids, 123, 221
hepatopancreas, 295, 297
herons, 41, 44
herring cale, 207, 208
Heterocephalus (*see* naked mole-rat)
hippopotamus, 78f, 79, 80f, 81f, 148, 176t, 190t, 203, 204, 293, 312f
hoatzin, 41, 43f, 126, 191t, 203, 222, 224, 227, 308, 309f
Homo (*see* humans)
hormones and paracrines, 301f
  aldosterone, 136, 244, 256, 263, 263f
  bombesin/gastrin-releasing polypeptide (GRP), 227t, 280f, 281t, 302
  cholecystokinin (CCK), 245, 277t, 288, 279f, 280–282, 285, 301, 302
  corticosterone, 184
  enkephalin, 275, 277t
  enterogastrone, 283
  enteroglucagons, 278, 281t, 283–285
  gastric inhibitory polypeptide (GIP), 278, 279f, 281t, 283, 285
  gastrin, 107, 115, 247, 274, 277t, 278–280, 280f, 281t, 285, 301, 302
  glucocorticoids, 256
  glucagon, 227t, 278, 280f, 281t
  5-hydroxytryptamine, 281t
  melatonin, 288

  motilin, 277t, 278, 281t, 285
  neuropeptide Y, 284
  neurotensin, 227t, 278, 280f, 281t
  pancreatic polypeptide (PP), 277t, 278, 280f, 281t, 284, 285
  polypeptide YY (PYY), 277t, 278, 284, 285
  renin, 263, 263f
  secretin, 245, 277t, 279f, 280f, 281t, 282, 283, 285, 301
  serotonin, 277t, 280f
  somatostatin, 277t, 278, 281t, 284
  thyroxin, 184
  vasoactive intestinal polypeptide (VIP), 185, 301
  vasopressin, 265
horned screamer, 44, 45
horses and ponies (*see also* donkeys; zebra), 12t, 183, 296, 315
  absorption in, 186t
  anatomy of, 17f, 20, 22f, 87f, 89, 89f, 96, 293, 306f, 311f
  coprophagy in, 216, 219f
  digesta transit in, 97, 98, 123t, 130, 321f, 132–134, 134f, 135, 147, 148, 150
  digestion (endogenous enzymes) in, 165t, 171, 173t, 176t, 183–185
  electrolyte secretion and absorption in, 233, 234f, 238–240, 244, 245, 250t, 251, 255, 256, 257, 260t, 261, 263f, 264f
  metabolism, 5t, 5f
  microbial fermentation and metabolism in, 191t, 192t, 199t, 205, 206f, 207, 209, 210t, 211–215, 224, 237f
  nitrogen metabolism in, 213, 214
  motility in, 111, 114f, 115
howler monkeys (*see also* monkeys), 73, 199t
Hucho (*see* salmon)
humans, 12t, 72, 183
  absorption in, 186t
  anatomy of, 17f, 76, 21, 70, 75f, 76
  body fluid compartments in, 230f, 231, 232f, 232, 259
  digesta transit in, 94, 148
  digestion (endogenous enzymes) in, 158t, 163, 164t, 168, 171, 174, 176t, 183, 185
  electrolyte secretion and absorption in, 234, 241, 242, 243f, 244, 247, 250t, 252, 254f, 254–258, 260, 261t, 266
  metabolism, 5t, 5f
  microbial fermentation and metabolism in, 192t, 199t, 208, 209, 211, 213–215, 225, 226
  motility in, 98, 100, 111, 112, 113f, 114, 114f, 115
  neuroendocrine control in, 269, 282–286
hummingbirds, 44, 121t, 126, 141, 163
Hydrochoerus (*see* capybara)
Hyemoschus (*see* chevrotain)
hyenas, 58
Hylobates (*see* gibbons)
hyrax, 9, 12t, 91f, 92, 129, 130f, 131, 191t, 193t, 198t, 205, 213, 223, 224, 293, 296, 313

iguanas (*see also* lizards), 7, 38, 39f, 40, 121t, 176t, 193t, 200t, 211, 215, 219t, 222, 231
indri (*see also* lower primates), 70, 72

invertebrates (*see also* protozoa), 2, 4, 10, 30, 51, 126, 144–147, 157t, 157, 167, 229, 288, 289, 291, 292, 294–298, 297f, 300–304, 301f
Isoodon (*see* bandicoots)

jackals, 58
javelines (*see* peccaries)
jerboa, 67

kangaroos, wallabies, and rat-kangaroos, 6, 9, 12t, 122t, 318–320
　absorption, 186t
　anatomy of, 62, 63f, 64f, 312f, 318f, 319f
　digesta transit in, 132, 133f, 147, 148
　digestion (endogenous enzymes) in, 162t, 163, 171, 173t, 175, 176t
　electrolyte absorption and secretion in, 241–245
　microbial fermentation and metabolism in, 190t, 198t, 203, 204
　metabolism, 6, 9, 198t
　motility, 101, 110, 122t
Kassina (*see* frogs)
kestrels, 44
ketone bodies, 212, 240
kingfishers, 44
knifefish, 29
koala, 12t
　metabolism, 6
　anatomy in, 62, 63f
　digesta transit in, 125t, 129, 139, 140
　digestion (endogenous enzymes), 162t, 163
　microbial fermentation and metabolism in, 193t, 199t, 205, 208, 215, 219t
kob, 176t
Kogia (*see* whales)
krill (*see also* crustaceans), 14, 204
Kyphosus (*see* sea chubs)

Lacerta (*see* lizards)
lactic acid, 196, 205, 224, 225, 234, 235f, 236f, 237f, 258, 266
Lagenorhynchus (*see* dolphins)
Lagopus (*see* ptarmigans)
Lagorchestes (*see* kangaroos and wallabies)
Lagothrix (*see* woolly monkeys)
laminarin, 160, 221
lampreys (*see also* cyclostomes), 12t, 13, 19, 24, 27f, 279
Lampropeltis (*see* snakes)
langur monkeys (*see also* monkeys), 72, 73f, 76, 148, 190t, 204, 293
Lasiorhinus (*see* wombats)
Latimeria (*see* hagfish)
lemmings, 70, 137, 137f, 138, 138f, 216, 219, 219t, 313, 315
lemurs, 12t, 70, 76, 77f, 158t, 216, 217
Lemmus (*see* lemmings)
Lepilemur (*see* lemurs)
Lepus (*see* hares)
Leuciscus (*see* chubs)
Liothrix (*see* nightingales)
lizards, 7, 12t, 296
　anatomy of, 36, 37f, 38, 39f, 40
　digesta transit in, 120t, 121t, 126, 148

digestion (endogenous enzymes) in, 159t, 176t
microbial fermentation and metabolism in, 193t, 200t, 211, 215, 219t, 222, 231
neuroendocrine control in, 274, 285
llama,
　anatomy in, 17f, 78f, 80f, 81, 82f, 83f
　digesta transit in, 122t
　microbial fermentation and metabolism in, 190, 190t, 202
　motility in, 107–109, 108f, 109f
　secretion and absorption of electrolytes in, 248
loons, 44
lorikeets, 44
lorises (*see also* lower primates) 70, 72, 158t, 164t
lower primates (*see also* aye-ayes; indries; lemurs; lorises; pottos; sifakas), 12t, 70
　anatomy in, 70, 71f, 72, 76, 77f
　coprophagy in, 216, 217
　digesta transit in, 128f, 129, 131
　digestion (endogenous enzymes) in, 158t, 164t
　microbial fermentation and metabolism in, 190t, 192t, 205, 206f, 207
Loxodonta (*see* elephants)
lungfish, 12t, 20, 25, 29, 172t, 176t
lutongs, 77f

Macaca (*see also* rhesus and pig-tailed monkeys; monkeys), 71f, 72, 96, 97, 164f, 176t
mackerel, 172
macropod marsupials (*see* kangaroos; wallabies; pademelons; quokka, and rat-kangaroos)
Macropus (*see also* kangaroos; wallabies), 63f, 64f, 122t
Madoqua (*see* dik dik)
manatees, 12t, 90, 92, 129, 196, 241
Manis (*see* anteaters), 50f, 51
mara, 124t
marmosets (*see also* monkeys), 7, 72, 76, 77f
marmot (*Marmota*), 67, 68f, 124t
marsupial mole, 60
martens (*Martes*), 158t
mastication, 14, 99, 222, 223, 227, 305, 306f, 308, 315, 319f
mean retention time (MRT) 119–126, 121t, 122t, 123t, 124t, 129, 132, 135, 139, 140, 141
Megaderma (*see* bats)
Meleagris (*see* turkeys)
Melopsittacus (*see* budgerigars)
menhaden, 26
Mesocrecetus (*see* hamsters)
metabolic rates, 3f
　basal, 2, 4, 4t, 6, 7
　field, 6, 7
　mass-specific, 4–6
　standard, 4, 4t, 5
methane (*see* gasses)
mice, 12t
　absorption by, 185, 186t
　anatomy of, 96, 97
　coprophagy in,
　digesta transit in, 125t
　digestion (endogenous enzymes) in, 158t, 160, 176t
　microbial fermentation and metabolism in, 211

396

metabolism, 5t, 5f
neuroendocrine control in, 286
Microtus (*see* voles)
microvilli (*see* brush border)
migrating myoelectric complexes (MMC), 110–114,
   113f, 114f
migrating spike bursts (MSB), 115, 116f
milk, 171, 183, 185
minnows, 25
mink (*Mustela*), 58, 59f, 60, 129
Mirounga, (*see* pinnipeds)
models
   of absorption and secretion, 239f, 247f, 253f,
      255f, 260f
   of digestion, 142f, 149f, 316
   of chemical reactors, 144f
moles, 12t, 53, 54f, 96, 97, 157, 158t, 291
Molossus (*see* bats)
mongoose, 58
monkeys (*see also* names of individual groups), 7,
   12t
   absorption, 186t
   anatomy of, 70, 71f, 72, 73, 74f, 76, 77f, 293, 312f
   digesta transit in, 128f, 129, 131, 148
   digestion in endogenous enzymes, 158t, 164t, 176t
   electrolytes, secretion and absorption in, 253
   microbial fermentation and metabolism in, 190,
      191t, 192t, 199t, 204, 205, 206f, 207
   motility in, 96, 97
   neuroendocrine control in, 272
moose, 197, 198t
mountain beaver, 23, 46, 216, 217
mouse deer (*see* chevrotains)
mud puppies, 173t, 176t
mullet (*Mugil*), 26, 31, 35f, 172, 174, 221
Mus (*see* mice)
musk ox, 86
muskrats, 70f
Mustela (*see* mink)
Myocastor (*see* nutria)
Myotis (*see* bats)
Myrmecobius (*see* numbat)
Myrmecophaga (*see* anteaters)
Myxinie (*see* hagfish)

naked mole rat (*see also* rats), 192, 199t, 216, 217
Natrix (*see* snakes)
Necturus (*see* mud puppies)
Nerodia (*see* snakes)
nervous system
   autonomic nervous system, 213, 269, 269f, 270,
      272, 272f, 273, 273f, 274
   enteric nervous system, 270, 271, 271f, 272–274
neurotransmitters and neuromodulators, 274–278,
   300, 301f
   acetylcholine, 244, 245, 270, 275, 276, 284
   adenosine diphosphate, 275
   adenosine monophosphate, 275
   adenosine triphosphate (ATP), 275, 276
   bombesin/gastrin-releasing polypeptide (GRP),
      275, 276, 277t, 278
   cholecystokinin (CCK), 276
   dopamine, 275
   histamine, 275

enkephalins, 275, 277t
5–hydroxytryptamine (5–HT), 275
neuropeptide Y/pancreatic polypeptide (PP), 275,
   277t
neurotensin, 276, 277t
nitric oxide, 275
norepinephrine, 244, 270, 275, 276
opioids (*see* enkephalins)
polypeptide YY, 277t
somatostatin, 276, 277t
substance P, 275, 276, 277t, 301
vasoactive intestinal polypeptide (VIP), 275–276,
   277t, 278
newts, 12t, 31
nightingales, 159t
night monkeys (*see also* monkeys), 73, 74f, 76, 77f
nitrogen recycling (*see also* urea) 189, 202, 203,
   212f, 213, 214
Noctillo (*see* bats)
Notopterus (*see* knifefish)
Notoryctes (*see* marsupial mole)
numbat, 60
nutria, 124t, 216, 219
Nycticebus (*see* bush baby)

Ochotona (*see* pika)
Octodon (*see* degu)
Odacids, 31, 221
Odobenus, 165
Odocolleus (*see* deer)
Oedipomidas, 164
omasum, 84, 85f, 101–105
Oncorhynchus (*see* salmon)
ontogeny, 183–186
Ondatra (*see* muskrat)
Opisthocomus (*see* hoatzin)
opossums, 60, 61f, 62, 115, 176t, 272
Opsanus, (*see* toadfish)
orangutans (*see also* apes), 75f, 76
Oreochromis (*see* tilapia)
Ornithorhynchus (*see* platypus)
Oryctolagus (*see* rabbits)
ostrich, 43f, 45, 121t, 126, 141, 222, 265, 308, 309f
otters, 58
Ovis (*see* sheep)
owls, 44, 98, 159t, 171, 274
ox (*see* cattle)

pacemakers (*see also* myoelectric activity), 100, 114,
   115, 115, 136
paddlefish, 12t, 25
pademelons (*see* kangaroos; wallabies; and rat-
   kangaroos)
pagolins (*see also* anteaters), 12t, 50f, 51
Pan (*see* chimpanzees)
pancreas, 295
   anatomy of, 20
   innervation of, 269, 270
   neuroendocrine control of,
   secretion by, 168, 244, 245, 259, 262, 294–296,
      301
panda, 58, 59, 220, 223, 313
panther (*Panthera*) 163, 164
Papio (*see* baboon)

parrotfish, 25, 26, 35f
parrots, 41, 44, 159t
partridge, 40
Passer (*see* sparrows)
peccaries (*Tayassu*), 78f, 79, 165, 190t, 203
pectin, 155, 156f, 189, 192, 196, 212, 225
pelicans, 41, 44
penguins, 44, 121
Perameles (*see* bandicoots)
perch (*Perca*), 26, 171, 280
Perodicticus (*see* lorises)
petrels, 41, 157, 167, 169
Petromyzon (*see* lampreys)
Petauroides (*see* greater glider)
pH of secretions or digesta, 178, 197, 217,
    235f–237f, 238, 243–251, 258
Phacochoerus (*see* warthog)
phagocytosis, 296
phalarope, 177
pharyngeal teeth (pharyngeal mill), 25, 26, 31, 35,
    221, 304
phascogales (*Phascogale*), 60, 61f
Phascolarctos (*see* wombats)
pheasants, 40, 121t
Phesianus (*see* pheasants)
Philodryas (*see* snakes)
Phoca (*see* seals)
Phoebetria (*see* albatross)
Phyllostomus (*see* bats)
Phyllotis (*see* mice)
Physeter (*see* whales)
phospholipids, 166, 168
pig deer, 78f, 79
pigeons, 41, 44, 159t, 160, 176t, 286
pigs, (*see also* bush pig; pig deer; warthog) 9, 10,
    12t, 79, 317
  absorption in, 186t
  anatomy of, 17f, 22f, 78f, 79, 80f, 96, 97, 293
  digesta transit in, 123t, 127, 128f, 129, 131, 143,
    148
  digestion (endogenous enzymes) in, 157, 158t,
    160, 165, 168, 169, 172, 174, 175, 176t, 185
  electrolyte secretion and absorption in, 234, 238,
    246, 247, 250t, 251, 256, 257, 258, 266
  microbial fermentation and metabolism in, 191t,
    192t, 199t, 205, 206f, 207, 209, 210t, 211, 213,
    214, 224, 236f
  motility in, 111, 112, 114f
  neuroendocrine control in, 116, 283–285
  nitrogen metabolism in, 214
pig-tailed monkeys (*see also* monkeys), 72
pika, 12t, 88, 137, 216, 217
pike, 26, 29f, 171
pinnipeds (*see also* seals and walruses), 163, 165t,
    166
piranhas, 25
Pizonyx (*see* bats)
plaice, 282
planigales, 60
Platanista (*see* dolphins)
Platessa (*see* plaice)
platypus, 12t, 49f, 50
Plecotus (*see* bats)
Pongo (*see* orangutans)

ponies (*see* horses and ponies)
porcupines, 67, 69f, 70, 171, 193t, 199t, 216, 315
porpoises, 12t, 51
possums, 6, 62, 124t, 125, 140, 141, 143, 160, 162t,
    163, 166, 193t, 199t, 213, 216, 217, 219, 272
Potamochoerus (*see* bush pig)
potoroo (*Potorus*), 122t, 320
Pottos, 70, 158t
Presbytis (*see* langur monkeys)
primates (*see also* lower primates; monkeys; apes;
    humans), 3, 70–77, 160, 164t, 183
proboscis monkey (*see also* monkeys), 76, 77f
Procavia (*see* hyrax)
prion, 169
Procolobus (*see also* monkeys), 76, 77f
Procyon (*see* racoons)
protein
  maintenance requirements, 9
  requirements of digestive system, 9, 10
  whole body turnover of, 9, 10
Proteles (*see* aardwolf)
Proropterus (*see* lungfish)
protozoa; gastrointestinal, 188, 190t–193t, 195, 196,
    203, 208, 226, 303
proventriculus (*see* gizzard)
Pseudemys (*see* turtles)
Pseudocheirus (*see* possums)
Psittacus (*see* parrots)
ptarmigan, 121t, 126, 131, 141, 193t, 200t, 214
Pterocnemia (*see* rhea)
Pteronotus (*see* bats)
Pteropus (*see* bats)
pyloric ceca, 28, 30f, 31f, 157, 169, 220

quail, 40, 115, 163, 264, 285
quoll, 60, 61f, 162
quokka, (*see* kangaroos, wallabies, and rat-
    kangaroos)
Q_{10} effect, 3, 4, 123

rabbits
  absorption in, 182, 186
  anatomy of, 87f, 88, 96, 311f
  coprophagy in, 216, 217, 218t, 219t
  digesta transit in, 125t, 129, 130f, 213, 135, 136,
    146
  digestion (endogenous enzymes) in, 158t, 160,
    163, 164t, 173t, 176t
  electrolyte secretion and absorption in, 234, 240,
    244, 251, 253, 256, 257
  microbial fermentation and metabolism in, 192t,
    199t, 209, 211, 212–214
  motility in, 112, 114f, 116
  neuroendocrine control in,
  nitrogen metabolism in, 213, 214, 217
raccoons, 58, 59f, 127f, 130, 183, 191t, 192t 205,
    206f, 207, 209
rails, 44
Raja (*see* rays)
Rana (*see* frogs)
Raniceps (*see* cod)
rat-kangaroos (*see* kangaroos, wallabies, and rat-
    kangaroos)

Printed in the United Kingdom
by Lightning Source UK Ltd.
105001UKS00002B/177-182

9 780521 617147